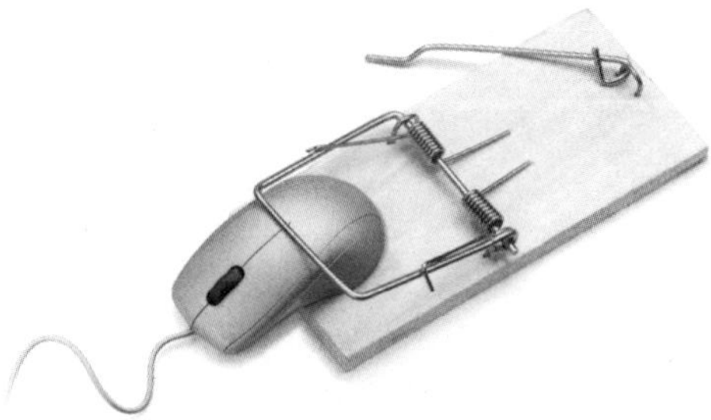

BEYOND THE KNOWLEDGE TRAP

Developing Asia's Knowledge-Based Economies

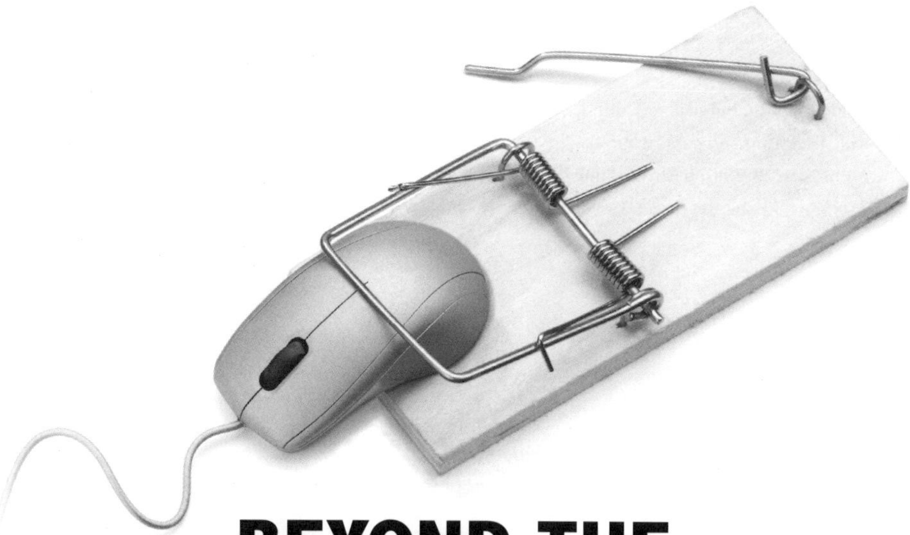

BEYOND THE KNOWLEDGE TRAP

Developing Asia's Knowledge-Based Economies

edited by

Thomas Menkhoff
Singapore Management University, Singapore

Hans-Dieter Evers
University of Bonn, Germany

Chay Yue Wah
SIM University, Singapore

Pang Eng Fong
Singapore Management University, Singapore

World Scientific

NEW JERSEY • LONDON • SINGAPORE • BEIJING • SHANGHAI • HONG KONG • TAIPEI • CHENNAI

Published by

World Scientific Publishing Co. Pte. Ltd.
5 Toh Tuck Link, Singapore 596224
USA office: 27 Warren Street, Suite 401-402, Hackensack, NJ 07601
UK office: 57 Shelton Street, Covent Garden, London WC2H 9HE

British Library Cataloguing-in-Publication Data
A catalogue record for this book is available from the British Library.

BEYOND THE KNOWLEDGE TRAP
Developing Asia's Knowledge-Based Economies

ISBN-13 978-981-4343-67-1
ISBN-10 981-4343-67-6

Typeset by Stallion Press
Email: enquiries@stallionpress.com

Printed in Singapore by B & Jo Enterprise Pte Ltd

Contents

List of Figures

Chapter 7

Chapter 9

Chapter 10

Chapter 11

Chapter 12

Chapter 14

Chapter 15

List of Tables

Chapter 4

Chapter 7

Chapter 8

Chapter 9

Chapter 11

Chapter 12

Chapter 16

List of Contributors

Tatjana BAUER is Junior Researcher at the Center for Development Research (ZEF), University of Bonn. She has done extensive field research on the Vietnamese community in Germany and on knowledge governance in Vietnam within the Vietnamese-German WISDOM project.

Yue Wah CHAY is Associate Professor of Psychology and Director of the Office of Graduate Studies, SIM University. Prior to joining SIM University, he was a faculty member of the Nanyang Technological University, the Singapore Management University and the National University of Singapore. His current research interests include knowledge systems, citizenship behavior, work commitment, assessment and executive careers. He is also co-editor with Thomas Menkhoff and Hans-Dieter Evers of *Governing and Managing Knowledge in Asia* (World Scientific 2010).

Derek H. C. CHEN is an Economist with the Growth and Competitiveness Practice of the World Bank Institute. A Singapore national, he joined the World Bank in 2000, after receiving his doctoral degree in Economics at the University of California, Davis. His fields of specialization include economic growth, international economics, poverty analysis and demographic economics. He leads WBI's research efforts on the effects of knowledge accumulation on long-term economic growth and employment, and he also oversees the initiative on the Knowledge Assessment Methodology (KAM — (www.worldbank.org/kam)), a web-based database and benchmarking tool that shows economies' progress towards the knowledge economy. Derek has written widely on topics such as knowledge accumulation and growth. Prior to joining the Bank, he taught undergraduate economics at UC-Davis, and before that, at the National University of Singapore.

Hans-Dieter EVERS is Emeritus Professor of Development Planning, Center for Development Research, University of Bonn, Germany and Senior Fellow, Centre for Development Policy and International Studies (CenPRIS), Universiti Sains Malaysia. He has taught at Bielefeld, Monash, Yale, Singapore, Universiti Kebangsaan Malaysia, University of Indonesia and Gadjah Mada University and was several times a Visiting Professor of Management at the Singapore Management University, a Distinguished Visiting Professor of Sociology, National University of Singapore and Eminent Visiting Professor, Universiti Brunei Darussalam. He has published widely on development planning, knowledge governance and urbanisation in Southeast Asia. He is also co-editor with Thomas Menkhoff and Chay Yue Wah of *Governing and Managing Knowledge in Asia* (World Scientific 2010).

Judith EHLERT is Junior Researcher at the Center for Development Research (ZEF), University of Bonn. She studied sociology at the University of Bielefeld and has carried out extensive field research in Vietnam on local knowledge to cope with floods in the Mekong Delta.

Solvay GERKE is a Professor of Development Research and Director of the Center for Development Research, University of Bonn, Germany. Prior to that she was Professor of Southeast Asian Studies at the University of Bonn, Germany and a Visiting Fellow at the Institute of Malay Civilization, National University of Malaysia.

She has carried out extensive research on family planning, migration, life styles and social stratification in Indonesia and Malaysia. She is also directing the work package "Knowledge Management" of the WISDOM project in Vietnam. She is the co-editor with Thomas Menkhoff of *Chinese Entrepreneurship and Asian Business Networks* (Routledge 2002). With Hans-Dieter Evers and Anna-Katharina Hornidge, she has edited a book on *The Straits of Malacca — Knowledge and Diversity* (LIT Verlag 2008).

Anna-Katharina HORNIDGE is Senior Researcher at the Department of Social and Cultural Change at the Center for Development Research, University of Bonn, Germany. She specializes in knowledge and development sociology and cultural studies, but is also interested in environmental sociology and development-oriented innovation creation and diffusion processes. She has extensively worked on processes of the social construction of 'knowledge society' in Germany and Singapore, and is currently working on farmer-led and -oriented knowledge creation and innovation development in Central Asia (Uzbekistan) and Southeast Asia (Indonesia). She is the author of *Knowledge Society. Vision and Social Construction of Reality in Germany and Singapore* (LIT Verlag 2007).

Kenneth HUANG is an Assistant Professor of Management at Singapore Management University (SMU), Lee Kong Chian School of Business, and has been a Research Affiliate at the Massachusetts Institute of Technology (MIT), Sloan School of Management. He has published widely in leading journals such as *Science*, *Academy of Management Journal*, and *Research Policy*. He also frequently presents in major conferences and research seminars around the world on topics related to technology management, intellectual property strategy, innovation policy and scientific entrepreneurship, especially in life sciences and biotechnology in the United States, Singapore and China contexts. He earned his Ph.D. (Technology Management and Policy) and M.S. (Technology and Policy) from MIT, and B.S. (with honors) in Biomedical Engineering and Electrical Engineering from the Johns Hopkins University.

Kong-Rae LEE is a Senior Research Fellow and Director at the Science and Technology Policy Institute in South Korea. He obtained his Ph.D. in Science and Technology Policy from SPRU, Sussex University (Brighton, UK) in 1993. He is Editor-in-chief of the *Asian Journal of Technology Innovation* and a member of the Korean Academy of Science and Technology (KAST).

Poornima LUTHRA is a Lecturer for Organizational Behaviour and Human Resources at the Lee Kong Chian School of Business, Singapore Management University. Her research interests include knowledge management, institutional theory, resource based view, leadership and teams. She has worked closely with the Asian Development Bank (ADB) in the context of knowledge management.

Lenny MARTINI is a Researcher at the School of Business and Management, Institut Teknologi Bandung (ITB), Indonesia. She holds a Bachelor degree in Industrial Engineering (2003) and a Master of Science in Management (MSM) degree from ITB (2009).

Thomas MENKHOFF is a Practice Associate Professor of Organizational Behavior and Human Resources in the Lee Kong Chian School of Business, Singapore Management University. He has formerly taught in the National University of Singapore, the University of Cologne and the University of Bielefeld in Germany. His current research interest is concentrated on knowledge management, change leadership and on Chinese business networks, on which he edited a triple issue for the Journal of Asian Business together with Pang Eng Fong and Hans-Dieter Evers entitled *The Power of Knowing — Studies in Chinese Business in Asia* (Ann Arbor, MI 2008).

Pamela NIENKEMPER is Geographer and Research Fellow at the Center for Development Research (ZEF), University of Bonn. She is specializing in GIS (geographical information system) and working on environmental topics in Malaysia and Indonesia in the ZEF research project on the Straits of Malacca.

Shan L PAN is a Associate Professor of Information Systems and the Coordinator of the Asian IT Case Series in the Department of Information Systems of the School of Computing at the National University of Singapore (NUS). His primary research focuses on the recursive interaction of organizations and information technology (enterprise systems), with particular emphasis on issues related to work practices, cultures and structures from a knowledge perspective.

PANG Eng Fong is a Professor of Management Practice at the Lee Kong Chian School of Business, Singapore Management University (SMU) and former Dean of the Lee Kong Chian School of Business (2006–2008). He earned his doctorate at the University of Illinois. Professor Pang was previously Ambassador of Singapore to the Republic of Korea and Mongolia, Belgium with concurrent accreditation to the European Communities, Luxembourg, the Holy See and the Netherlands and Permanent Representative to the Organisation for the Prohibition of Chemical Weapons. His last ambassadorial posting was as High Commissioner to the United Kingdom with concurrent appointment as Ambassador to the Republic of Ireland. Pang Eng Fong is the author/editor of several books, including *Competing for Global Talent* with Christiane Kuptsch (Geneva 2006) and *Foreign Investment and Industrialisation in Malaysia, Singapore, Taiwan and Thailand* with Linda Lim (Paris 1991).

Suparb PASONG is a Researcher and Member of Thailand Creative and Design Center (http://www.tcdc.or.th/), one of the Thai government's knowledge management organizations. He holds a doctoral degree in Sociology from Bielefeld University, Germany, and conducts research on the human dimensions of global environmental change, in particular, institutions and environmental governance. He is associated with an international research group known as Institutional Dimensions of Global Environmental Change (IDGEC), based at the University of California at Santa Barbara, California. Until recently, he taught environment and development at several universities in Thailand, including Prince of Songkla and Walailak. He also occasionally serves as a consultant with UNDP and ESCAP in areas such as water governance and knowledge management for development.

Benjamin SCHRAVEN holds a masters degree in political science and a Ph.D. degree from the University of Bonn, Germany. He is a Research Fellow at the Center for Development Research (ZEF), University of Bonn. He has done extensive research on early warning capacities in Sri Lanka, water storage solutions in the context of climate change adaptation in sub-Saharan Africa as well as livelihood adaptation strategies of peasant households in Northern Ghana.

Jasmin P. SUMINSTRADO is the Knowledge Officer of ILO Microinsurance Innovation Facility in Geneva. As one of the pioneers of KM in the Philippines, she facilitated the sharing of best practices for the UNDP (Small Grants Programme) Global Environment Fund, Philippines (UNDP-SGP-GEF, Philippines). She is the co-author of the e-book entitled *Community Wealth Rediscovered: Knowledge for Poverty Alleviation* and has written several papers on KM and development economics commissioned by the USAID, IDRC, the Poverty and Economic Policy Network (PEP), the Asian Development Bank etc. Ms. Suministrado holds a Master's degree in Business Administration from De La Salle University and a Bachelor of Science degree in Business Economics from the University of the Philippines.

Serafin D. TALISAYON is a Professor (retired) at the University of the Philippines, where he continues to teach a graduate course in knowledge management. He is a co-founder of CCLFI, a non-profit foundation, which advocates and provides technical assistance in knowledge-based development, knowledge for poverty alleviation and knowledge management Two of his latest edited books include *Knowledge Management in Asia: Experiences and Lessons* (Asian Productivity Organisation, 2008), and *Knowledge Management: From Brain to Business* (Asian Productivity Organisation, 2008). He served as KM consultant for numerous multi-lateral development agencies such as the World Bank (Manila), Asian Development Bank and Asian Productivity Organization (Tokyo).

Jann Hidajat TJAKRAATMADJA is a Professor of Knowledge Management and Director of Center of Knowledge for Business Competitiveness at the School of Business and Management, Institut Teknologi Bandung, Indonesia (SBM — ITB), where he was also the Vice Dean for Academic Affairs between 2003–2009. He obtained his Ph.D. (2002) as well as his Bachelor Degree (1978) in Industrial Engineering from ITB. He also holds a MSIE (1982) in Industrial Engineering from the Catholic University of Leuven in Belgium and was awarded a Visiting Fellowship at Rensselaer Polytechnic Institute (RPI) in 1994. He has written numerous books and journal articles in the areas of technology management, human resource management and knowledge management. Prof. Jann is a member of the Indonesian Production and Operations Management Society (IPOMS), and currently the President of the Knowledge Management Society Indonesia (KMSI) which he co-founded in 2007.

Caleb WALL studied at Massey University, New Zealand and received his Ph.D. degree from the Bonn Graduate School of Development Research, University of Bonn. He did extensive field research on local knowledge and economic and ecological restructuring of land and water use in Khorezm, Uzbekistan (UNESCO-ZEF Project). His book, *Argorods of Western Uzbekistan. Knowledge Control and Agriculture in Khorezm* (Berlin, Münster, London: LIT) was published in 2008. He now works as a consultant on social issues for Environmental Resources Management.

Peter WEINGART is Emeritus Professor of sociology and former director of the Institute for Science and Technology Studies (IWT), University of Bielefeld, Germany. He is a Visiting Professor at the University of Stellenbosch, South Africa, and a fellow of the Stellenbosch Institute of Advanced Studies (STIAS). Since 2008 he is editor of Minerva. His research interests include science advice to politics, science — media interrelation, and science communication. He has published widely in the field of sociology of science and science policy. One of his book publications (with S. Maasen) is *Metaphors and the Dynamics of Knowledge* (Routledge 2000).

Agung WICAKSONO is an Assistant to the Head of President's Delivery Unit for Development Monitoring and Oversight (UKP-PPP), Government of Indonesia, and Assistant Professor at the School of Business and Management, Institut Teknologi Bandung, Indonesia (SBM — ITB). He received his Ph.D. degree from the University of St. Gallen, Switzerland. Initially trained as an engineer at ITB with a Bachelor Degree (2000) in Industrial Engineering, he also obtained a double degree of M.Sc. and MBA in International Management from the Technical University Hamburg-Harburg (TUHH), Germany (2002). His research areas include corporate governance, cross-cultural management and political economy of state-owned enterprises in Southeast Asia.

Acknowledgements

The editors gratefully acknowledge the support they received from Singapore Management University (SMU) with regard to the set-up and scholarly activities of the research interest group 'The Knowledge Force' as well as the generous funding of several SMU research projects (e.g., *Antecedents of Knowledge Sharing in Knowledge-Intensive Teams: A Cross-Cultural Perspective/07-C207-SMU-010*) and international conferences (e.g., the Fourth International Research Conference on Asian Business entitled *"Knowledge Architectures for Development"*) jointly organised by the Lee Kong Chian School of Business and the Wee Kim Wee Centre, Singapore Management University in March 2008 which inspired the production of this review volume.

Introduction: Strategic Aspects of Developing Asia's Knowledge-Based Economies

Thomas Menkhoff, Hans-Dieter Evers, Chay Yue Wah and Pang Eng Fong

The increasing importance of knowledge as a new factor of production is a significant consequence of the process of globalization (Drucker, 1994; Evers, 2003; Foss, 2005). Actionable knowledge is crucial to deal with very diverse issues such as creating robust knowledge clusters in emerging markets (Evers *et al.* in this volume) or the need to develop innovations in service-related industries (Kogut and Zander, 1992; Chan and Mauborgne, 1997; Wirtz *et al.*, 2008). The creation and effective *governance* of knowledge represents a key driver of sustainable knowledge economies in an era of rapid globalization and latent knowledge conflicts (Tallman *et al.*, 2004; Neo and Chen, 2007; Willke, 2007; Menkhoff, Evers and Chay, 2010). This represents new opportunities, challenges and threats for policy makers, leaders, managers and employees in both public and private sector organizations in Asia and beyond tasked with managing knowledge effectively (Talisayon, 2008). Several broad research questions and issues are of interest:

- What is the role of knowledge as an engine of growth for Asia's increasingly knowledge-centric development?
- How fit are Asian countries for the knowledge economy in terms of their knowledge architectures and national information and communication technology frameworks?

- Which role do governments and the state play in related transformation processes in terms of knowledge governance, technology and innovation management, science and technology, e-governance, institutional arrangements and so forth so as to leverage on *glocal* knowledge flows and to ensure sustainable development?
- What happens in multi-cultural (diverse) knowledge contexts when members of different (ethnic) groups interact and collaborate (e.g., in knowledge-intensive teams) aimed at achieving development progress, e.g., by creating new knowledge and innovative products?
- How effective are present knowledge governance systems in Asia in terms of managing knowledge traps and to ensure development progress? How can public policy contribute in leveraging on knowledge for development (K4D)?
- Where are the Asian role models who have succeeded in building up sustainable knowledge architectures? What can Asian societies learn from each other in this respect? What development lessons can be drawn from Asian case studies?

These were some of the issues which were intensively discussed during the Fourth International Research Conference on Asian Business entitled "Knowledge Architectures for Development" jointly organised by the Lee Kong Chian School of Business and the Wee Kim Wee Centre, Singapore Management University, in March 2008. This book features selected papers presented at the conference which is a follow-up of three earlier international conference projects on business and governance issues conducted in Germany and Singapore (Menkhoff and Gerke, 2002; Menkhoff, Pang and Evers, 2008).[1]

[1] The 3rd international conference was held in Singapore in March 2006. It was funded by both the Lee Kong Chian School of Business and the Wee Kim Wee Centre, Singapore Management University and chaired by Pang Eng Fong and Thomas Menkhoff. The various contributions were published as a Special Triple Issue of the Journal of Asian Business in 2008.

A key objective of the 2008 meeting was to bring different groups of scholars (representing various disciplines from management to area studies) together to advance both relevant conceptual frameworks and empirical knowledge about the specifics of sustainable knowledge governance in Asian societies in a globalised world. Additional papers written by staff members of the Centre for Development Research (ZEF) at Bonn University, Germany, were also included. ZEF has been at the forefront of critical reflections about the rise of knowledge as evidenced by the innovative works of its research group "Culture, Knowledge and Development".

The current volume is a sequel to the earlier book *Governing and Managing Knowledge in Asia*, whose 2nd edition was published in 2010 (Menkhoff, Evers and Chay, 2010). Whereas the earlier book discussed conceptual issues and theories, the current volume leans more heavily on empirical research and detailed case studies. This clearly shows the advance in research on knowledge-based economies and societies.

Topics related to knowledge such as knowledge policy, access to knowledge, dynamic governance, innovation systems, comparative advantages of local knowledge or learning clusters remain high on the agenda of policy-makers and business leaders in Asia-Pacific and elsewhere (Evers, 2003; Pan, 2004; ADB, 2007; Neo and Chen, 2007). Reflective, empirical-theoretical material on these issues is still rare. Therefore, this monograph is timely as it attempts to advance both conceptual frameworks and empirical knowledge about the specifics of moving towards sustainable knowledge governance in Asian business and society. We believe that the various chapters will offer valuable insights into the varied experiences of Asian countries such as Singapore, South Korea, Malaysia, Thailand, Indonesia, Vietnam and Uzbekistan and their attempts to make knowledge work. We hope that the monograph will be useful for leaders in development and that it will be used as required reading material in courses on Asian Business, Public Policy, Knowledge and Innovation Management (TIM), Knowledge Economy Development etc.

Challenges Ahead[2]

> "Opening up access to knowledge is a demand of global justice; it is both a human rights issue and a crucial factor in spurring economic development and technological innovation" (Yale Law School Professor Jack Balkin, founder and director of the Information Society Project at Yale Law School). http://www.law.yale.edu/news/6191.htm

Knowledge[3] is deemed relevant for meeting the Millennium Development Goals (MDG) in three areas: education, internet access and water management (BMZ, 2005). It has been obvious for a long time, however, that knowledge itself has become an essential development factor, if not the most important one of all (Hornidge in this volume: Chapter 3). In a knowledge-based economy, the creation of wealth from immaterial production (services, computer-assisted production, etc.) exceeds that from material production (manufacturing). Levels of education are relatively high in a knowledge society, knowledge workers make up a large part of the workforce, and a considerable portion of gross domestic product is invested in education,

[2] This part draws on: Evers, Gerke and Menkhoff (2006). Little understood knowledge trap. *Development and Cooperation*, 33(6):246–247.

[3] *Knowledge management* refers to the totality of organisational strategies aimed at creating a smart organisation, which is able to leverage upon its various IC assets, to learn from past experiences, whether successful or unsuccessful, and to create new value through knowledge. At the people level, KM puts emphasis on the competencies, education and learning abilities of organisational members. At the organisational level, KM is concerned about the creation, utilization and development of the collective intelligence of an organisation. Technologically, effective KM requires the efficient organisation of a suitable communication and information infrastructure (e.g., intranet) based on relevant taxonomies and knowledge repositories (Zeleny, 2010). *Knowledge governance* is both an administrative process and a structure of authority relations; it involves the channelling of resources in building up knowledge management capabilities and improving the competitive advantage of a country in the world market by utilizing knowledge as a factor of production. At the national level, one of the challenges is to *govern* knowledge effectively by providing governmental support for the development of a knowledge infrastructure via a conducive legal framework, re-organising the educational system and setting up institutions to support research and development activities (Evers, 2003).

research and development. The World Develop-ment Report 1998–1999 has drawn attention to these facts (World Bank, 1999). The case of South Korea's successful knowledge-based development experience (Suh and Chen, 2007) offers many valuable lessons for developing economies in this regard (Chen in this volume: Chapter 2; Lee in this volume: Chapter 7).

According to conventional wisdom, knowledge can be imported quickly and at minimal transaction costs by use of the latest information technology. Governments in countries with few natural resources are thus tempted to try to leap into the post-industrial age. They might even succeed, were it not for the "knowledge trap" (Evers, Gerke and Menkhoff, 2006).

Knowledge is being created on an ever-increasing scale. It is estimated that the volume of knowledge at our disposal (including "junk knowledge") is currently doubling every five years (Stehr, 2001). However, the creation of new knowledge always goes along with the creation of new unknowns. The more one knows, the better one understands what one does not know. In this regard, the frontiers of science as a knowledge-producing institution are indeed endless (Weingart in this volume: Chapter 5). Anyone who wants to do relevant research, therefore, needs to know about the unknowns. They are important because they provide the stimuli to search for adequate answers. Sometimes, however, even experts get trapped because they simply do not know how to resolve certain problems such as risks associated with the explosion of nuclear power plants (Weingart in this volume: Chapter 5) or because the desire for recognition surpasses the endeavor and will to advance knowledge (Huang in this volume: Chapter 6). In certain situations, even 'experts' are unable to make sense of unknowns. This becomes even more critical if entire country development strategies fail to take off due to the inability to troubleshoot hi-tech manufacturing equipment purchased in fully developed knowledge economies or problems to effectively market innovative products and services due to insufficient know how.

The "knowledge trap" lies in the fact that data, information and knowledge are often taken over without any understanding of the corresponding unknowns. This is particularly so when the people acquiring such knowledge simply copy solutions. Failing to import an understanding of the unknowns consequently leads to bad investments and stagnation. Desired results are not achieved, and a "knowledge economy" will certainly not be created. Visionary and capable knowledge elites can prevent this from happening provided they manage to effectively leverage on both local and global knowledge depending on their needs (Pasong in this volume: Chapter 8; Menkhoff *et al.* in this volume: Chapter 11).

Since the late 1990s, it has been debated how to bridge the digital divide, both at national and the international levels. It mostly boils down to development strategies that supposedly allow countries to "catch up": developing countries should emulate the developed knowledge societies, ensure computer literacy, technical infrastructures and facilities, and increase the number of higher education graduates in each age group. Conventional development strategies therefore include:

— interventions in primary and secondary education,
— the promotion of hardware and software infrastructures,
— the digitisation of governments (E-Government) and
— legal reforms (for instance, copyright protection).

In addition, there are also attempts to use modern information technology in various business sectors, for example, in agriculture.

Not a single one of these approaches takes into account the significance of what is unknown. This is problematic. In the initial stages, at least, the strategies mentioned will widen, rather than close, the gulf between established knowledge societies and the societies which are supposed to catch up. Users in poor countries have less and less understanding of how technologies function, and how they might be further developed to serve their interests.

Steps Forward

What is needed is a strategy that takes the relevance of unknowns into account (Evers *et al.* in this volume: Chapter 1; Chen in this volume: Chapter 2). Such a strategy would have to be directed at several things, namely:

— knowledge hubs, competence centres and centres of excellence,
— knowledge clusters as "learning regions",
— the transfer of knowledge through global production networks, and
— the use of comparative advantages of local knowledge.

Knowledge hubs: Various developing countries have attempted to pursue strategies of localising knowledge. In the early 1980s, Indonesia declared four of its universities to be "centres of excellence", and gave them particular financial support. The results were not convincing. Only isolated competence centres were established, and they were not adequately networked with other knowledge and production hubs (Tjakraatmadja *et al.* in this volume: Chapter 10). Knowledge clusters were not created. Malaysia has followed suit by declaring its Universiti Sains Malaysia (USM) in 2010 an "APEX university", i.e., a centre of excellence and several others, like Universiti Kebagsaan Malaysia (UKM) as "research universities". It remains to be seen, whether this policy leads to the formation of vibrant knowledge clusters (Evers *et al.* in this volume: Chapter 4).

Knowledge clusters: Silicon Valley near San Francisco and Silicon Plateau around Bangalore are examples of cluster formation, as is the knowledge region of Munich. There are also the first indications of cluster formation in the Multimedia Super Corridor in Malaysia. Economic research emphasizes the close inter-connectedness of innovation, local economic growth and cluster formation (Porter, 2000; Pinch *et al.*, 2003). "Clusters" are regional concentrations of companies, manufacturing subsidiaries, research institutes, universities and other institutions which are connected to knowledge utilization and production. What is important is the diversity of the players involved (Kuptsch and Pang, 2006; Chen, Sun and McQueen, 2010). They may complement one another, be in competition, or cooperate (Menkhoff *et al.* in this volume: Chapter 11). Our research in Asian knowledge cities such as Singapore has demonstrated the relevance of these structures (Pang, 2006; Chay *et al.*, 2009; Menkhoff *et al.*, 2010). Geographical clustering without knowledge sharing tends to reduce the effectiveness of knowledge production and knowledge output (Evers and Bauer in this volume: Chapter 9; Menkhoff *et al.* in this volume: Chapter 11).

Global production networks: Today, the production and transfer of knowledge take place against the background of globalization (Mowery and Silverman, 1996). Global production networks (GPNs) are expanding rapidly, and, under certain conditions, ensure the cross-border transfers of knowledge. Typically, knowledge clusters with a high level of diversity are particularly closely involved in such networks. However, the trend towards GPNs is sometimes double-edged. While larger suppliers are furnished with knowledge, even if only to satisfy the quality standards of potential customers, small and medium-sized enterprises at the end of the knowledge chain often become marginalized (Menkhoff *et al.*, 2010; Menkhoff, 2010). They also often lack the skill and the expertise to absorb global knowledge. Given the speed with which knowledge is produced in the already operational knowledge economies, it is hardly possible to catch up simply by closing "knowledge gaps". Moreover, transnational corporations mainly impart "packaged knowledge", a type of "fast food" for rapid consumption, which is disassociated from any understanding of the unknowns relevant for enhancing local capacities.

Local knowledge: The transfer of global knowledge into the local context always requires the use of local knowledge (Evers and Caleb in this volume: Chapter 13; Gerke and Ehlert in this volume: Chapter 14). This provides special opportunities for developing countries which open up to the world market in the course of globalisation, either through choice or because they are forced to do so. Globalisation of local knowledge and localisation of global knowledge are the prerequisites for the use of knowledge as an engine for growth. Poor knowledge governance and knowledge management as well as failure to learn lead to further underdevelopment and knowledge loss (Foss and Michailova, 2009; Madsen and Desai, 2010; Chen *et al.*, 2010) which can be prevented through the development of sustainable knowledge clusters and good knowledge-based management (Talisayon and Suminstrado in this volume: Chapter 12; Luthra and Pan in this volume: Chapter 16).

Contents

Conceptualizing and strategizing knowledge as an engine for growth

In Chapter 1, "Knowledge Hubs and Knowledge Clusters: A Knowledge Architecture for Development", Hans-Dieter Evers, Solvay Gerke and Thomas Menkhoff analyse the internal dynamics of knowledge clusters and knowledge hubs and show why clustering takes place despite globalisation and the rapid growth of ICT. They define a knowledge cluster as a local innovation system which is organized around universities, research institutions and firms which successfully drive innovations and create new industries. Knowledge hubs are localities with high internal and external networking and knowledge sharing capabilities. Both form a new knowledge architecture within an epistemic landscape of knowledge creation and dissemination, structured by knowledge gaps and areas of low knowledge intensity. The basic argument that firms and their delivery chains attempt to reduce transport (transaction) costs by choosing the same location is still valid for most industrial economies, but knowledge hubs have different dynamics relating to externalities produced from knowledge sharing and research and development outputs. With globalisation and knowledge-based production, firms may cooperate on a global scale, outsource parts of their administrative or productive units and negate location altogether. As the authors point out, the extremely low transaction costs of data, information and knowledge seem to invalidate the theory of agglomeration and the spatial clustering of firms, going back to the classical work by Alfred Weber (1868–1958) and Alfred Marshall (1842–1924), who emphasized the microeconomic benefits of industrial collocation. The paper, however, argues against this view and shows why the growth of knowledge societies will rather increase than decrease the relevance of location by creating knowledge clusters and knowledge hubs. The essay draws on empirical data derived from ongoing research in the Lee Kong Chian School of Business, Singapore Management University and in the Center for Development Research (ZEF), University of Bonn, supported by the German Aeronautics and Space Agency (DLR).

In "Using Knowledge as an Engine for Growth — The Case of South Korea" (Chapter 2), Derek H.C. Chen reconstructs the rise of South Korea's knowledge-based economy (KBE) which is attributed to effective knowledge accumulation and application. Both represent key determinants of total factor productivity and the "new engine of growth" for many economies. Korea has been experiencing rapid, and more importantly, sustained economic growth since the 1960s. This has resulted in its real GDP per capita increasing rapidly enabling the once low-income country to join the ranks of high-income industrialized nations within a short time span of four and half decades. The majority of this growth can be attributed to knowledge, rather than to the accumulation of traditional factors of production of capital and labor. As Derek Chen shows, Korea achieved this knowledge-based growth by investing heavily in education and training, boosting innovation through intensive research and development, developing a modern and accessible information infrastructure, all coupled with a stable economic and conducive institutional regime that enabled the knowledge-related investments to flourish. Due to this, Korea has ably made its transition to a knowledge economy, that is, an economy that uses knowledge as the key engine of growth. Its successful knowledge-based development experience offers many valuable lessons for developing economies. Conceptually, the paper is based on the Knowledge Economy approach of the Knowledge for Development Program (www.worldbank.org/wbi/k4d) of the World Bank Institute (WBI) which aims at assisting client countries in accessing and using knowledge to enhance their long-term economic growth.

In Chapter 3 "'Knowledge Society' as Academic Concept and Stage of Development — A Conceptual and Historical Review", Anna-Katharina Hornidge reconstructs the scientific and political "construction" of the idea of a 'knowledge society' (in short: k-society). Inspired by rapid developments in the information and communication technologies, a wide range of conceptual approaches to 'knowledge society', 'information society' and 'knowledge-based economy' have emerged during the last few decades. Eventually, despite terminological overlaps and the lack of clear-cut definitions, these notions were taken up by governments around the globe for either legitimising ongoing development programmes or pushing future economy and technology oriented activities. Singapore, for example, moved (while economically motivated) within a very short time frame from a purely technological definition to a definition integrating technological, economic aspects with areas such as culture, the arts and creativity. A key question that arises from her paper is this: If knowledge societies are socially constructed and not merely the result of technological and economic developments, which 'actors' are the dominant drivers of such processes and how do they achieve legitimacy? According to the author, key driving forces include the scientific community which constructs k-society concepts, political elites who create a vision (sometimes without the active contributions of scientists) of an arguably 'self-emerging' k-society and the powerful subsystem state which constructs a k-society as a stage of socio-economic development. According to her analysis, political activities, legitimised by this vision, then actually construct what is commonly believed to emerge by itself. Conceptions of k-society are also strongly influenced by structural realities such as the political and legal system, historical experience, and economic imperatives.

In Chapter 4 "Measuring Spatial Density: Knowledge Clusters in Malaysia", Hans-Dieter Evers, Pamela Nienkemper and Benjamin Schraven make a case for the construction of knowledge maps to reveal the spatial distribution of knowledge. Knowledge assets, knowledge producing and disseminating organisations are referenced to spatial objects and integrated into a geographic information system (GIS). Knowledge-relevant assets and organisational units are represented in thematic maps and in 3-D perspective graphs. Special attention is given to mapping and measuring knowledge clusters.

Statistical procedures to measure the degree of spatial density and knowledge clustering are discussed and ways are indicated to compare and determine the emergence of knowledge clusters. Data on Malaysia are used to illustrate the distribution of knowledge clusters in Peninsular Malaysia and within the state of Selangor. The authors conclude that the construction of knowledge maps is useful to show the complexity of epistemic landscapes which will enhance the chances of government agencies, companies and civic organisations to understand and use knowledge for development. This paper is meant as a guideline for GIS-based knowledge mapping analyses at country or regional level.

In "The Moment of Truth for Science — The Consequences of the Knowledge Society for Society and Science" (Chapter 5), Peter Weingart reflects on the fastest growing enterprise in our society, science.[4] Since the 17th Century, every doubling of the population has seen a tripling in the number of scientists. This means that, over the last 300 years, Western societies have invested a major part of their resources into the production, revision and verification of knowledge. It is a process that perpetuates itself if sufficiently nourished — the more resources go into science, the more scientists are educated, the more results they produce and the more questions arise. In this regard, the frontiers of science are indeed endless. The exponential growth of science is seen as the main factor in the shift from an industrialised society to a knowledge society which implies multiple new challenges and risks as evidenced by nuclear catastrophes such as Three Mile Island and Chernobyl, symbols of the loss of public confidence in scientific authority. Nevertheless, science as a knowledge producing institution remains irreplaceable because of the growing need for "reliable knowledge" in business and society (Weingart, 2002).

---

[4] This is a reprint of an artcile which first appeared in EMBO (European Molecular Biology Organization) Reports: Weingart, P. (2002). The Moment of Truth for Science. EMBO Reports, 3(8):703–706. The editors gratefully acknowledge EMBO's permission to reprint the essay in this edition. Any future references should refer to the original publication.

Similar to the importance of reliability and validity in statistics, scientific knowledge works are guided by customary rules of behavior and standards of conduct that coordinate interactions with others. Sometimes scientists transgress the ethical boundaries associated with their work as exemplified by cases of data suppression or plagiarism. How scientific norms evolve and are interpreted by knowledge workers is a fascinating area of research with immense practical implications. Against this background, the 6th Chapter by Kenneth G. Huang entitled "The Conflicting Norms of Science in the Institutionalization of Scientific Knowledge" examines the power of scientific norms as conceptualised by sociologist Robert K. Merton in the 1960s. As the author argues, the idealised conception of the norms or "ethos" of science as originally proposed are not necessarily always adhered to in modern science disciplines such as the life sciences due to the evolving nature of science and the increasing tension between pure scientific inquiry and motivation for commercialization. This makes it necessary to examine more closely the socio-political context in which the scientific community is embedded, including the inter- and intra-group behaviors and dynamics amongst scientists before accurate and meaningful inferences can be drawn to fully depict modern science.

Governing and managing knowledge as a strategic resource: Asian experiences

The importance to create a condusive national knowledge governance infrastructure is taken up in Chapter 7 written by Kong-Rae Lee entitled "Korea's Public Research System as a National Knowledge Infrastructure". Kong-Rae Lee investigates Korea's public research system as a case study of a national knowledge infrastructure. It reviews the reform of governance system of government-sponsored research institutes (GRIs) which caused the separation of all GRIs from the ministries to which they belonged in 1999. It was believed that R&D performance and efficiency of GRIs was low because ministries intervene too much in their research activities. As a consequence, the separation of GRIs from their subordinate ministries was seen as a viable policy option to activate and increase performance and efficiency of GRIs. R&D performance of government research institutes (GRIs) has been regularly evaluated by social science research institutes like STEPI or private consulting companies based on a research contract with a high level administration body. It is found that GRIs under research councils have recently improved their R&D performance in terms of paper publications and patent applications. The results of the comparison appeared to be similar when compared using the index of publication of SCI papers per 100 million wons. This finding implies that public research institutes in Korea have successfully built up their research capability within a relatively short development history as a result of skillful and dynamic adjustments of the national governance mechanisms.

In "Practicing Knowledge Management — Thailand's Experience" (Chapter 8), Suparb Pasong examines how knowledge of various kinds and sources plays a role in the development of the Thai economy and society. The study traces the recent history of knowledge management as a development practice in Thailand. It describes how KM has evolved in the public and private sectors, among the state and the civil society actors. Experiences of two organizations representing different approaches towards the practice of knowledge management are documented and analyzed. Both cases offer a comparative analysis of the architecture underlying each practice, people involved and their interactions as well as the mechanisms and institutional arrangements driving the dynamics and the application of knowledge as development inputs. The paper also assesses the outcomes of knowledge-driven development and identifies challenges for the state, society and the corporate world in managing knowledge for development. As the author concludes, to create value in a creative, high power and competitive economy, the strategy is simple: switch on the power of liberal arts and technology.

The objective of Chapter 9 ("Emerging Epistemic Landscapes: Knowledge Clusters in Ho Chi Minh City and the Mekong Delta") by Hans-Dieter Evers and Tatjana Bauer is to study the effects of knowledge clusters in Ho Chi Minh City and the Mekong Delta *vis-à-vis* Vietnam's path towards a knowledge-based economy. As the paper suggests, clustering appears to have a positive effect not only on the increase of knowledge output, but also on the economic growth of these regions. Using a GIS-based mapping method, the authors identify two major knowledge clusters — Ho Chi Minh City and Can Tho City and their internal structure in terms of knowledge clustering. Both areas create hubs in the south of Vietnam, with favourable conditions for knowledge production and a large pool of skilled people and an advanced infrastructure. Their own survey data as well as an analysis of databases and economic statistics show that productivity is higher and innovation in terms of knowledge spillovers and cooperation are more likely to take place in knowledge clusters. On the other hand, geographical clustering without knowledge sharing has tended to reduce the effectiveness of knowledge production and knowledge output in the south of Vietnam. This preliminary result is currently further pursued in a larger research project on scientific knowledge management systems in Vietnam to determine the extent to which proximity or clustering have led to inter-organisational networking, knowledge sharing and tangible knowledge outcomes.

Chapter 10 ("Knowledge Sharing in the Indonesian Context: Institut Teknologi Bandung (ITB) as Potential Knowledge Hub To Create Value from Academia, Business and Government Linkages") by Jann Hidajat Tjakraatmadja, Agung Wicaksono and Lenny Martini examines the knowledge driver potential of ITB Bandung (ITB) which is Indonesia's oldest and most renowned technology school. As the authors argue, organizational knowledge will grow and develop if the organization has the ability to learn continuously, irregardless of whether it is an academic, business or government (ABG) organization. Their research in this area was triggered by the desire to shed light on the antecedents and outcomes of systematic knowledge acquisition, sharing and creation between academia, business and government. To do so, empirical research in nine Indonesian companies was conducted and enriched with empirical findings from a case study of knowledge sharing activities within faculties of SBM ITB, a business school at ITB Bandung. The conceptual framework which resulted from their research is used as reference point to discuss whether ITB plays an important role as a knowledge transformation hub in bridging ABG sectors to create value through knowledge sharing among participating actors. In view of the track record of Institut Teknologi Bandung (it was here where the idea of creating the so-called 'Bandung High-Tech Valley' was launched decades ago), the authors believe that ITB possesses huge potentials to drive the success of knowledge governance in Indonesia. They also briefly highlight the role of the new Knowledge Management Society Indonesia (KMSI) which is aimed at facilitating the implementation of the above ABG linkages.

In "Achieving Knowledge Economy Status through Good Knowledge Governance: The 'Singapore KBE story' Revisited" (Chapter 11), Thomas Menkhoff , Hans-Dieter Evers, Chay Yue Wah and Solvay Gerke identify some of the critical success factors of Singapore's successful transition to a knowledge-based economy. Singapore, an island city-state located on the southern tip of the Malay Peninsula, south of the Malaysian state of Johor, and north of the Indonesian Riau Islands, has a population of about 5 million people. It is the smallest country in Southeast Asia. When Singapore became fully independent in 1965, having no natural resources, it was faced with severe problems such as unemployment, labor unrest, poverty, poor housing conditions etc. As a consequence, the government, under the People's Action Party (PAP), embarked on a strategy of export-led growth driven largely by foreign multinational firms (MNCs) and government-linked companies (GLCs). During the last 40 years, fast-paced government-led industrialisation and cluster-based growth has created a robust and increasingly knowledge-based economy with a well-diversified manufacturing industry based on electronics, chemicals, mechanical engineering and biomedical sciences as well as a sophisticated service sector. Important success factors for Singapore's economic success include efficient and visionary knowledge elites, dynamic knowledge leadership and governance, systematic utilisation of foreign know how and talents, continuous investments into human capital and supportive cultural values. Effective diversity management and the capability to create national value through the absorption of foreign knowledge assets represent other competitive weapons of Singapore.

Chapter 12 ("Knowledge for Poverty Alleviation: A Framework for Developing Intangible Assets of Low-Income Communities") by Serafin Talisayon and Ms. Jasmin P. Suminstrado describes a framework that brings together two powerful and recent development paradigms — knowledge-based management and sustainable development and presents a new perspective for developing low-income communities. It proposes the Knowledge for Poverty Alleviation (KPA) framework, a new model that can be used in the design and evaluation of anti-poverty projects at the community level to increase chances of project success and sustainability. KPA emphasizes the importance of recognizing a community's intangible assets, and proposes four criteria for the success of community-based anti-poverty projects: leveraging community intangible assets, neutralization of community intangible liabilities, using its available tangible assets, and the build-up of sustainable structural and stakeholder capital. Sustainability is pursued along the three domains of sustainable development: social empowerment, environmental protection, and economic sustainability. Initial applications of the KPA framework in the analysis of the most successful among several hundred poverty alleviation projects in the Philippines provide initial evidence of its explanatory and diagnostic usefulness. Some directions are indicated for development of program and project management toolkits to deliver better results.

In "Knowledge Loss: Managing Local Knowledge in Rural Uzbekistan" (Chapter 13), Hans-Dieter Evers and Caleb Wall critically discuss an important aspect of the discussion about knowledge, namely the loss of knowledge with special reference to indigenous and local knowledge in rural Uzbekistan. Indigenenous knowledge is passed on from generation to generation and is firmly grounded in the tradition of a group, community or society. Local knowledge is acquired through learning and adoption to local conditions. Knowledge that has been brought back from studying abroad, from reading, viewing or listening to mass media or adopted through other channels has become local knowledge, as soon as it has been adopted to local social and cultural conditions and is ready to be applied locally. Local knowledge constitutes a 'milieu' within a network of social interaction. Indigenous as well as local knowledge may be lost. The dialectics of knowledge production leads both to knowing and not-knowing. The more knowledge is produced the more we know what we don't know. A particular form of creating ignorance is knowledge loss. This process does not indicate that old knowledge is up-dated or replaced by new knowledge but the attrition of a stock of available and usable knowledge. In their paper the authors examine the root causes of knowledge loss and how knowledge seapage has occurred in rural Uzbekistan. Based on empirical field research on livestock production and post-harvest processing, the paper conceptualises knowledge loss, for example with regard to cattle farming know how, as evidence of poor knowledge management. As they stress, in its own right it is a failure of management and governance to allow knowledge resources, expensively produced within the community, to be lost. On a wider level, it evidences a lack of knowledge reproduction and retention, which can be seen as a result of excessive transaction costs and risks to knowledge sharing.

Chapter 14 by Solvay Gerke and Judith Ehlert entitled "Local Knowledge as Strategic Resource: Fishery in the Seasonal Floodplains of the Mekong Delta, Vietnam" puts emphasis on the crucial role of water in Vietnam's rural Mekong Delta in the pursuit of diverse live-lihood activities. The chapter examines the social interaction of paddy farmers and landless people during the annual flood season from a local knowledge perspective. It is argued that local knowledge turns into a strategic resource in the context of growing competition for decreasing natural water products. Knowledge is not like "light"; hence the strategic moment as well as the tacit component of local knowledge makes its sharing a technically limited form of interaction which is based on a delicate process of trust-building. As the authors stress, the current transformation process of agricultural modernization as pushed forward in the Mekong Delta will probably aggravate livelihood insecurities in the future. Local knowledge as adaptive asset in this process of changing society-water relationships will continue to be of strategic importance for survival and sustained development.

In "Integrating Natural and Social Sciences: Developing a Water Information System for the Mekong Delta" (Chapter 15), Thomas Menkhoff, Solvay Gerke and Hans-Dieter Evers describe how the WISDOM project, a joint programme between Vietnam and Germany, is seeking to devise sustainable solutions to life in the Mekong Delta by combining integrated watershed and knowledge management approaches.

Chapter 16 "Implementing Knowledge Management in a Geographically Dispersed Organization: The Case of an NGO" by Poornima Luthra and Shan L. Pan explores ongoing attempts by a non-government cultural relations organization to integrate knowledge across geographical boundaries to yield a competitive advantage. Particular emphasis is put on the internal realignment of knowledge vision and wider organizational structure as well as the challenges of creating a robust knowledge sharing culture so as to leverage on intellectual capital assets.

Conclusion

Knowledge has been widely recognised by economists as the most important factor of production in a "new economy". The production and utilisation of knowledge is therefore essential for development and the introduction of information and communication technology (ICT) is a precondition for developing a knowledge society. Countries, regions and populations are, however, divided, in terms of access to ICT. Our analyses suggest that the existing global digital divide and the knowledge gap are widening between developing countries and the industrial countries and within individual nations. Access to primary education and the acquisition of reading and writing skills is a basic human right and an internal digital divide between those that have access to further knowledge and others without access is unjust and not acceptable. Furthermore a civilisation needs "meta-narratives" as a common ground, an anchorage for basic cultural values, which have to be disseminated, known and accepted by all members of a society to avoid violent conflict, fundamentalisms of various kinds and alienation.

Some countries such as South Korea, Singapore and Malaysia among others have embarked on an ambitious plan to close the digital divide and to use knowledge as a base for economic development, by-passing earlier stages of industrialisation. Some commentators have, in contrast, asserted "that it is doubtful that the knowledge revolution will let developing countries leapfrog to higher levels of development" as "the knowledge economy will actually expand the gap between rich and poor" (Persaud, 2001). We argue that the digital divide, the knowledge gap and hidden knowledge traps are not natural phenomena. Global knowledge has to be localized and local knowledge utilized in developing a knowledge society.

It can be argued that knowledge gaps are a precondition to development and innovation, and that a knowledge gap will always be found between and within countries. As is the case with any unequal distribution of resources, development ethics and human rights issues have to be considered. The right to education and information should be safeguarded. A global standardisation or total commercialisation of knowledge under the guise of "relevance" is counterproductive to development.

Devaluation of local knowledge by globally operating experts as well as marketing strategies of large corporations are as much responsible for the widening knowledge gap as other factors of global development and governance. A comparison within world regions or with comparable countries will, however, reveal the effective advantage that can be gained by high investments in ICT combined with local knowledge production and dissemination.

Policy-makers in both developing and emerging markets should consider giving greater thought to the development of knowledge clusters. For that to happen, they would have to give up their predominantly economy-oriented approach and adopt a more comprehensive one. After all, a complex environment with government institutions, scientific institutions and civil society organisations is necessary to structure local knowledge development successfully. Above all, however, development policy would have to begin dealing once again with higher education, research, science and technology and related issues such as good knowledge governance.

References

Asian Development Bank (ADB) (2007). Moving toward knowledge-based economies. Technical Notes.

BMZ (2005). Der Beitrag Deutschlands zur Umsetzung der Millenniums-Entwicklungsziele. Bonn, Federal Ministry for Economic Cooperation and Development (BMZ), Materialien No. 140.

Chan Kim, W. and R. Mauborgne (1997). Value innovation: The strategic logic of high growth. *Harvard Business Review*, (January–February):103–112 (Reprint Number: 97108).

Chay, Y.W., T. Menkhoff, B. Loh and H.-D. Evers (2009). Social capital and knowledge sharing in knowledge-based organizations: An empirical study. In *Knowledge Management, Organizational Memory and Transfer Behavior: Global Approaches and Advancements*, M. E. Jennex (ed.)., Hershey, PA, USA: IGI Global.

Chen, J., P. Y. T. Sun and R. J. McQueen (2010). The impact of national cultures on structured knowledge transfer. *Journal of Knowledge Management*, 14(2):228–242.

Drucker, P. F. (1994). *Postcapitalist Society*. New York: Harper Business.

Ernst, D. (2003). Placing the networks on the internet: Challenges and opportunities for knowledge creation in developing Asia. In *Knowledge Creation in the Learning Economy*, B. A. Lundvall and K. Smith (Eds.). Cheltenham: Edward Elgar.

Evers, H.-D. (2003). Transition towards a knowledge society: Indonesia and Malaysia in comparative perspective, *Comparative Sociology*, 2(2):355–373.

Evers, H.-D. and S. Gerke (2005). Closing the digital divide: Southeast Asia's path towards a knowledge society. Working Paper No 1. Bonn, Centre for Development Research, University of Bonn.

Evers, H.-D., S. Gerke, and T. Menkhoff (2006). Little understood knowledge trap, *Development and Cooperation*, 33(6):246–247 (German version: "Unerkannte wissensfalle", *Entwicklung and Zusammenarbeit*, 33(6):246–247.

Foss, N. J. (2005). *Strategy, Economic Organization, and the Knowledge Economy — The Coordination of Firms and Resources*. New York: Oxford University Press.

Foss, N. J. and S. Michailova (2009). Knowledge governance: What have we learned? And where are we heading? In *Knowledge Governance: Perspectives, Processes and Problems*, N. J. Foss and S. Michailova (Eds.), pp. 272–288. Oxford: Oxford University Press.

Hornidge, A.-K. (2007). *Knowledge Society: Vision and Social Construction of Reality in Germany and Singapore*. Berlin and London: LIT Verlag.

Kogut, B. and U. Zander (1992). Knowledge of the firm, combination capabilities, and the replication of technology. *Organization Science*, 3:383–397.

Kuptsch, C. and E. F. Pang (2006). Competing for global talent. Geneva: International Labor Office: International Institute for Labour Studies.

Madsen, P.M. and V. Desai (2010). Failing to learn? The effects of failure and success on organizational learning in the global orbital launch vehicle industry. *The Academy of Management Journal (AMJ)*, 53(3):451–476.

Menkhoff, T. and S. Gerke (Eds.) (2002). *Chinese Entrepreneurship and Asian Business Networks.* Richmond: RoutledgeCurzon.

Menkhoff, T., Pang, E. F. and Evers, H.-D. (2008). The power of knowing — Studies of Chinese business in Asia. *Journal of Asian Studies*, 22–23(1–3), Special Issue.

Menkhoff, T. (2010). Creating value through knowledge management. Business Times (Singapore), June 15, p. 24.

Menkhoff, T., H.-D. Evers, and Y. W. Chay (Eds.) (2010). *Governing and Managing Knowledge in Asia*. New Jersey: World Scientific Publishing (2nd edition).

Menkhoff, T., Y. W. Chay, and B. Loh (2004). Notes from an 'Intelligent Island': Towards strategic knowledge management in Singapore's small business sector. *International Quarterly for Asian Studies*, 35(1–2):85–99.

Mowery, D.C., Oxley, J. and B. Silverman (1996). Strategic alliances and interfirm knowledge transfer. *Strategic Management Journal*, 17:77–91.

Neo Boon Siong and Chen, G. (2007). *Dynamic Governance — Embedding Culture, Capabilities and Change in Singapore.* New Jersey: World Scientific Publishing.

Pang, E. F. (2006). Foreign talent and development in Singapore. In *Competing for Global Talent*, C. Kuptsch and E. F. Pang (Eds.), pp. 155–170. Geneva: International Labor Office: International Institute for Labour Studies.

Pan Shan-Ling ed. (2004). *Managing Strategic Enterprise Systems and E-Government Initiatives in Asia — A Casebook.* New Jersey: World Scientific Publishing.

Persaud, A. (2001). The knowledge gap. *Foreign Affairs*, 80(2):107–117.

Pinch, S., N. Henry, M. Jenkins and S. Tallman (2003). From 'Industrial Districts' to 'Knowledge Clusters': A model of knowledge dissemination and competitive advantage in industrial agglomerations. *Journal of Economic Geography*, 3:373–388.

Porter, M. (2000). Location, competition, and economic development: Local clusters in a global economy. *Economic Development Quarterly*, 14:15–34.

Stehr, N. (2001). A world made of knowledge. *Society*, 39(1):89–92.

Stehr, N. (ed.) (2004). *The Governance of Knowledge.* Transaction Publishers.

Straits Times May 11 (2010). Bringing life to a high-tech ghost town (p. A14).

Suh, J. and D. H. C. Chen (2007). Korea as a knowledge economy: Evolutionary process and lessons learned. The World Bank, Washington D.C.

Talisayon, S. D. (2008). Knowledge management in Asia: Experience and Lessons. Report of the APO Survey on the Status of Knowledge Management in Member Countries. Tokyo: APO.

Tallman, S., M. Jenkins, N. Henry and S. Pinch (2004). Knowledge clusters and competitive advantage. *Academy of Management Review*, 29:258–71.

Tsang, Eric W. K. (2002). Learning from overseas venturing experience: The case of Chinese family business. *Journal of Business Venturing*, 17:21–40.

Toh, M. H. *et al.* (2002). Mapping Singapore's knowledge-based economy. *Economic Survey of Singapore*, Singapore, pp. 56–75.

Weingart, P. (2002). The moment of truth for science. *EMBO Reports*, 3(8): 703–706.

Willke, H. (2007). *Smart Governance. Governing the Global Knowledge Society.* Frankfurt, New York: Campus.

Wirtz, J., L. Heracleous, and T. Menkhoff (2008). Value creation through strategic knowledge management: The case of Singapore Airlines. In Menkhoff, T., E. F. Pang, and H.-D. Evers (Eds.) *The Power of Knowing: Studies of Chinese Business in Asia*, *Journal of Asian Business*, 22/23(1–3):249–263.

World Bank (1999). *Knowledge for Development: World Development Report 1998/1999.* New York: Oxford University Press.

Yli-Renko, H., E. Autio, and H. J. Sapienza (2001). Social capital, knowledge acquisition, and knowledge exploitation in young technology-based firms. *Strategic Management Journal*, 22:587–613.

Zeleny, M. (2010). Knowledge of enterprise: KM or knowledge technology? In *Governing and Managing Knowledge in Asia, Series on Innovation and Knowledge Management*, T. Menkhoff *et al.* (Eds.), Vol. 3, pp. 23–57. New Jersey: World Scientific Publishing.

Part 1

Conceptualizing and Strategizing Knowledge as an Engine for Growth

Chapter

1

Knowledge Hubs and Knowledge Clusters

A Knowledge Architecture for Development

Hans-Dieter Evers, Solvay Gerke and Thomas Menkhoff

1. Introduction: The Devaluation of Space and the End of Industrial Agglomeration?

With the advent of globalisation and knowledge-based production, firms now cooperate on a global scale, outsource parts of their administrative or productive units and negate location altogether. Geographical space has been theoretically downgraded and proximity or distance devalued (Brown and Duguid, 2002). Furthermore, rapid advances in ICT have enabled the emergence of global production networks (Coe *et al.*, 2004), outsourcing, just-in-time production, high-level manpower migration (Fallick *et al.*, 2006), and global "head hunting" for managers and engineers.

Globalisation theorists such as Saskia Sassen (Sassen, 1991) have proclaimed the existence of a "global city", consisting of CBDs (central business districts) in major cities worldwide, amalgamated into a huge global city that is welded together by intense electronic communication, a shared common language and the universal corporate culture of a capitalist world economy. The extremely low transaction costs of data, information and knowledge seem to invalidate the theory of agglomeration and the spatial clustering of firms (James, 2005), going back to the classical work by Alfred Weber and Alfred Marshall, who emphasised the microeconomic benefits of industrial collocation (Weber, 1909).

Despite this compelling theoretical argument, empirical reality demonstrates a different picture; industries well versed in ICT, outsourcing and cooperation via the internet still tend to cluster and form industrial agglomerations. "Contrary to the views of some economists, most categories of knowledge and information are not a ubiquitously available public good. Rather, they display large spatial disparities which show a remarkable historical persistence" (Meusburger, 2000:352). More recently, Carillo, Erkazakis and others have emphasised the concentration of knowledge-intensive institutions in urban areas and coined the concept of the "knowledge city" (Carrillo, 2006; Ergakakis *et al.*, 2006), a notion which, according to Carillo, is still in a "pre-paradigmatic stage" and in need of further refinement (Carrillo, 2004:33). Proximity increases a company's innovative capacity when it can share ideas, products and services with other firms operating in the same field, examples of which are Silicon Valley, the Hyderabad IT cluster, the Munich high-tech zone and the ABC (Aachen-Bonn-Cologne) cluster in Germany, the MSC in Malaysia, Biopolis and adjacent areas in Singapore, and many others. In short, it is exactly innovative non-material production, applied research and knowledge-based manufacturing that tend to cluster in specific locations. Was Durkheim wrong when he asserted that "the truths of science are independent of any local context" (Durkheim, 1972:88)? In a modern global context the question then arises, why do knowledge-based industries form clusters rather than making use of ICT to connect diverse locations worldwide?

Following the recent trend in recognising knowledge as a factor of production, cluster research has increasingly turned away from an emphasis on agglomeration economics and the minimisation of transaction costs.

Michael Porter, in his well-known study *The Competitive Advantage of Nations*, produced a "diamond of advantage" to explain why clusters develop (Porter, 1990).

This diamond consisted of the following elements:

- *Factor conditions* — a region's endowment of production factors including human, physical, knowledge, capital resources, and infrastructure, which make it more conducive to success in a given industry.
- *Demand conditions* — the nature of home demand for a given product or service, which can pressurise local firms to innovate faster.
- *Related and supporting industries* — networks of buyers and suppliers transacting in close proximity to foster active information exchange, collective learning and supply chain innovation.
- *Firm strategy, structure, and rivalry* — a climate that combines both intense competition among localised producers, with cooperation and collective action on shared needs, making it fertile for innovation and regional competitive advantage (Porter, 2000, 1990).

His widely accepted view was recently challenged by Henry and Pinch (2006). They argued that more important are "the competitive advantages secured by firms through gaining rapid access to knowledge concerning the innovations, techniques and strategies of competitor firms" (Henry and Pinch, 2006:114). In view of the high ICT capabilities of high-tech firms, this argument reveals only half the truth. Why is rapid access to knowledge not gained through video conferencing, networking with other technical staff through the World Wide Web, accessing data banks that could be located anywhere on the globe, via chat rooms on the internet, or just using old-fashioned telephone connections? All these modern means of communication are used to negate geographical distance by allowing ad hoc communication within seconds; yet, still, high-tech firms and knowledge-based industries show an avid tendency to cluster in geographical space. Why should this be the case?

2. Types of Knowledge: A Revised Nonaka Thesis

To answer this question we have to go back to the basics of knowledge management.

In their much cited work, Nonaka and Takeuchi distinguish between tacit and explicit knowledge (Nonaka and Takeuchi, 1995). Tacit knowledge is, in essence, experience gained through action, and explicit knowledge refers to knowledge stored and made available in books, databanks or other types of media. Maintaining competence within an organisation, despite a high turnover of employees through either retirement or retrenchment, poses a major management challenge, as tacit knowledge is lost. Michel Polanyi, in an earlier work, emphasised that tacit knowledge is based primarily on doing rather than cognition. A person can therefore "do" more than he or she "knows" (Polanyi, 1967). Moreover, Botkin and Seeley estimate that eighty per cent of knowledge is tacit (Botkin and Seeley, 2001). One of the most difficult tasks of knowledge management is therefore to facilitate the transfer of tacit knowledge into explicit knowledge or to transfer personal into organisational knowledge, e.g., turning a firm or government agency into an intelligent learning organisation.

The conversion of tacit to explicit knowledge is difficult and provides an essential challenge to the practice of knowledge management. The best way to transmit tacit knowledge or experience is still by observation, face-to-face contacts and learning from doing. Routine work can easily be outsourced, but innovative, knowledge-based work needs teamwork and the existence of communities of practice, frequent social interaction and capacity building by direct face-to-face learning. This line of argument eventually leads to the hypothesis that:

> "the transfer of tacit knowledge is a major factor in the emergence of knowledge clusters. The more important tacit knowledge is for production, the more localised production is likely to be" (knowledge transfer hypothesis).

There is, up to now, only some empirical evidence to support our "knowledge transfer hypothesis", but the fact remains that clusters are still emerging and keep going by banking on their competitive advantage. We believe that our hypothesis holds both for pre-industrial handicraft manufacturing as well as for modern research and development work and knowledge-based production. Pre-modern handicraft production tended to be clustered in special quarters or streets (Enright, 2003:100). The craft quarters in medieval European cities or the Hang (merchandise) streets in the Hoan Kiem district of Hanoi were, indeed, knowledge clusters driven by the transfer of the expertise and experience of master craftsmen to their apprentices, as well as through keen observation of practices in neighbouring shops. Imitation of successful competitors and early access to crucial information is conducive to clustering (Meusburger, 2000:259); observations of competitors' practices rather than blind market forces of supply and demand appear to be the most salient factors driving economic processes in this context. This insight has also been used to argue for a sociological theory of markets and prices (Evers and Gerke, 2007; Fligstein, 2002; White, 1981).

A fair number of relevant studies have provided empirical evidence that proximity and face-to-face interaction indeed facilitate the transfer of tacit knowledge and together form a decisive asset in the emergence of knowledge hubs. As Zook argues, "Marshallian interaction and observation retain their relevance even within the decidedly global financial and internet industries" (Zook, 2004). One study in modern Italy, for example, examined the approaches used in determining communication and innovation in Italian technological districts in order to identify their distinctive features and provide a framework for empirical analysis (Antonelli, 2000). The study found that clusters cannot rely solely on agglomeration for their success, but develop differently due to different knowledge sharing and research and development chances.

This view is contested by Håkanson, who raises doubts that privileged access to "tacit knowledge" alone provides competitive advantages that cause the growth and development of both firms and regions (Håkanson, 2005). His point is indeed acceptable, insofar as tacit knowledge is always embedded in cultural and social contexts that need to be taken into account together with market conditions.

Menkhoff *et al.* (2010) studied knowledge in science parks and found that intense ethnically-based interaction played a decisive role in the dynamics of knowledge hubs. Similarly, close interaction in socially diverse communities of practice was more productive than homogeneous knowledge hubs.

A study on rural areas in the US emphasises the importance of local actors and argues that "rural knowledge clusters are specialized networks of innovative, interrelated firms… deriving competitive advantages primarily through accumulated, embedded and imported knowledge among local actors about highly specific technologies, processes and markets" (Munnich *et al.*, 2002). Another US-wide study concludes that tacit knowledge is an important factor in creating innovation (Audretsch and Feldman, 1996).

In a different social arena, empirical studies by Karin Knorr-Cetina in high-tech research laboratories have shown that face-to-face interactions between scientists, inside and outside the laboratory, have a decisive impact on the "manufacture" of knowledge (Knorr Cetina, 1981). Knowledge production is always a social process that requires interaction, and may take place, to a certain extent, in cyberspace; however, innovation and discovery are also driven by emotions, fun and anger, excitement and frustration, which are projected at persons in direct interaction. Emotions are a less studied, but nevertheless important, enabler (or hindrance) of knowledge sharing (Chay *et al.*, 2010).

From these studies we can conclude that, whereas industrial clusters have gained their competitive advantage primarily from a reduction of transaction costs (Iammarino and McCann, 2006), knowledge clusters emerge primarily through a direct transfer of tacit knowledge.

3. Knowledge Architecture

The marshalling of tacit knowledge and the use of proximity (Boschma, 2005) for competitive gains needs a specific institutional frame, a specific "knowledge architecture" (Evers *et al.*, 2003). In a social science context, Fligstein uses the term 'architecture' to describe the interrelation between markets and governments (Fligstein, 2002). In ICT research, the term 'architecture' "typically describes how the system or program is constructed, how it fits together, and the protocols and interfaces used for communication and cooperation among modules or components of the system" (www.courts.state.ny.us/ad4/LIB/gloss.html). "IT architecture is a design for the arrangement and interoperation of technical components that together provide an organisation of its information and communication infrastructure" (http://www.ichnet.org/glossary.htm). An effective ICT architecture is by now the backbone of knowledge clusters in knowledge-based societies, but the impact of different architectures or ICT regimes on knowledge flows is unknown, except for the fact that ICT speeds up communication. "Thus, clusters are made up not only of physical flows of inputs and outputs, but also include the intense exchange of business information, know-how and technological expertise, both in traded and untraded forms" (Sölvell, 2009:15).

Pinch *et al.* have drawn attention to the fact that "agglomerations may develop a cluster-specific form of architectural knowledge that facilitates the rapid dissemination of knowledge throughout the cluster by increasing the learning capacity of proximate firms and thereby conferring cluster-specific competitive advantages" (Pinch *et al.*, 2003:373). In line with this argument, we define the ***knowledge architecture*** of a knowledge cluster as:

> "the institutions of communication and the type and intensity of knowledge flows (knowledge sharing), based on the formal and informal interaction between persons and organisations."

Steven Pinch has described the characteristics of architectural knowledge, which "tend[s] to be specific to, or embedded in, particular organisations within which it evolves endogenously over time in a complex trajectory... architectural knowledge is highly path dependent... and tacit in character... Crucially, architectural knowledge is also essential in determining the capacity of organisations to acquire, assimilate and adopt new knowledge" (Henry and Pinch, 2006). What holds true for individual organisations can also be applied to a knowledge hub within a large corporation, or consisting of several smaller organisations. In short, the knowledge architecture is a crucial determinant for the innovative capacity of firms, knowledge hubs and, indeed, the whole knowledge cluster.

As the knowledge architecture is basically "tacit" in character, tacit knowledge transfer is an essential factor in the emergence of knowledge hubs, as we have argued in the "knowledge transfer hypothesis" above. A knowledge architecture emerges on the basis of knowledge (Chay *et al.*, 2010; Chay *et al.*, 2007), and knowledge about the knowledge architecture within a cluster or within a firm provides a competitive advantage for persons in the know, as well as for intelligent firms in comparison to organisations outside a cluster. Architectural knowledge, nevertheless, must be distinguished from "component knowledge", which is "normally tied to the technology of the industry, is relatively coherent and definable, and is usually acontextual" (Tallman *et al.*, 2004:264). Component knowledge can easily be shared with experts in the same field or transmitted to organisations. Architectural knowledge, like organisational or managerial processes is, however, more difficult to pass on, as it evolves as an inseparable part of a firm and is therefore contextualised (Tallman *et al.*, 2004:265).

Knowledge flows and knowledge depositories constitute the knowledge architecture of an organisation or a cluster of organisations. Knowledge architecture is therefore a property of an organisation or cluster, the argument for which may be supported from the vantage point of sociological systems theory (Luhmann, 1984). As Helmut Willke has argued, the intelligence of an organisation is more than the sum of the knowledge of its members. The knowledge of organisations is, indeed, different from personal knowledge, because "organisational or institutional knowledge resides in de-personalized, anonymous rule systems" (Willke, 2007:113) and, we would argue, its knowledge architecture. In a modern knowledge society, Willke argues, large organisations tend to be more knowledgeable, more intelligent than individuals. No single individual is capable of building a modern airplane (Willke, 2007:114); it needs organisational intelligence to accomplish this task and, additionally from our point of view, industrial clusters and knowledge hubs as well.

4. K-Clusters and K-Hubs

Most of the current literature does not draw a distinction between knowledge clusters and knowledge hubs. Policy statements in particular use both terms arbitrarily. We feel that turning these terms into different analytical concepts would enhance our understanding of spatial processes. The most general concept would be 'agglomeration', i.e., clusters are agglomerations with 'proximity' as a crucial variable. Henry and Pinch use the terms 'agglomeration' and 'cluster' synonymously "to refer to geographical groupings of firms (both large and small but often SMEs), broadly in the same sector, but extending beyond to incorporate greater parts of the value chain" (Henry and Pinch, 2006:117). This definition is in line with that used by the Cluster Mapping Project of the Harvard Center for Strategy and Competitiveness: "Clusters are geographically concentrated groups of interconnected companies, universities, and related institutions that arise out of linkages or externalities across industries" (https://secure.hbs.edu/isc/index.jsp).

The cluster concept emphasises the organisational aspect of agglomerations, while the term 'hub' refers to the knowledge sharing and dissemination aspect. A more precise definition reads as follows:

> Knowledge clusters are agglomerations of organisations that are production-oriented. Their production is primarily directed to knowledge as output or input. Knowledge clusters have the organisational capability to drive innovations and create new industries. They are central places within an epistemic landscape, i.e. in a wider structure of knowledge production and dissemination. Examples of organisations in knowledge clusters are universities and colleges, research institutions, think-tanks, government research agencies, and knowledge-intensive firms.

Knowledge hubs may exist in the same locations as knowledge clusters and may be nested within them:

> Knowledge hubs are local innovation systems, nodes in networks of knowledge production and knowledge sharing. They are characterised by high connectedness and high internal and external networking and knowledge sharing capabilities. As meeting points of communities of knowledge and interest, knowledge hubs fulfill three major functions: to generate knowledge, to transfer knowledge to sites of application, and to transmit knowledge to other people through education and training.

Knowledge hubs are always nodes in networks of knowledge dissemination and knowledge sharing within and beyond clusters. Their knowledge architecture shows specific characteristics that can be made apparent in empirical studies. As a study of the wine industries in Italy and Chile has shown, firms with a strong knowledge base are more likely to exchange innovation-related knowledge with other firms. However, this is considered to occur only among firms whose cognitive distance is not too high. "This may explain the formation of densely connected cohesive subgroups and the emergence of local knowledge communities" (Giuliani, 2007:163), or, in our terminology, to the formation of knowledge hubs.

With the development of the World Wide Web, a new architecture was introduced by leaving core resources of the internet in a "commons", which was "built into the very architecture of the original network" and was decisive for the innovation and creativity spurned by the internet (Lessig, 2004:227–228). Despite the wide use of common knowledge on the internet, communication is still concentrated within organisations and knowledge hubs (see Table 1). E-mail communication is supplemented by attendance at formal meetings, discussion groups, and informal chats in coffee rooms or canteens, mostly within an organisation, but occasionally also at conferences. It is characteristic of knowledge hubs that other knowledge hubs are also accessed and knowledge is shared throughout a knowledge network. Further, the resilience and

Table 1: Concepts

Concept	Short Definition	Measurement (Examples)
K-cluster	Agglomerations of organisations emphasising knowledge as output or input	Number of organisations per location
K-hub	Local innovation systems that are nodes in networks of knowledge production and knowledge sharing	Number of knowledge workers and their products (patents, papers, software)
K-architecture	The structures and institutions of communication and the related type and intensity of knowledge flow	ICT governance regimes, regular meetings, k-sharing incentives
Epistemic landscape	Areas of high or low knowledge intensity	Regional R&D expenditure, location of k-clusters and k-hubs

strength of a knowledge hub seems to rest in its connectivity, based on strong internal and external ties. As one always needs existing knowledge to acquire and use new knowledge, organisations with a low level of knowledge assets will seek consultancy services elsewhere, rather than joining an emerging knowledge hub and engaging in knowledge sharing.

To visualise a complex matter in simple terms, we may say that clusters are most visible as an agglomeration of organisations, buildings and hubs as a community of knowledge sharing and knowledge producing people.

The concepts discussed above are summarised in Table 1.

Knowledge clusters and knowledge hubs show distinctive *knowledge architectures*. Countries or regions exhibit *epistemic landscapes* of knowledge assets, structured by knowledge clusters, knowledge hubs, knowledge gaps and areas of high or low knowledge intensity. The emergence of epistemic landscapes will be demonstrated in the following section.

5. Epistemic Landscapes

Epistemic landscapes develop over long periods of time. They are seldom shaped by individual actors, but more often by the collective action of strategic groups (Evers and Gerke, 2009). Firms connected by a common interest to capitalise on the competitive advantage of clustering have an impact on epistemic landscapes through their location decisions. Moreover, government strategies to develop knowledge-based societies and economies have often been decisive in shaping epistemic landscapes — relevant development policies have been assessed in detail elsewhere for Malaysia and Indonesia (Evers, 2003), the Straits of Malacca region (Evers and Hornidge, 2008), Vietnam (Evers and Bauer, 2010), and Singapore and Germany (Hornidge, 2007). Developing industrial regions, clusters or knowledge hubs are, indeed, standard practice in many regional planning departments around the world.

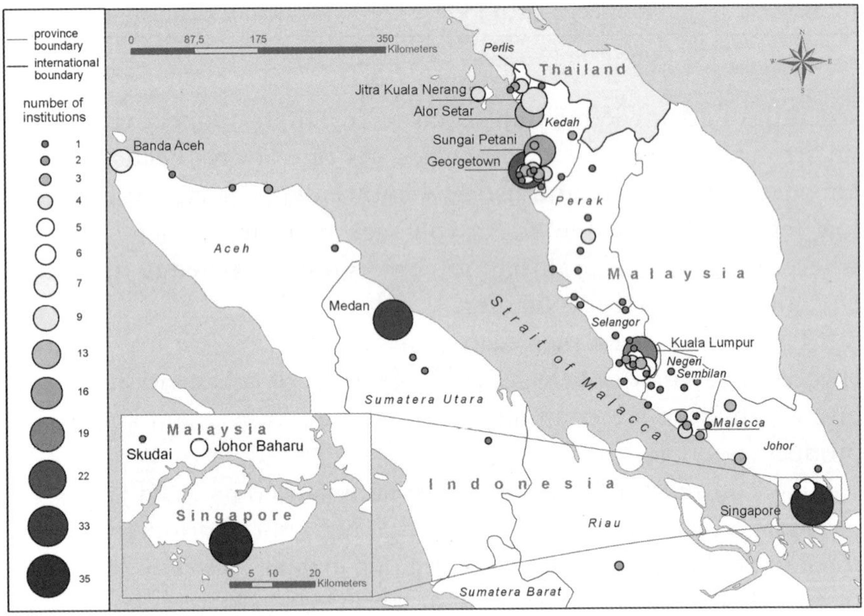

Figure 1: Epistemic landscape — The straits of Malacca region

Source: Own data, showing the location of knowledge producing organisations such as universities and research institutes (Evers and Hornidge, 2007).

In this context, we define epistemic landscapes in a geographical sense, i.e., we refer to the spatial distribution of knowledge assets within a predefined region. The term is not yet standard scientific terminology, although it has been used in different contexts. One line of argument refers back to Bacon and 18th century 'encyclopaedism', and defines an epistemic landscape as depicting a synthesis of knowledge (Wernick, 2006). In Weisberg and Muldoon's study, a single epistemic landscape corresponds to the research topic that engages a group of scientists. Agent-based modelling with NetLogo software is used to model the changing epistemic landscape according to the research strategies of participating scientists (Weisberg and Muldoon, 2007). In our study, we follow a slightly different path and focus on the development strategies of governments, strategic groups, firms, and research institutes and their success in shaping the epistemic landscape of a region.[1] The allocation of human and financial resources creates knowledge assets, which can be measured, mapped and made to depict the contours of an epistemic landscape.

6. Towards a New Architecture of Knowledge for Development

Asian governments, as well as international development agencies, are increasingly relying on knowledge as a factor of production (ADB, 2005; Gerke and Evers, 2006:2–3; Gerke *et al.*, 2005; Hornidge, 2007:4–10, 62–65). In 2003, the Asian Development Bank identified knowledge as the most important resource in maintaining the region's competitiveness, given the rapid rate of change created by globalisation and technological innovation. Besides banking on an increased transfer of knowledge through FDI, as well as increased investment in education and R&D, experts are advocating the creation

[1] This refers to ongoing research on knowledge management and knowledge governance in the water sector of the Mekong Delta (WISDOM project http://www.zef.de/1052.0.html), carried out jointly by the Center for Development Research (ZEF), University of Bonn, the Southern Institute of Social Sciences, HCMC and The Mekong Development Research Institute of Can Tho University.

of knowledge hubs as the incubators of future economic development. Japan's Ministry of Education, Culture, Sports, Science and Technology (MEXT) launched a programme in 2003 to set up knowledge clusters throughout Japan. In their terms, knowledge clusters are described as follows: "A 'Knowledge Cluster' is a local innovation system organised around universities, research institutions and firms which have unique R&D themes and potentialities".[2]

In 2006, the Asian Development Bank announced a programme to develop knowledge hubs in selected developing countries throughout the Asian and Pacific region to support and strengthen research and disseminate new development concepts and technologies (ADB, 2005). Since 2006, ADB has supported Tsinghua University in Beijing in establishing a regional knowledge hub on climate change. The knowledge hub is being established under an ADB grant and the contribution of a great deal of expertise, which together are enabling the setting-up of centres of excellence in the region to support and strengthen research and disseminate new and emerging concepts and technologies. Other centres are planned in Thailand and India, strengthening and supplementing the already existing knowledge hubs.

> "These knowledge hubs should aim to mainstream new concepts in innovation, science, technology, management development, and related fields for the region. They should also promote improved exchange of data, information, and knowledge; and increase the capabilities of institutions and organisations in the region. Initiatives have created a wealth of knowledge base and expertise throughout the region. However, the capabilities of regional organisations and institutes in disseminating and sharing their findings are limited. Information is not enriched through regional cooperation, and information and expertise bases largely remain scattered around the region and fail to provide the multiplier effect that could be achieved if it were nurtured with more support for regional knowledge exchange. As the knowledge hub will focus on new development topics, experience and lessons learned from ADB knowledge sharing initiatives such as the Consultative Group on International Agricultural Research (CGIAR)

[2] See http://www.mext.go.jp/a_menu/kagaku/chiiki/cluster/h16_pamphlet_e/01.pdf.

centre of excellence will be applied in the establishment of the knowledge hubs" (ADB, 2005:2).

Singapore and Malaysia have followed a similar policy of designating specific areas to house knowledge clusters and identifying special areas of research and development to set up knowledge hubs. We have analysed elsewhere the strategies to develop knowledge clusters in the Straits of Malacca region in greater detail (Evers *et al.*, 2008), in Indonesia (Evers, 2003), Malaysia (Evers, 2003; Evers, 2004a; Evers, 2004b) and Singapore (Evers, 2003; Hornidge, 2007; Menkhoff *et al.*, 2009). So far, these development policies have been fairly successful, although it should be noted, however, that the emergence of knowledge clusters and knowledge hubs has been embedded in a wider epistemic landscape. Knowledge capital was created by supporting colleges, universities, research institutes and centres of applied research and development, and tacit knowledge imported through the immigration of foreign talent and overseas training schemes. This facilitated the leverage of an important principle of knowledge management, namely that knowledge is required to use and create more knowledge. This also entails deleting barriers to knowledge flows, building an ICT backbone, increasing knowledge assets, closing knowledge gaps, and developing a legal infrastructure that allows and encourages creative and diverse knowledge production. Without the thorough implementation of a knowledge architecture, as well as an epistemic landscape, the successful development of a knowledge-based economy and society will be highly unlikely.

7. Conclusions

The combination of geographical knowledge mapping and the design of epistemic landscapes culminate in a tool that can visualise the distribution of knowledge assets. A look at an epistemic landscape will show us the knowledge clusters, the gaps, valleys and heights of knowledge assets within a predefined region. As in poverty mapping, it will allow a more precise targeting of development measures. In this sense, knowledge mapping is a planning tool, as it will also prove helpful to assess the impact of development measures in the fields of education, research and development and communication. If information or decision support systems are installed, epistemic landscapes will highlight the availability of certain areas to receive information and implement development programmes. We also suggest that the mapping of epistemic landscapes is a precondition for the successful implementation of a sustainable knowledge architecture for development.

References

ADB, Asian Development Bank (2005). Technical assistance. Establishment of regional knowledge hubs. Manila: ADB, Asian Development Bank.

Antonelli, C. (2000). Collective knowledge communication and innovation: The evidence of technological districts. *Regional Studies*, 34:535–547.

Audretsch, D. and M. P. Feldman (1996). R&D Spillovers and the geography of innovation and production. *American Economic Review*, 86:630–640.

Boschma, R. (2005). Role of proximity in interaction and performance: Conceptual and empirical challenges. *Regional Studies*, 39:41–45.

Botkin, J. and C. Seeley (2001). The knowledge management manifesto: Why KM requires community-building. *Knowledge Management Review*, 3.

Brown, J. S. and P. Duguid (2002). Local knowledge — Innovation in the networked age. *Management Learning*, 33:427–437.

Carrillo, F. J. (2004). Capital cities: A taxonomy of capital accounts for knowledge cities. *Journal of Knowledge Management*, 8:28–46.

Carrillo, F. J. (2006). *Knowledge Cities: Approaches, Experiences and Perspectives.* Butterworth-Heinemann.

Chay, Y. W., T. Menkhoff, B. Loh, and H.-D. Evers (2010). What makes knowledge sharing in organisations tick? An empirical study. In *Governing and Managing Knowledge in Asia*, T. Menkhoff, H.-D. Evers, and Y. W. Chay (Eds.), pp. 301–325. Singapore: World Scientific Publishing.

——. (2007). Social capital and knowledge sharing in knowledge-based organisations: An empirical study. *International Journal of Knowledge Management*, 3:37–56.

Coe, N., M. Hess, H. W. C. Yeung, P. Dicken, and J. Henderson (2004). 'Globalizing' regional development: A global production networks perspective. *Transactions of the Institute of British Geographers*, New Series, 29:468–484.

Durkheim, E. (1972). *Selected Writings, Translated and Edited by Anthony Giddens.* Cambridge: Cambridge University Press.

Enright, M. J. (2003). Regional clusters: What we know and what we should know. In *Innovation Clusters and Interregional Competition*, J. Bröcker, D. Dohse, and R. Soltwedel (Eds.), pp. 99–129. Berlin, Heidelberg, New York: Springer-Verlag.

Ergakakis, K., K. Metaxiotis, J. Psarras, and D. Askounis (2006). A unified methodological approach for the development of knowledge cities. *Journal of Knowledge Management*, 10:65–78.

Evers, H.-D. (2003). Transition towards a knowledge society: Malaysia and Indonesia in comparative perspective. *Comparative Sociology*, 2:355–373.

——. (2004a). Knowledge society and the knowledge gap. In *Globalisation, Culture and Inequalities*, Abdul Rahman Embong (ed.), pp. 301–316. Bangi: Penerbit Universiti Kebangsaan Malaysia.

——. (2004b). The path towards a Malaysian knowledge society. In *Public Policy, Culture and the Impact of Globalisation in Malaysia*, Mohd Hazim Shah and Kai Lit Phua (Eds.), pp. 21–39. Kuala Lumpur: Persatuan Sains Sosial Malaysia (Malaysian Social Science Association).

Evers, H.-D. and S. Gerke (2007). Social and cultural dimensions of market expansion. *ZEF Working Paper Series No. 24*. Bonn: Department of political and cultural change, center for development research, University of Bonn.

——. (2009). Strategic group analysis. *ZEF Working Paper Series*, 34.

Evers, H.-D., S. Gerke, and A.-K. Hornidge (eds.) (2008). *The Straits of Malacca: Knowledge and Diversity*. Berlin and London: LIT Verlag.

Evers, H.-D. and A.-K. Hornidge (2007). Knowledge hubs in the straits of malacca. *Asia Europe Journal*, 5:417–433.

——. (2008). The straits of malacca as a knowledge corridor. In *The Straits of Malacca: Knowledge and Diversity. ZEF Development Studies Vol. 8*, S. Gerke, H.-D. Evers, and A.-K. Hornidge (Eds.), pp. 41–62. Berlin, London, Penang: LIT Verlag/Straits G.T.

Evers, H.-D., M. Kaiser, and C. Müller (2003). Entwicklung durch Wissen: Eine neue globale Wissensarchitektur. *Soziale Welt*, 54:49–70.

Fallick, B., C. A. Fleischman and J. B. Rebitzer (2006). Job-hopping in silicon valley: Some evidence concerning the microfoundations of a high-technology cluster. *Review of Economics and Statistics*, 88:472–481.

Fligstein, N. (2002). *The Architecture of Markets: An Economic Sociology of Twenty-First-Century Capitalist Societies*. Princeton and Oxford: Princeton University Press.

Gerke, S. and H.-D. Evers (2006). Globalizing local knowledge: Social science research on Southeast Asia, 1970–2000. *SOJOURN: Journal of Social Issues in Southeast Asia*, 21:1–21.

Gerke, S., H.-D. Evers, and R. Schweisshelm (2005). Wissen als Produktionsfaktor: Südostasiens Aufbruch zur Wissensgesellschaft. *Soziale Welt*, 55:39–52.

Giuliani, E. (2007). The selective nature of knowledge networks in clusters: Evidence from the wine industry. *Journal of Economic Geography*, 7:139–168.

Håkanson, L. (2005). Epistemic communities and cluster dynamics: On the role of knowledge in industrial districts. *Industry & Innovation*, 12:433–463.

Henry, N. and S. Pinch (2006). Knowledge and clusters. In *Clusters and Globalisation*, C. Pitelis, R. Sugden and J. R. Wilson (Eds.), pp. 114–132. Cheltenham: Edgar Elgar.

Hornidge, A.-K. (2007). *Knowledge Society: Vision and Social Construction of Reality in Germany and Singapore*. Berlin and London: LIT Verlag.

Iammarino, S. and P. McCann (2006). The structure and evolution of industrial clusters: Transactions, technology and knowledge spillovers. *Research Policy*, 35:1018–1036.

James, A. (2005). Demystifying the role of culture in innovative regional economies. *Regional Studies*, 39:1197–1216.

Knorr Cetina, K. (1981). *The Manufacture of Knowledge: An Essay on the Constructivist and Contextual Nature of Science*. Oxford: Pergamon Press.

Lessig, L. (2004). The innovation commons. In *The Governance of Knowledge*, N. Stehr (ed.), pp. 227–239. London and New Brunswick: Transaction Publishers.

Luhmann, N. (1984). *Soziale Systeme. Grundriß Einer Allgemeinen Theorie*. Frankfurt am Main: Suhrkamp.

Menkhoff, T., H.-D. Evers, M. W. Meyer and L. M. H. Lim (2010). Building vibrant science and technology parks with knowledge management: Trends in Singapore. In *Governing and Managing Knowledge in Asia*, T. Menkhoff, H.-D. Evers, and Y. W. Chay (Eds.), pp. 187–220. London and Singapore: World Scientific Publishing.

Menkhoff, T., S. Gerke, H.-D. Evers, and Y. W. Chay (2009). Wissen und Entwicklung in Singapur. *ASIEN — The German Journal on Contemporary Asia*, 112:31–52.

Meusburger, P. (2000). The spatial concentration of knowledge. Some theoretical considerations. *Erdkunde*, 54:352–364.

Munnich, Lee W. Jr., G. Schrock, and K. Cook (2002). Rural knowledge clusters: The challenge of rural economic prosperity. *Reviews of Economic Development Literature and Practice, University of Minnesota*, 12.

Nonaka, I. and H. Takeuchi (1995). *The Knowledge-Creating Company: How Japanese Companies Create the Dynamics of Innovation*. New York: Oxford University Press.

Pinch, S., N. Henry, M. Jenkins, and S. Tallman (2003). From 'Industrial Districts' to 'Knowledge Clusters': A model of knowledge dissemination and competitive advantage in industrial agglomerations. *Journal of Economic Geography*, 3:373–388.

Polanyi, M. (1967). *The Tacit Dimension*. New York: Anchor Books.

Porter, M. (2000). Location, competition, and economic development: Local clusters in a global economy. *Economic Development Quarterly*, 14:15–34.

Porter, M. (1990). *The Competitive Advantage of Nations.* New York: The Free Press.

Sassen, S. (1991). *The Global City.* Princeton, NJ: Princeton University Press.

Sölvell, Ö. (2009). *Clusters — Balancing Evolutionary and Constructive Forces.* Stockholm: Ivory Tower Publishers.

Tallman, S., M. Jenkins, N. Henry, and S. Pinch (2004). Knowledge clusters and competitive advantage. *Academy of Management Review*, 29:258–271.

Weber, A. *Über den Standort der Industrien. Erster Teil: Reine Theorie des Standorts.* Tübingen: Verlag JCB Mohr.

Weisberg, M. and R. Muldoon (2007). Epistemic landscapes and the division of cognitive labor. Unpublished manuscript Pittsburgh. Penn: University of Pennsylvania.

White, H. C. (1981). Where do markets come from? *American Journal of Sociology*, 87:517–547.

Willke, H. (2007). *Smart Governance. Governing the Global Knowledge Society.* Frankfurt, New York: Campus.

Zook, M. A. (2004). The knowledge brokers: Venture capitalists, tacit knowledge and regional development. *International Journal of Urban and Regional Research*, 28:621–641.

Chapter

2

Using Knowledge as an Engine of Growth

The Case of South Korea*

Derek H. C. Chen

1. Introduction

Korea has experienced rapid and sustained economic growth over the past five decades. In the aftermath of World War II, South Korea's GDP per capita was comparable to levels in the poorer countries in Africa (Figure 1). Then the Korean War, from which South Korea emerged in 1953, made conditions even worse; South Korea was considered by many to be a hopeless case after four years of mass destruction. However, in a relatively short span of about 55 years,

* The paper is essentially an updated summary of Suh and Chen (2007), *Korea as a Knowledge Economy: Evolutionary Process and Lessons Learned*, and therefore borrows heavily from the original report.

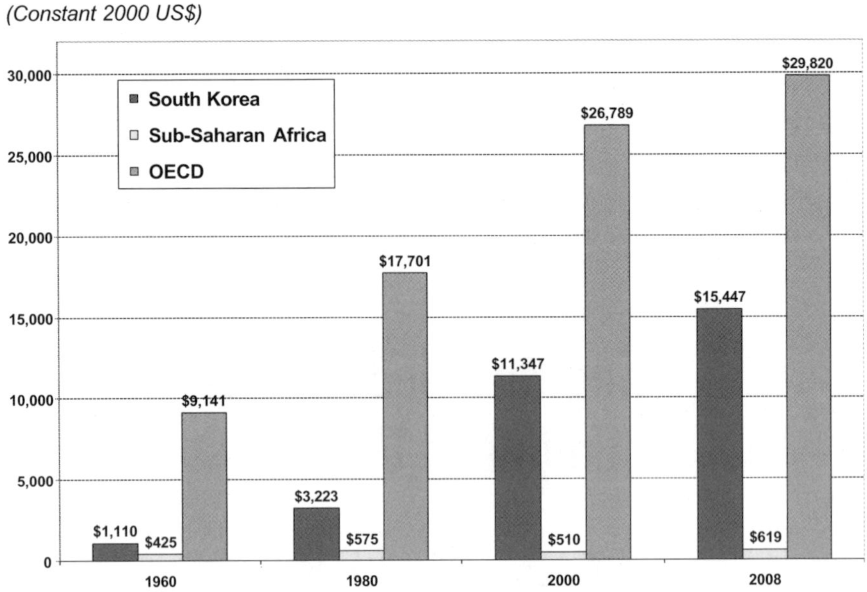

Figure 1: Real GDP per capita: South Korea, Sub-Saharan Africa, and the OECD
Source: Author's construction based on data obtained from World Bank database.

Korea's GDP per capita has increased nearly 14-fold, to over US$15,000, which is on par with the medium economies of the European Union.

Figure 2 presents the decomposition of South Korea's economic growth over the past four decades and clearly shows that the contribution of knowledge, represented here by total factor productivity, was a key factor to South Korea's miracle of rapid economic growth. By comparison, Mexico's GDP per capita in 1960 was about 2.5 times larger than that of Korea; however, by 2005 Korea's GDP per capita was more than twice Mexico's. Without the contribution of knowledge, Korea's real GDP per capita would still be below that of Mexico. It is this rapid and sustained knowledge-led economic growth that makes the South Korean case particularly interesting to analyze. In addition, its journey toward the knowledge economy offers valuable policy lessons for other developing economies that are seeking to make that transition.

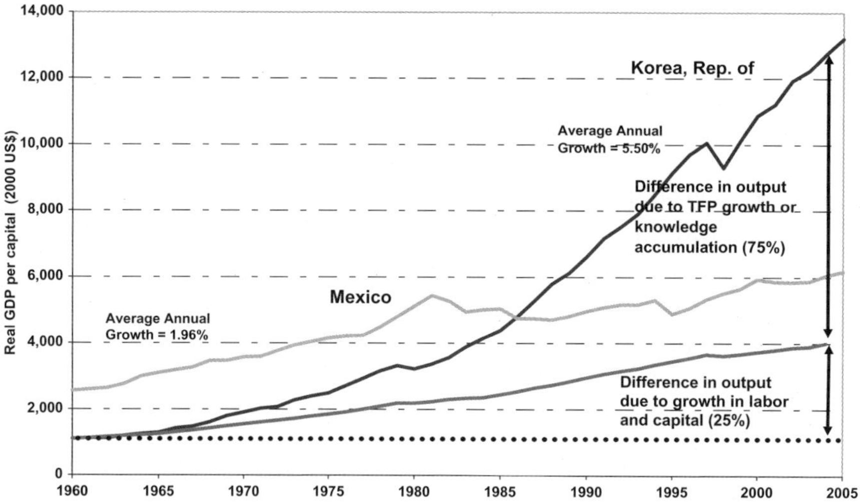

Figure 2: The contribution of knowledge to South Korea's GDP growth

Source: Author's construction based on data obtained from World Bank database.

2. The Knowledge Economy Framework

An economy becomes a "knowledge economy" when the sustained use and creation of knowledge are at the center of its economic development process. A knowledge economy is one that uses knowledge as the key engine of economic growth. It is an economy in which knowledge is acquired, created, disseminated, and used effectively to enhance economic development.[1] The successful transition to a knowledge economy typically involves elements such as making long-term investments in education, developing innovation capability, modernizing the information infrastructure, and having an economic environment that is conducive to market transactions. The

[1] Contrary to some beliefs, the concept of the knowledge economy does not necessarily revolve around high technology or information technology. For example, the application of new techniques to subsistence farming can increase yields significantly, or the use of modern logistical services can enable traditional craft sectors to serve broader markets than before.

World Bank has termed these elements the pillars of the knowledge economy, and together they constitute the knowledge economy (KE) framework.

Specifically, the four pillars of the KE framework are the following:

- ***An economic incentive and institutional regime*** that provides good economic policies and institutions, which promote efficient allocation of resources and stimulate creativity and incentives for the efficient creation, dissemination, and use of existing knowledge.
- ***An educated and skilled labor force*** that continuously upgrades and adapts skills to efficiently create and use knowledge.
- ***An effective innovation system*** of firms, research centers, universities, consultants, and other organizations that keeps up with the knowledge revolution, taps into the growing stock of global knowledge, and assimilates and adapts new knowledge to local needs.
- ***A modern, adequate and accessible information infrastructure*** that facilitates the effective communication, dissemination, and processing of information and knowledge.

The KE framework thus asserts that investments and interactions among these four pillars are necessary for the sustained creation, adoption, adaptation, and use of knowledge in domestic economic production. The result will be competitively-priced higher-value-added goods and services, which increases the probability of economic success in the current highly competitive and globalized world economy.

3. The Korean Development Strategy

3.1 *The Catch-up Period: 1950–1997*

South Korea's rapid and sustained economic growth from the time when it was starting out as a low-income country was an outcome of the knowledge economy approach, eventhough an explicit knowledge economy development strategy was not laid out. During this time from 1950 to 1997, Korea's economic development hinged on the critical interactions among the four pillars of the knowledge economy, which have evolved together with the various stages of economic development (see Table 1). In particular, the pragmatic development strategies focused on achieving sustained productivity growth by consistently increasing the value added of output. These strategies involved intensive learning processes consisting of active technological capability building and complementary human resources development. At the same time, the Korean government assumed the very necessary proactive leadership role of supporting the market and providing an environment that would foster and sustain the transformation.

In the 1960s, Korea embarked on the promotion of both export- and import-substitution industries, starting with subsistence agriculture (rice) and labor-intensive light manufacturing sectors (textiles and bicycles). Considerable capital accumulation and investment in primary education during this period allowed a gradual shift up the value-added chain toward more sophisticated commodities. Key to this shift was also the use of technologies obtained through foreign licensing and adapted for domestic production. In the mid-1970s, the government's use of a well-targeted industrial policy resulted in a major shift to the development of heavy industries (for example, chemicals, shipbuilding). Along with industrial targeting, policies were enacted to further improve technological capabilities, together with improving access to and quality of technical and vocational training.

In the 1980s, Korea undertook efforts to ensure a market-conducive environment by deregulating various sectors and liberalizing trade. Concurrently, it expanded higher education while investing in indigenous research and development through the establishment of the National Research and Development Program. Korea continued to pursue high-value-added manufacturing in the 1990s by promoting indigenous high-technology innovation. Domestic wage hikes and the appreciation of the Korean won had resulted in chronic current account deficits, which sparked a series of reforms, including the reform of the financial market. Together with the setting up of a modern and accessible information infrastructure, there was continued expansion of research and development capabilities in Korean

Table 1: South Korea's economic development strategies, 1960–Present

	Development Goals	Major Policy Directions	Macroeconomic Policy Framework	Human Resources Development	Science and Technology (S&T)
1960s	• Build a production base for export-oriented industrialization	• Expand export-oriented light industries • Mobilize domestic and foreign capital	• Prepare legal and institutional bases to support industrialization	• Decrease illiteracy • Establish a national infrastructure	• Build scientific institutions' legal and administrative frameworks
1970s	• Build a self-reliant growth base	• Promote heavy and chemical industries • Build social overhead capital	• Maximize growth by expanding concessionary loans from the government • Increase government's intervention into the markets	• Increase vocational training • Increase engineering-major college graduates	• Set up a scientific infrastructure: specialized S&T institutions, Daeduck Science Town

(*Continued*)

Table 1: (*Continued*)

	Development Goals	Major Policy Directions	Macroeconomic Policy Framework	Human Resources Development	Science and Technology (S&T)
1980s	• Expand technology-intensive industries	• Conduct industrial rationalization • Decrease export subsidy and expanding import liberalization	• Stabilize macroeconomy • Enhance private autonomy and competition	• Expand the higher education system • Develop semiskilled human resources	• Promote R&D and private research centers • Develop national R&D programs
1990s	• Promote high-technology innovation	• Support technology development • Build an information infrastructure	• Liberalize trade and foreign direct investment • Reform financial markets and restructure the economy	• Develop highly skilled human resources in strategic fields: information technology, biotechnology, etc. • Develop lifelong learning systems	• Promote R&D programs on highly advanced technologies

Source: Reproduced from Suh and Chen (2007).

industries, which drew on the skilled labor force that had resulted from the government's aggressive expansion of the higher education system.

3.2 *The Financial Crisis of 1997*

There were limitations of the Korean model, especially in the recent era of globalization and rapid technological changes. The mechanism of resource allocation that the government used to wield discretionary power over the market had been effective when the economy was burgeoning. However, it approached its limits as the economy developed and became larger and more complex. The financial crisis of 1997 manifested the limitations of discretionary resource allocation and underscored the urgent need for widespread economic reform. The old policy framework and institutions that had led Korea in the early high-growth era turned out to be bottlenecks for sustained economic growth in the new economic environment. As such, Korea, with a clear national consensus, implemented a comprehensive restructuring of ailing corporate and financial sectors and undertook reforms in the public sector and the labor market, all of which contributed to overcoming the crisis and ensuring a rapid economic recovery.

3.3 *Post-Crisis Period: 1998 to Present*

In the aftermath of the financial crisis in 1998, South Korea officially launched a national campaign to make the transition to an advanced knowledge-based economy in which domestic innovation would thrive, thereby enhancing overall productivity and sustaining economic growth. The initial impetus came from the *Maeil Business Newspaper*, which presented in 1996 a more coherent vision of the future for the Korean economy. Subsequently, economywide action plans were drawn up and implemented through the coordinated efforts of a number of government agencies. Key to the success of the national knowledge economy strategy was the strong political leadership from the Kim Dae-jung government and the national consensus with the private sector and civil society.

Korea's new knowledge-based development strategy was based on the KE framework developed by the World Bank.[2] In particular, the strategy was based on the four pillars of the knowledge economy. Korea's transition toward a knowledge economy has been relatively successful. For example, the World Bank's Knowledge Assessment Methodology (KAM) Basic Scorecard,[3] which benchmarks countries' overall readiness to use knowledge for economic development, shows that Korea has evolved into a relatively mature knowledge-based economy, performing well above the average country in East Asia and on par with average high-income countries (Figure 3).

The OECD has documented Korea's active investment in knowledge,[4] which in 2002 was 5.8 percent of GDP, the fourth

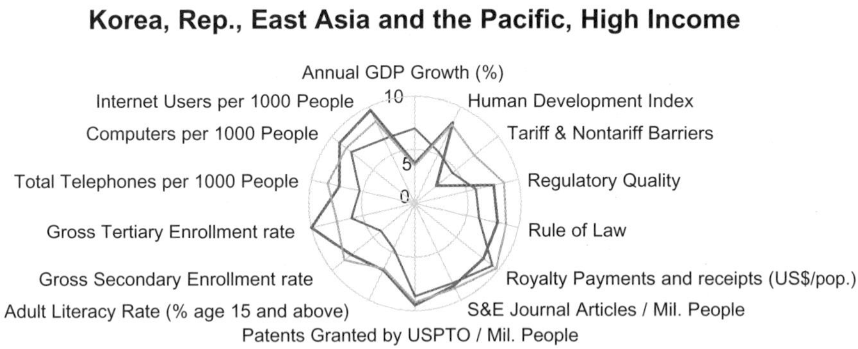

Figure 3: KAM basic scorecard for Korea, East Asia, and high-income country average
Source: Knowledge Assessment Methodology (KAM) July 2009 (www.worldbank.org/kam).

---

[2] The report *Korea and the Knowledge-based Economy: Making the Transition*, a joint publication of the World Bank and the OECD, outlined concrete steps for reforms in the various policy domains.

[3] The KAM Basic Scorecard uses 12 knowledge indicators and two performance indicators to illustrate a country's overall readiness for the knowledge economy. All variables are normalized or rescaled onto a 0 to 10 interval. The center of the illustrated spider chart denotes the minimum normalized value of 0, and the outer perimeter of the chart denotes the maximum normalized value of 10. Thus, a "bigger" or "fuller" spider chart implies that the country or region is better positioned in terms of the knowledge economy.

[4] The OECD defined investment in knowledge as spending on R&D, software, and higher education.

highest among the OECD countries (Figure 4). Moreover, Korea's investment pattern has been changing significantly in recent years, from investment in physical capital such as machinery and equipment to knowledge inputs. Among the OECD countries in 2002, Korea still had the highest ratio of spending on machinery and equipment to GDP, but the ratio is rapidly decreasing. From 1995 to 2002 the ratio decreased by 3.7 percentage points. For the same period, in contrast, Korea's investment in knowledge increased by 1 percentage point.

3.4 *Meeting Skill and Human Resources Requirements*

Education and human resources have been key factors in Korea's rapid economic growth over the past four decades.

Expanding enrollment rates

Korea's educational system has expanded rapidly over the past 40 years (Figure 5). Universal primary education was adopted as early as the 1950s, which was instrumental for full-fledged industrialization in subsequent years. The secondary enrollment rate was relatively low in 1970, at less than 40 percent. However, subsequent sharp increases led to secondary enrollment being more than 90 percent in 2004. Similarly, enrollment for higher education has experienced steep increases since 1990 and surpassed 60 percent in 2004, which was the highest level among the OECD countries.

Enhancing educational quality

Korea's education system has also achieved qualitative improvement in tandem with quantitative expansion. Korea ranks very high in various international tests administered in recent years. For example, Korea ranked third for mathematics and tenth for science out of 57 countries included in the OECD's Programme for International Student Assessment (PISA) 2006 survey (Figure 6).

Figure 4: Investments in knowledge and in machinery and equipment in OECD countries, 2002

Source: Reproduced from Suh and Chen (2007).

a. 1994–2001 for Greece and Italy. 1995–2002 for Korea. The EU figure excludes Belgium, Greece, and Italy. OECD figure excludes Belgium, Greece, Italy, and New Zealand.

b. Excludes Greece and Italy.

c. 2001 data.

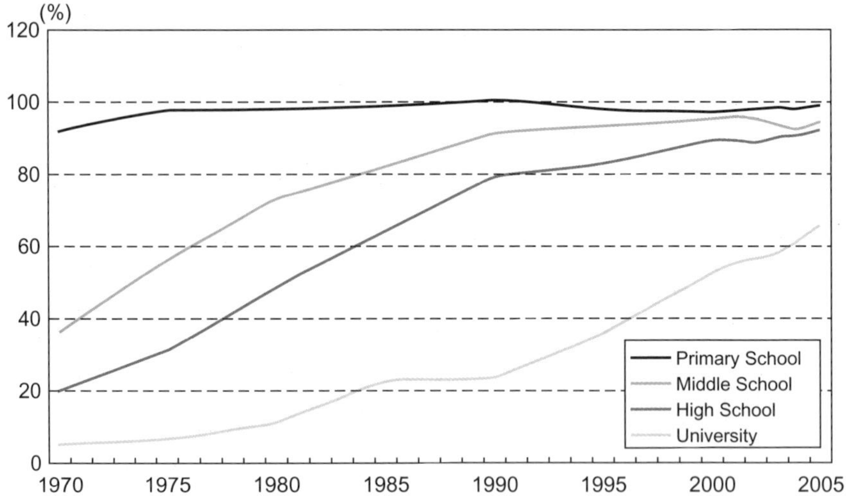

Figure 5: Educational expansion in Korea, gross enrollment rates
Source: Reproduced from Suh and Chen (2007).

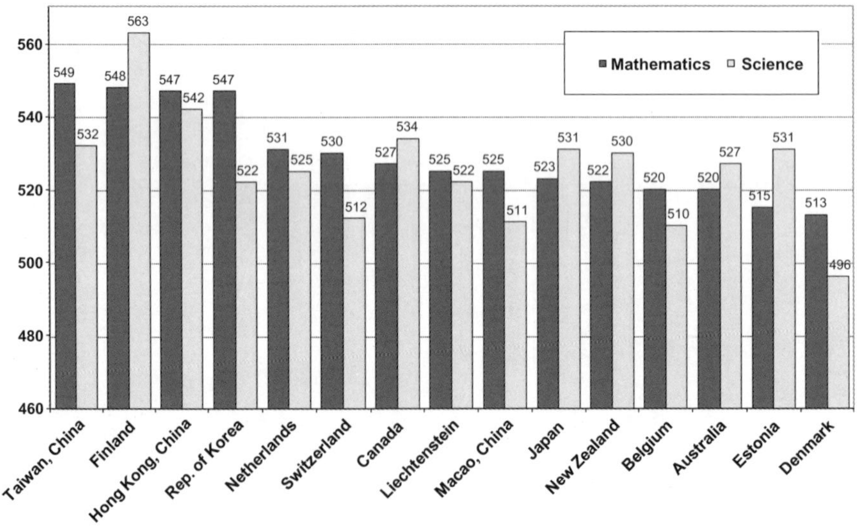

Figure 6: PISA mathematics and science scores, selected countries, 2006
Source: Constructed using data from Baldi *et al.* (2007).

Expanding the education system in tandem with economic development

Since the 1950s, Korea's educational policy and planning have complemented the government-led economic development plans. Specifically, the education system was nurtured and expanded according to the manpower needs of the economy (Table 2). Hence, in the 1950s and 1960s education policies focused on the expansion of primary and secondary education, which was critical to supply at least a literate workforce to the industries. Vocational high schools were also established in the 1960s to provide training in craft skills for the growing labor-intensive light manufacturing industries. Junior vocational colleges were set up in the 1970s to supply technicians for the heavy and chemical industries. In the 1980s, the higher-education expansion policies adopted by the government were effective in supplying high-quality white-collar workers and R&D personnel that were required as Korea began enhancing domestic innovation.

Public and private expenditure on education

Korea's emphasis on education and developing a skilled workforce can be seen from the economy's total expenditure on education. In 2006, 7.3 percent of GDP was spent on education, a level much higher than the OECD average of 5.7 percent (Figure 7). Among the OECD countries, only Iceland (8.0 percent), and the United States (7.4 percent) had education expenditure shares that surpassed Korea. The government's priority on education can be seen from the growth rates of the public education budget, which have outpaced those of GDP over the past three decades. In particular, public spending on education as a share of GDP was 4.3 percent in 2006, compared to 2.1 percent in 1975. Another characteristic of the Korean education system is the unusually large component of private financing. Private expenditures on education account for 3.0 percent of GDP, which is the highest among the OECD countries and far above the OECD average of 0.9 percent.

Table 2: Evolution of the Korean education system, 1948–present

Government Focus	1948–1960	1961–1980	1981–2000	2001–present
Challenges at national level	Establishing a national infrastructure	Planning education for economic development	Enhancing lifelong learning	Increasing human resources innovation
Strategy	Using a government-initiated approach	Focusing on traditional institutions of higher education Continuing the government-initiated approach	Reaching out to the nontraditional education sector Using a government-led, partial market approach	Tightening up the loosely connected system of HR development Using a coordinated govt.-market approach (market influence increased)
Primary tasks and activities	Building elementary schools Developing vocational schools Developing HR in medicine, engineering, agriculture, teacher education	Improving teaching quality (elementary and secondary education) Increasing college graduates with engineering major Developing medium-skilled HR	Developing high-skilled HR in national strategic fields (IT, BT, S&T etc.) Developing a system of lifelong learning	Improving quality or relevance of university education Increasing research productivity Enhancing the efficiency of the HRD system Enhancing regional development and innovation

(*Continued*)

Table 2: (*Continued*)

Government Focus	1948–1960	1961–1980	1981–2000	2001–present
Resources (tools)	Using foreign assistance (UNKRA, OEC, USOM etc.)	Increasing educational period for new elem. and second. (2 to 4 years) Creating vocational colleges Mobilizing private resources for expansion of educational sector	Increasing research funds in science and technology Creating diverse types of higher education institutions Introducing credit-bank system	Restructuring at government, system, and institutional levels Using massive financial support from government (Brain Korea 21/Post-BK 21, NURI educational Internet network)

Note: UNKRA = United Nations Korean Reconstruction Agency; OEC = Office of Economic Coordination; USOM = United States Operation Missions.

Source: Reproduced from Suh and Chen (2007).

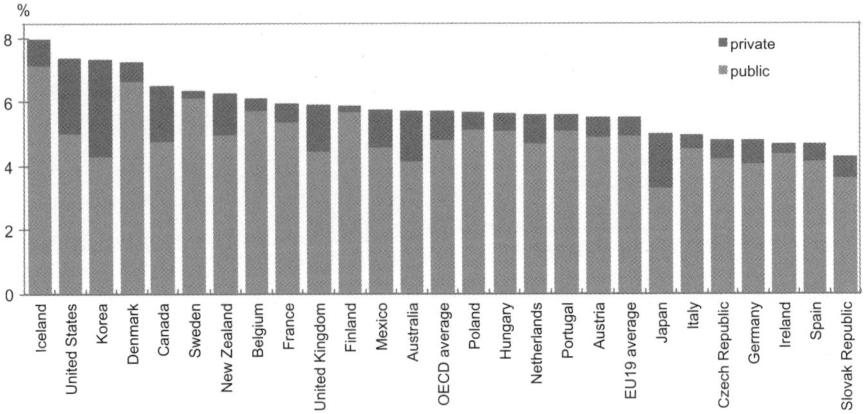

Figure 7: Education spending as a percentage of GDP, 2006

Source: Author's construction based on data from OECD 2009.

The Korean government has been successful in encouraging the private sector, either households or private foundations, to bear a significant portion of total education costs. In particular, private foundations have established a number of secondary schools and higher education institutions, in which expenses are paid for by user fees. At the secondary level, enrollment at private institutions accounts for more than 40 percent of total secondary enrollment, whereas private enrollment for tertiary education is over 70 percent (Figure 8). On the other hand, primary education in Korea has been treated as a public good and has been mostly publicly funded, with around 99 percent of primary school students in 2005 being enrolled in public schools.

The heavy reliance on private funding in secondary and higher education has important policy implications. Inducing the private sector to play a more active role in providing secondary and higher education services has had a leveraging effect where scarce government resources have been spent on key priority areas. More specifically, by leaving higher education to the private sector and targeting public resources for primary education, Korea was able to address one of the main equity issues, basic education for all.

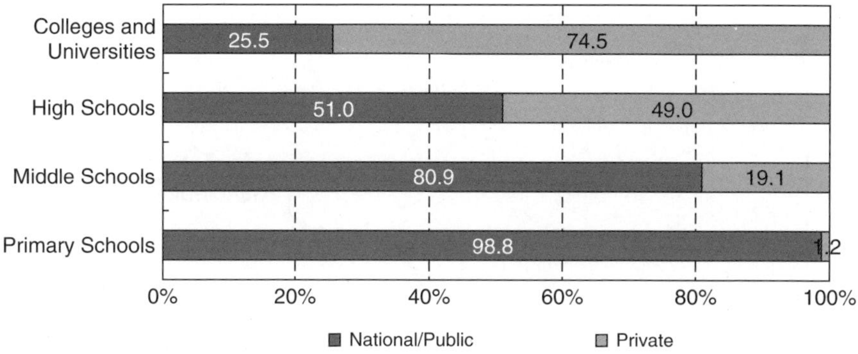

Figure 8: Ratio of enrollment in private schools, 2005

Source: Reproduced from Suh and Chen (2007).

The cultural factor

One factor that is contributing to the high propensity for private spending in education is the intrinsic social values that literacy and education have in Korean culture. Indeed, educating children has always been the overriding family task for Korean parents. Coupled with the rising demand for a more educated workforce, a natural consequence of the industrialization process, this cultural factor has greatly contributed to Korean families' willingness to pay for educating their children, and consequently to the rapid increases in school enrollment rates.

3.5 *Harnessing the Potential of Science and Technology*

Rapid development of indigenous research and development capabilities

Korea has recognized the importance of developing indigenous capabilities in science and technology for successful industrialization and therefore has made building such capabilities one of the key priority policy areas over the past four decades. As can be seen in Figure 9, Korea's gross expenditure on research and development (GERD) has grown both in size and as a share of GDP, which increased from 0.25 percent in 1963 to 2.84 percent in 2004. Along with the rapid

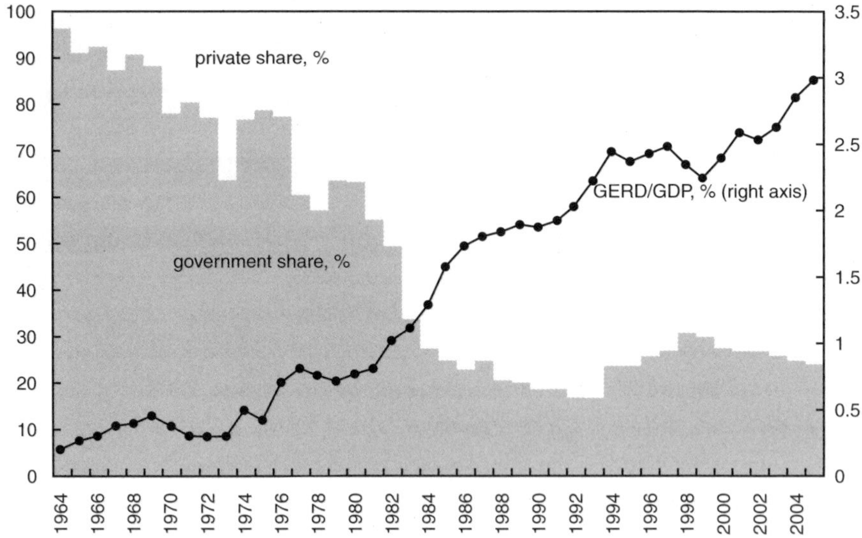

Figure 9: Gross expenditure on research and development in Korea, 1964–2004
Source: Reproduced from Suh and Chen (2007).

increase in R&D investment, the number of researchers has also increased more than 100-fold during the same period, from 1,900 to 209,979 researchers. The rapid increases in total R&D expenditure have been possible because of active expansion of investment by the private sector. During the earlier years of industrialization, private sector R&D spending was negligible, but rapid economic growth has called for commensurate investment in technology development, resulting in continuous increases in R&D spending by private enterprises over the past four decades. Consequently, the funding sources have also greatly changed: the government's share of GERD has been continuously reduced, with only one-fourth of GERD coming from the government in recent years.

The rapid increase in R&D investments has led to a corresponding increase in indigenous innovation and adoption of foreign technologies. This can be seen from Korea's improved performance in various international indicators of innovation and technology adoption, such as the number of scientific and technical journal

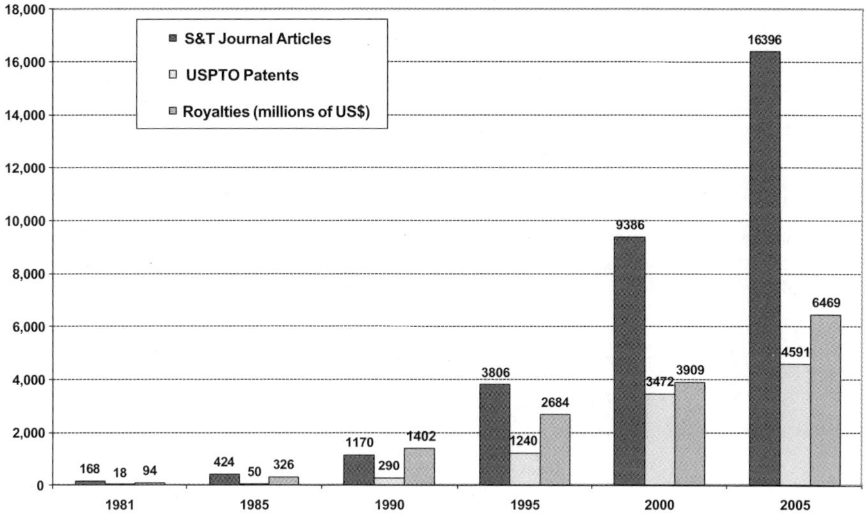

Figure 10: R&D capabilities and technology adoption in Korea, 1981–2005

Source: Author's constructed based on data from U.S. Patent and Trademark Office's website and World Bank database.

articles written and published by Korean authors,[5] the number of patents granted to Korean inventors by the United States Patent and Trademark Office (USPTO), and the amount of royalties and license fees paid and received by Korea (Figure 10).[6] According to the KAM, South Korea is currently on par with a typically high-income country in many aspects related to innovation and technological adoption (Figure 11).

[5] Refers to the number of scientific and engineering articles published in the following fields: physics, biology, chemistry, mathematics, clinical medicine, biomedical research, engineering and technology, and earth and space sciences.

[6] Royalty and license fees are payments and receipts between residents and nonresidents for the authorized use of intangible, nonproduced, nonfinancial assets and proprietary rights (such as patents, copyrights, trademarks, industrial processes, and franchises) and for the use, through licensing agreements, of produced originals of prototypes (such as films and manuscripts). Data are in current U.S. dollars.

Figure 11: KAM: Innovation and technological adoption indicators

Source: Knowledge Assessment Methodology (KAM), July 2009 (www.worldbank.org/wbi/kam). See footnote 3 for explanation of chart.

Evolution of technological assimilation and domestic innovation capabilities

When Korea launched its industrialization drive in the early 1960s, it had to rely almost completely on imported foreign technologies. By doing so, Korea pursued two key objectives. The first was to promote the inward transfer of foreign technologies, the second was to develop domestic absorptive capacity to digest, assimilate, and improve upon the transferred technologies and to adapt them to domestic production. The second objective required a relatively skilled labor force, which Korea had because of its concurrent aggressive educational policies. The technology assimilation strategy used various channels, such as original equipment manufacturing, foreign licensing–based production, reverse engineering of imported capital goods, and learning from the building of turnkey plants. These channels of informal technology assimilation enabled Korea to minimize its dependence on FDI, which had become more prominent since the 1997 financial crisis, and to maintain independence from multinational corporations (Table 3). The strategy proved to be a success and Korean firms were able to assimilate technologies rapidly enough to undertake subsequent expansion and improvement with little assistance from foreign suppliers.

Table 3: Channels of foreign technology transfer to Korea, 1962–2001

	(US$ millions)							
	1962–66	1967–71	1972–76	1977–81	1982–86	1987–91	1992–96	1997–2001
FDI	45.4	218.6	879.4	720.6	1767.7	5635.9	8405.2	57,850.8
Foreign licensing	0.8	16.3	96.6	451.4	1184.9	4359.4	7317.8	13,194.0
Capital goods imports	316	2,541	8,841	27,978	50,978	120,952	220,843	252,034
Sum of foreign licensing and capital goods imports	317	2,557	8,938	2,8429	52,163	125,311	228,161	265,228

Source: Reproduced from Suh and Chen (2007).

Changes in economic environments in the early 1980s induced Korea to embark on serious investments in indigenous R&D. On the one hand, Korean industrial development had reached the stage at which domestic industries found it more difficult to be competitive in the international market because they were reliant on imported technologies and employed domestic labor that was becoming more and more expensive. On the other hand, Korean industries had grown to become potential competitors in the international market, making foreign companies increasingly reluctant to transfer technologies to Korea; thus, it was inevitable that Korea would have to develop an indigenous base for research and innovation. Meeting the challenge required highly trained scientists and engineers as well as financial resources to support R&D activities, which are by nature uncertain and risky.

The government and innovation in Korea

The role of the government in the efforts of developing indigenous innovation capabilities evolved together with the phases of industrial

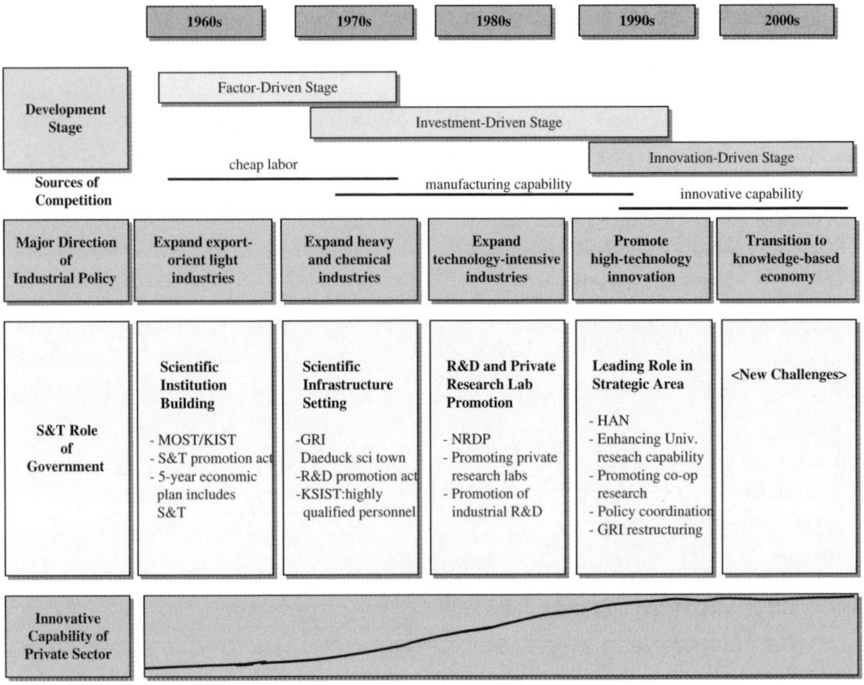

Figure 12: Five decades of Korean innovation policy

Source: Reproduced from Suh and Chen (2007).

development (Figure 12). For example, to enable firms to finance the massive importing of capital goods and the building of turnkey plants in line with its technological assimilation strategy in the 1960s, the Korean government brought in large-scale, long-term foreign loans and allocated them to selected industries. In the 1970s, as the economy had developed and was moving into heavy industries, the government created government research institutes (GRIs) in the fields of heavy machinery and chemicals to compensate domestic industries for their technological weakness. These GRIs, such as the Korea Institute of Machinery and Metals (KIMM) and the Electronics and Telecommunications Research Institute (ETRI), worked with private industry to some extent to enhance technological capabilities for further industrial development.

The Korean government has also contributed significantly to the rapid rise in private sector investments in research and development. First, the government's outward-looking or export-driven development strategy forced domestic industries into international markets, exposing them to intense global competition. For these firms to remain competitive, they had to keep pace with technological changes by investing heavily in R&D. Second, the government's industrial policy that favored large firms gave birth to a unique business organization in Korea, the chaebol. Chaebols enjoy greater financial affluence owing to the economies of both scale and scope of their business operations. Chaebols, which are usually big multinational firms, were able to engage in risky and expensive R&D projects. The top 20 firms account for about 57 percent of the total industrial R&D investments in Korea (KITA, 2004). Most important, Korea has been able to increase R&D investments at such a rapid rate because it has an abundant pool of highly educated manpower that could meet the increasing demand for R&D services in both private and public sectors.

However, Korea's efforts to build an indigenous innovative service base are not without challenges. For instance, most of the research and development is being performed by the chaebols, and little being conducted by the universities. This disproportionate distribution of R&D investment is not sustainable for the economy as a whole — insufficient resources are being devoted to R&D in basic sciences, which is necessary for the long-term development of science and technology. Indeed, by using the USPTO count of patents granted as an indicator of R&D productivity, Figure 13 shows how Korea has been overtaken by other economies with strong innovation systems.

3.6 *Building Information Infrastructure*

In the early 1970s, Korea's information infrastructure was inadequate and the provision of information and communication technologies (ICT) services was insufficient and too inefficient to meet the telecommunication demands associated with rapid economic growth. In 1975, only 3 percent of Koreans had a telephone. To improve

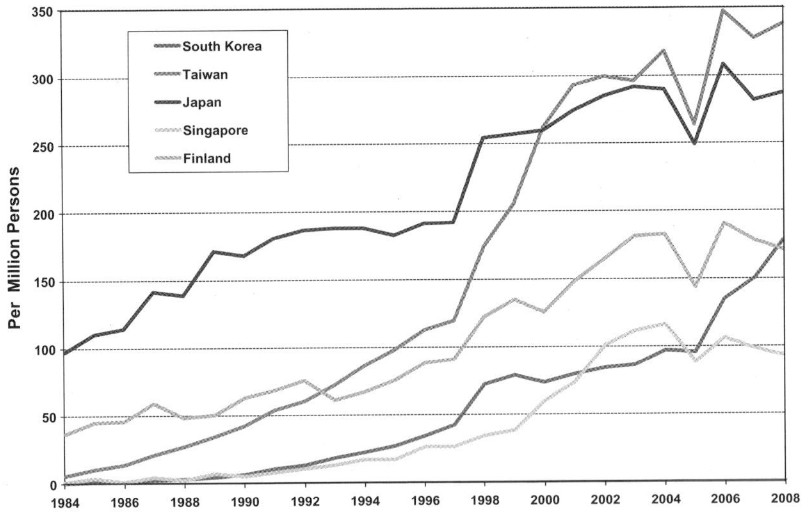

Figure 13: Number of USPTO patents granted, various countries

Source: Author's construction based on data from U.S. Patent and Trademark Office's website and World Bank database.

efficiency in the provision of telecommunication services, the Korean government decided to rely on the invisible hands of the price mechanism and thus focused on introducing competition into the ICT infrastructure sector. As such, a series of sequential but rapid policy measures were implemented for the deregulation and liberalization of the ICT services sector, along with privatizing the government-owned telecom operators (Table 4). The Korea Telecommunication Authority was established in 1981 to spearhead this effort.

The reform of the ICT infrastructure sector resulted in tremendous improvements in terms of ICT penetration rates. Most noteworthy are the recent penetration rates of cell phones and the Internet. From 1995 to 2008, the proportion of Koreans having a cell phone increased by nearly 26 times, to 94 percent, while the proportion of Internet users increased by a whopping 95 times, to 77 percent (Figure 14). The most recent data from the World Bank's KAM shows that Korea has succeeded in joining the ranks of high-income countries in terms of its informatization level (Figure 15).

Table 4: Introducing competition in telecommunication services

1981	Korea Telecommunication Authority established
1990	Competition introduced in value-added services Dacom enters international market
1992	Competition introduced in paging services (10 new operators)
1994	Second cellular license issued (Shinsegi Telecom)
1995	Competition introduced in long-distance market (Dacom)
1996	27 new licenses granted: three for personal communication services (PCSs), six for trunked radio systems (TRSs), 11 for second-generation cordless telephones (CT-2), etc.
1997	10 new licenses granted: one local operator (Hanaro), one long-distance operator (Onse Corporation), six TRSs, one paging operator
1997	Resale-based competition introduced

Source: Reproduced from Suh and Chen (2007).

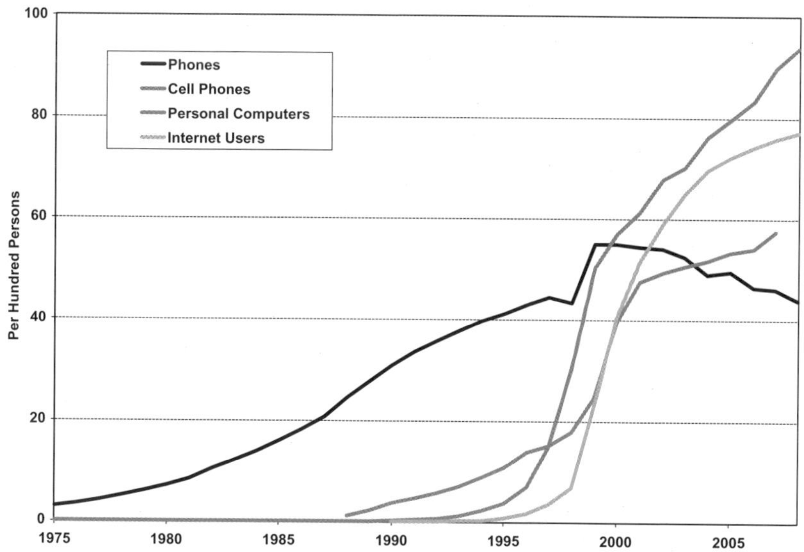

Figure 14: ICT density in South Korea, 1975–2008

Source: Author's construction based on data from World Bank database.

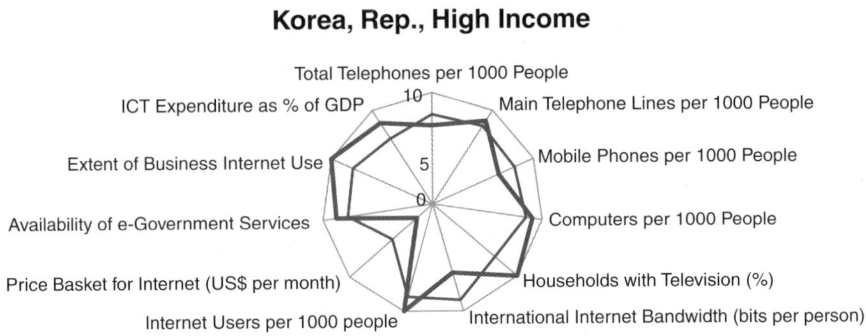

Figure 15: Information and communications technologies indicators
Source: Knowledge Assessment Methodology 2009 (www.worldbank.org/wbi/kam).

Currently, Korea has constructed ICT networks connecting all 114 areas of the country, and ICTs are used extensively in numerous economic and social activities. The number of individuals using ICT-related services is also constantly rising: the number of subscribers to Internet banking services reached 22.58 million as of March 2005, and e-commerce has rapidly increased from 50 billion won in 1998 to 314 billion won in 2004, which is equivalent to 40 percent of GDP. Led by an e-government initiative, the public sector is also extensively using ICTs. In 2004, about 97 percent of documents were dealt with through the e-approval system in the government agencies, compared with only 21 percent in 1998.

The rapid development of the Korean information infrastructure hinged on key government organizations that were responsible for the informatization strategy. These organizations were restructured in the 1990s. They included the Informatization Promotion Committee, chaired by the prime minister; the Informatization Strategy Meeting, chaired by the president; and the Ministry of Information and Communication. Concurrently, the Korean government established three master plans for the development of the information society. In 1995, the Informatization Promotion Act was enacted, and the first master plan for promoting informatization was formulated a year later. In 1998 in the second master plan, Cyber

Table 5: Master plans for promotion of informatization

Plan	Period	Vision
Basic informatization promotion plan	1996–2000	Attain world-class informatization levels by 2010
Cyber Korea 21	1999–2002	Build a leading knowledge-based society
e-Korea Vision 2007	2002–2007	Build e-Korea as the global leader

Korea 21 was established to cope with the changing environment that resulted from the Asian financial crisis. And in 2002, when most of the policy goals set up by Cyber Korea 21 had been achieved ahead of the original schedule, the third blueprint, e-Korea Vision 2007 was laid out (Table 5).

To build the infrastructure efficiently and economically while actively responding to the technological development and changing demands, diverse implementation methods have been used. Networks in commercial and densely populated areas have been built with optical cables; networks that extend to subscriber premises have been built partly with optical cables and partly by digitizing and enhancing the speed of existing telephone lines or CATV networks or by building new wireless local loops.

In order to finance the investment for the rapid deployment of Korea's information infrastructure, the Informatization Promotion Fund was established as a special vehicle to overcome short-term budgetary constraints. From 1993 to 2002, the fund reached a total of US$7.78 billion. About 40 percent of that was from government budgetary contributions, 46 percent came from private enterprises (of which licensing fees for new communication services composed major portions), and the remaining 14 percent came from miscellaneous profits and interests. The funds were allocated in a way that would balance development in ICT-related activities: 38 percent for technology development programs (the best known is CDMA wireless technology), 18 percent for human resources development, and the remaining 44 percent for building of infrastructure and diffusion, including standardization. A particularly important purpose for the fund was to narrow the digital divide. Socially alienated groups, such as the aged and rural villagers, have continuously received various supports and assistance, including computer-literacy education.

3.7 *Designing a New Economic Framework*

The 1997 financial crisis required major economic reforms and also presented an opportunity for the reform of the economic regime. The reforms, which began in the wake of the crisis, have been extensive and substantial, covering most of the areas in public and private domains. The economic reforms since the 1997 crisis had three major objectives:

1. To transform Korea into a market-oriented economy by deregulating across the sectors and thereby promoting competition and entrepreneurship. At the same time, a modern regulatory framework would be set up to support the efficient and equitable functioning of the markets.
2. To improve the institutional regime by improving the rule of law and by having greater transparency, disclosure of information, and accountability on the part of the government as well as the private sector.
3. To continue the transition to the knowledge-based economy by developing a relevant and modern legal and institutional infrastructure in such areas as intellectual property rights, valuation of intangible assets, and laws to cover privacy and security in digital transactions.

Specifically, major structural reforms included improving the efficiency and soundness of the financial system, creating stable corporate environments to support the entry of new players into the economy, enhancing the flexibility of the labor market, and redefining the role of the government by strengthening the basic institutional infrastructure and fortifying the rule of law (Figure 16). The Korean government also undertook efforts to develop venture business firms that were expected to spearhead innovation and generate employment.

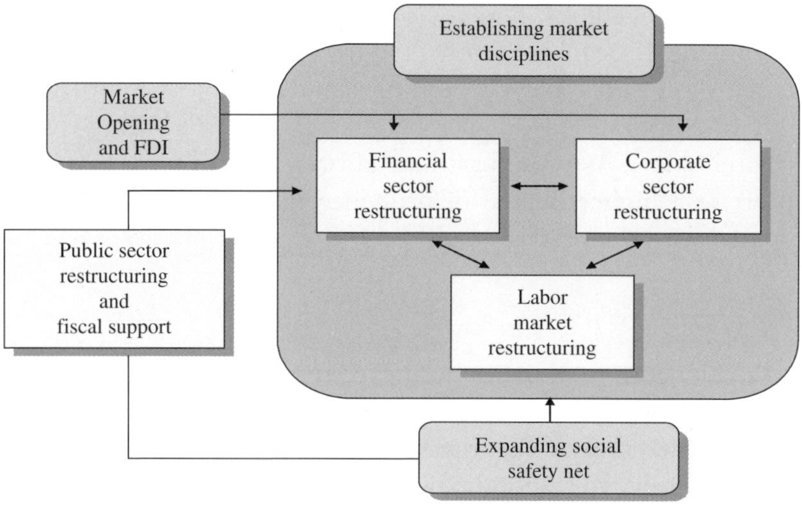

Figure 16: Economic reforms after the financial crisis

Source: Reproduced from Suh and Chen (2007).

Financial sector reform

Restoring confidence in Korea's financial system was a top priority for the government in the wake of the crisis. In order to rehabilitate the financial system, the government liquidated troubled institutions, removed nonperforming loans, and recapitalized promi-sing financial institutions by injecting public funds amounting to 108 trillion won. Responding to the government's initiatives, financial institutions also adopted stricter standards. These have contributed to increasing the health and profitability of the financial industry. For example, the share of bad loans out of total loans sharply decreased from 11.3 percent in 1999 to 3.9 percent in 2002.

Corporate sector reform

Reforms in the corporate sector have focused on eliminating inefficiencies and encouraging firms to reorganize themselves based on the strength of their core competencies. Initiatives have been implemented to improve corporate governance systems, revise bankruptcy procedures, and remove anticompetitive regulations. In addition, the Korean government introduced measures to improve financial disclosures and accounting standards and to enhance the quality of corporate governance by guaranteeing the voting rights of the shareholders. As a result of these efforts, Korean firms have begun to exercise a higher degree of prudence in investment decisions as compared with the precrisis period.

Labor market reform

The Korean labor market has been excessively rigid in the past. Labor legislation, labor relations, and industrial and market structure have contributed to labor market rigidities. Such rigidities hamper the ability of the country to respond rapidly to changing market needs and to opportunities, both those resulting from restructuring pressures and those brought about by the "knowledge revolution". A number of initiatives have been implemented since the crisis to create more flexible labor markets. For example, labor laws have been revised to legalize layoffs, and a legal framework for manpower-leasing services has been introduced.

Enhancement of social safety nets

Besides enacting measures to enhance labor market flexibility, Korea has also begun to strengthen its social safety nets. Unemployment insurance, a well-functioning pension system, and properly targeted poverty programs form part of an overall insurance package. Since the crisis, the benefits structure and coverage of Korea's unemployment insurance program have significantly improved. Expanded social safety nets have helped in mitigating negative effects of economic restructuring and promoting innovation and entrepreneurship by encouraging risk taking by individuals.

Nurturing entrepreneurship and cultivating venture businesses

The development of venture businesses requires a stable supply of concomitant risky capital, which was absent from Korea until the early 1980s. Accordingly, the government actively promoted the formation of a venture capital market, which has grown rapidly since the late 1990s. Korea is now among the leading OECD countries in terms of venture capital investment as a share of GDP. The government recently revised its venture business policy to avoid any direct intervention into the market and thus to improve the self-sustaining capability of venture business. The government now focuses more on the establishment of regulations to maintain a level playing field and on monitoring and supervision to make sure the regulations are observed.

The new role of the Korean government

Since the 1997 crisis, the most important change in the role of government was to allow the market mechanism to work and to let the private sector take the lead in spurring economic activity. Through a stream of institutional reforms, the government has since focused on fostering both domestic and foreign competition, encouraging and maintaining market- and innovation-conducive macroeconomic environments, and providing a well-developed framework for a knowledge economy, consisting of a highly educated workforce and a well-established and modern information infrastructure.

The institutional reforms include deregulation across all sectors, removal of institutions not conducive to market competition, further reductions in tariff and nontariff barriers, and liberalization of foreign investment. The recent liberalization of trade and foreign direct investment, together with the many current regulatory reforms, has significantly increased the potential for foreign competition across the board. It is hoped that these reforms will foster competition in domestic markets, thereby forcing Korean industries to be even more efficient, innovative, and competitive.

4. Lessons Learned

Korea is an example for developing countries in many ways. Admittedly, there are unique aspects of the country that limit the direct applicability of lessons to be derived:[7] ethnic and cultural homogeneity, a strong Confucian tradition that places high value on education, achievement and loyalty to the nation, a security threat, and political leadership. In addition, some features of the country are typical of the Asian model of development and growth, which cannot be easily replicated in other cultures — among them, a strong state involvement in the orientation and management of the economy, very high saving and investment rates, and an industrial organization well-fitted to mass production. Nevertheless, there are a number of key features of the Korean experience that illustrate how a country can gradually build a KBE with approaches to follow and pitfalls to avoid.

4.1. *Synchronizing the KE Pillars with Economic Development*

Korea has been exemplary in developing each of the pillars of the knowledge economy at the level, type, and pace that complement the other KE pillars, and together the KE pillars complemented the various stages of the economy's development. For example, in the 1960s, when the economy was starting out with subsistence agriculture and light manufacturing, the main focus of education was on providing universal primary and secondary education. At the same time, the universal primary and secondary education of the labor force was also critical to the technology assimilation efforts of the manufacturing industries. Similarly, when Korea was deploying world-class industries

[7] Contrasting Korea-specific contexts and characteristics with other countries or country classifications will be useful for delineating the applicability of the Korean experiences. Readers can refer to some earlier works. For example, using six factors of classifying countries, Ernst and O'Connor (1989) distinguish five developing country groups, among which Korea belongs to first-tier Asian newly industrializing economies. Focusing on the role that FDI played in building technological capabilities of various nations, Lall (2003) contrasts the capability-building strategy of Korea and Taiwan (China) with the FDI-dependence strategy of other countries, including China and Malaysia.

in communications, IT, and advanced electronics in the 1990s and beyond, there was the necessary expansion of the higher education system to ensure an ample supply of workers with tertiary education for the intensive R&D programs. At the same time, the modern and high-technology information infrastructure was being established to facilitate the timely and prompt dissemination of information and knowledge from abroad and around the country.

To reiterate, Korea developed the KE pillars by adopting a gradual, step-by-step approach that emphasized coordination and complementarities among the various KE pillars, and also with the country's stage of economic development and industrialization. In the early years, these complementarities took place even without an explicit KE agenda. The KE approach is thus important to minimizing mismatches in terms of quantity and type in any of the KE pillars and hence reducing misallocations and wastage of the extremely scarce resources of any developing country.

4.2. *Education, Human Resource Development, and Lifelong Learning*

Education builds a nation's ability to absorb new knowledge and technology, because it gives rise to individuals' basic competence, which is an essential building block in technological learning. Therefore, the government should assume full responsibility for taking the measures necessary to promote human resource development. For example, investing in primary and secondary education in advance, as Korea did in the 1960s and 1970s, is essential in laying a foundation for industrial development. In addition, to help workers cope with changes in technology, the government should also provide vocational and technical training and constant retraining or take steps to promote such training at workplaces.

As an economy becomes more advanced, technological competence becomes a critical factor. Building of that competence requires high-caliber scientists and engineers who are familiar with the developments on the frontiers of S&T. More specifically, advanced education in S&T should come first in preparing for entrance into a developed world. In the case of Korea, education and industrialization were mutually complementary in sustaining and accelerating mutual development. Education made technological learning and therefore industrialization possible, and industrialization enhanced the rate of return on investment in education, further promoting demand for education.

4.3. *Technological Adoption and Innovation*

Korea's industrialization evolved from imitation to innovation. At the initial stage, Korean industries attained technological capability through informal channels for technology transfer, such as production arrangements for OEM, reverse engineering of imported machines, and technical training as part of turnkey plant importation. These practices were necessarily coupled with strong reliance on the technological absorptive capacity of the workers. Thus, contrary to the experiences of other developing countries, FDI played a modest role in technological learning in the course of development in Korea. These channels of informal technology transfer enabled Koreans to acquire technology at lower cost and maintain independence in business operations. However, this independence probably came at a significant price because Korea had to abandon many of the technological opportunities that foreign direct investors would have offered.

Today Korea continues to import a significant amount of technology and know-how from abroad,[8] but it has developed a strong indigenous R&D platform and spends almost 3 percent of its GDP on R&D. These approaches for education and technology illustrate both the pragmatism and the discipline that should inspire developing countries in their knowledge efforts. As a whole, it is estimated that for Korea, TFP growth, including knowledge-related contributions, accounted for more than two-thirds of the growth of GDP per capita over the period 1960–2005 (see Figure 1.3). This demonstrates

[8] Until recently, Korea has recorded sizable deficits in TBOP (technology balance of payments): US$2.4 billion in 2003. Korea's TBOP payments in 2003 were US$3.2 billion, which amounts to 20.2 percent of GERD (OECD 2005c).

the need for accumulating not only physical capital but also intellectual capital for successful takeoff and growth.

4.4. *Market-Oriented Role of the State*

One of the most important lessons from Korea's economic development process is the necessity of a market-oriented approach for the transformation to a knowledge economy. Korea's economic success was not based on policies that substituted for the market. Instead, the success of Korea's development policies up to 1997 was the result of policies that complemented and facilitated the functioning of the market. However, a contributing factor in the 1997 financial crisis was the continued discretionary resource allocation on the part of the government in the early 1990s. This had resulted in a very weak financial sector that was saddled with large debts from the bankrupt chaebols. The financial crisis thus showed the limitations of the government's market intervention and underscored the importance of a market-based approach for sustained economic development. A market-based approach requires the existence of competitive forces and, as such, policies that in one way or another foster competition. Transparency of financial markets, accountability of the government, a level playing field for all market participants, liberalized trade, and foreign investment regimes are all crucial characteristics of a market-based knowledge economy.

4.5. *Export-Led Industrialization*

By adopting an outward-looking development strategy, the government exposed Korean industries to competition in the global market, which compelled them to invest heavily in technological assimilation and innovation to remain competitive. This was true in the 1970s and 1980s, when Korea developed its heavy industries. This outward-looking development strategy was complemented by intensive development of government R&D programs and strong fiscal incentives for the firms in the early 1980s. More recently, stimulated by the same strategy, the resulting increases in competitiveness have enabled Korean firms to gain significant shares in global markets in high-technology areas such as telecommunications, semiconductor memory chips, and LCDs. Protectionist policies may be effective in creating initial market opportunities for domestic industries, but they eventually lead to complacency in terms of innovation because of the lack of exposure to competitive forces.

4.6. *The Pivotal Role of the Government*

The Korean government has played a very significant role throughout the entire development process. Since the beginning of the industrialization process, Korea's visionary government provided effective leadership that ensured a stable and conducive macroeconomic environment, providing the mass education and training of the population, encouraging the assimilation of foreign technologies and developing a domestic R&D initiative, and establishing an accessible and modern information infrastructure. As the economy developed and became larger and more complex, it was best to leave economic activities to market forces, and the government consequently adopted a less direct interventionist approach and changed its role to that of an architect and regulator.

One characteristic of the Korean government that deserves special recognition is its long-term fiscal prudence, which allowed the Korean government to implement a series of post-1997 economic reforms even though it had incurred astronomical costs. Initiatives such as removing NPLs and recapitalizing financial institutions, establishing social programs such as unemployment insurance, and providing financial support for low-income families required public funds and contributed to enormous fiscal pressures. However, because of its long history of fiscal prudence and financial credibility, the Korean government was not saddled with public debt and was able to issue new bonds to finance the necessary reforms.

4.7. *Top-down and Bottom-up Approaches to Build a National Consensus*

Because the KE approach is broad, in that it permeates many areas of the economy and involves many actors, it is important to use both top-down and bottom-up approaches to build a national consensus and achieve a set of coherent strategies across the different of parts of the government, the private sector, and civil society. Top-down and bottom-up approaches are also key for successful reform processes. In the case of Korea, the sensitization of the society as a whole, including opinion leaders and politicians, to the notion of the knowledge economy had begun before the 1997 crisis as a bottom-up initiative of the *Maeil Business Newspaper*, the principal business newspaper in Korea. Before the crisis, the paper's management had organized policy forums that introduced the concept of the knowledge economy and highlighted its benefits, and the public was informed of the basics of the KE approach. The 1997 crisis may have actually energized existing KE awareness campaigns, which led to the government finally initiating its top-down approach — the government sought policy evaluations and advice from international organizations on the KE approach. Thus, in 2000, a report was jointly prepared by the World Bank and the OECD: *Korea and the Knowledge-Based Economy: Making the Transition*. The report inspired the subsequent government plans. In a typical Korean manner, all concerned government departments were mobilized to elaborate new policy measures with quantified objectives and budget estimates for implementing those measures. Meanwhile, to maintain and enlarge the KBE mobilization, the *Maeil Business Newspaper* has pursued its awareness actions in the direction of all sorts of target groups: business, civil servants, households, children, and even prisoners.

Indeed, without the resulting national consensus, efforts to reform and restructure the Korean economy would likely have been unsuccessful, because institutional reforms inevitably invoke resistance from vested interests, which frequently retards the reform process. For example, the legalizing of layoffs was made possible only through dialogues among the government, the private sector, and the labor unions, all of whom shouldered a portion of the costs associated with labor reform. No doubt there are specific features of Korean society that have facilitated its rapid adhesion to the KE concept. However, the general concept of a simultaneous top-down, bottom-up approach is one that warrants the consideration of policy makers implementing economy-wide strategies.

5. Conclusion

Korea's design and implementation of knowledge-based development strategies and the resulting rapid and sustained knowledge-led economic growth over the past four decades provide a wealth of valuable policy lessons for other developing economies. First, and particularly important, are the coordinated and complementary expansion of the four pillars of the KE framework — economic incentive and institutional regimes, educated and skilled workers, an effective innovation system, and modern and accessible information infrastructure — that evolved in tandem with the economy's various stages of development. These pillars provided the economy with the necessary means to effectively acquire and use knowledge to improve productivity and enhance long-term economic growth. Second, the strong and effective leadership provided by the government, which led to the coordinated development of the education, innovation, and ICTs pillars, was particularly important during the earlier stages of industrialization, when appropriate institutions to coordinate an economy-wide development agenda were not yet sufficiently established. The role of Korean government has appropriately mellowed in recent times to allow the market to further spur economic activity. Third, the economy-wide reforms and the array of policy measures that were implemented after the 1997 crisis serve as good examples of making the best use of opportunities to improve economic conditions. The government initiated the formalization of the action plan, which was orchestrated by MOFE, but also sought support from the civil society and some media groups. One outcome of those measures was the successful wiring of the Korean economy and the public, resulting in a first-class information infrastructure. And fourth were the exemplary ways in which the educational basis has been built and gradually expanded and the way technology has been gradually mastered and upgraded throughout the stages of industrialization.

Although Korea has made these advances, it needs to continue and increase efforts in reforming its higher education and innovation systems. These pillars have not sufficiently evolved in recent years to meet the demands of the current global economy, in part as a result of several pockets of resistance inherent in the Korean society and culture. In addition, a more proactive policy response is required to achieve more balanced economic development across the different sectors and niches of the economy. Concrete steps to resolve these issues will eventually become critical to Korea's continued transition to the knowledge economy and further sustained economic growth. To sum up these four decades of change, it has been easier to invest in hard infrastructure than to change mentalities, traditions, and institutional behaviors rooted in culture and history. This is not a new insight, and most, if not all, societies must endure this painful experience. Only by seeking a better understanding of a country's unique features and inner functioning can cultures progress in this matter.

References

Baldi, S., Y. Jin, M. Skemer, P. J. Green and D. Herget (2007). *Highlights from PISA 2006: Performance of 15-Year-Old Students in Science and Mathematics Literacy in an International Context* (NCES 2008–016). National Center for Education Statistics, Institute of Education Sciences, U.S. Department of Education, Washington DC.

IMD (International Institute for Management Development) (2005). The World Competitiveness Report 2005. Lausanne, Switzerland: IMD.

KEDI/MOE (2005). Statistical Yearbook of Education. Seoul: Korea Educational Development Institute.

KITA (Korea Industrial Technology Association) (2004). Major R&D Indicators. Seoul: KITA.

Knowledge Assessment Methodology (KAM). March 2006 (www.worldbank.org/kam).

Ministry of Education, various years. Statistical Yearbook of Education. Seoul: Ministry of Education.

Mitchell, G. R. (1997). Korea's strategy for leadership in research and development, Washington DC: US Department of Commerce.

Nam, D.-W. (1997). Korea's economic growth in a changing world. Seoul: Samsung Economic Research Institute.

OECD (2004). Learning for tomorrow's world — First results from PISA 2003. Paris, France.

OECD (2005a). OECD Science, Technology and Industry Scoreboard. Paris: OECD.

OECD (2005b). Education at a Glance. Paris: OECD.

OECD (2005c). Promoting Adult Learning. Paris: OECD.

OECD (2009). *Education at a Glance 2009: OECD Indicators.* Paris: OECD.

Suh, J. and D. H. C. Chen (2007). *Korea as a Knowledge Economy: Evolutionary Process and Lessons Learned.* The World Bank, Washington D.C.

The Government of the Republic of Korea (1999). DJnomics: A new foundation for the Korean Economy. Published for the Ministry of Finance and Economy. Seoul: Korea Development Institute.

Chapter

3

'Knowledge Society' as Academic Concept and Stage of Development

A Conceptual and Historical Review

Anna-Katharina Hornidge

1. Introduction[1]

In the past twenty to thirty years, visionary terms which have been incorporated into public speeches, academic writings, and day-to-day journalism such as 'knowledge society', 'information society' and 'knowledge-based economy'[2] announced a future in which social and economic development is increasingly based on knowledge. While

[1] This book chapter is largely based on Hornidge (2007).

[2] For reasons of terminological clarity, this paper subsumes the wide range of terms including 'knowledge society', 'information society', 'knowledge-based economy' under the term 'knowledge society'. The remaining terms are merely addressed in the sections specifically devoted to them.

the concepts 'knowledge society' and 'information society' were mainly developed by academics from Japan, the USA and Europe, the concept 'knowledge-based economy' was proposed somewhat later by international organizations such as the OECD.[3] From there — although far from complete — all three concepts entered the national politics of many countries which aimed at the active creation of better futures. Governments worldwide adopted the general idea of knowledge society as well as the manifold terminology originating from the scientific community. However, the theoretical concepts and definitions of knowledge society supporting this vision were hardly taken into account. Diagram 1 illustrates the overall focus and line of argument of this chapter: the conceptual and actual construction of knowledge society. While the creation of the concepts was mainly driven by the international scientific community, the construction of knowledge societies as stages of development has been pushed by national governments as well as (although not forming the focus of this chapter) actors from the private sector, the media and civil society groups. In redrawing this process of conceptual and actual construction, I follow Berger and Luckmann's approach of the social construction

---

[3] The following scholars can be mentioned: Machlup (1962); Umesao (1963); Lane (1966); Drucker (1969, 1993a, 1993b); Touraine (1969); Bell (1973, 1987); Porat (1976); Nora and Minc (1979); Böhme and Stehr (1986); Kreibich (1986); Castells (1989, 1996, 1997, 1998); Gibbons *et al.* (1994); Stehr (1994); and Willke (1998). They were later scrutinised and their concepts of knowledge society developed further by Kumar (1978); Gershuny (1978); Collins (1981); Lyon (1988, 1996); Dordick and Wang (1993); Stehr (1994, 1999, 2001a, 2001b); Webster (1995); Willke (1998, 1999); Maasen (1999); Dunning (2000); Evers (2000, 2002a, 2002b, 2003, 2005); Evers *et al.* (2000); Hofmann (2001); Steinbicker (2001); David and Foray (2002); Lloyd and Payne (2002); Evers and Menkhoff (2003); Mattelart (2003); Evers and Gerke (2005); Knoblauch (2004, 2005); Kübler (2005); Tänzler, Knoblauch and Soeffner (2006) and Hornidge (2007) to name a few.

Few scholars i.e., Lyon (1988, 1996); Webster (1995); Lloyd/Payne (2002); Mattelart (2003); Evers (2003); Knoblauch (2004, 2005); Tänzler, Knoblauch and Soeffner (2006); Kübler (2005); Evers and Hornidge (2007) and Hornidge (2007) point to the aspect of knowledge societies being constructed by social actors. The remaining scholars implicitly subscribe to the notion of knowledge societies emerging due to technological, economic and social developments taking place.

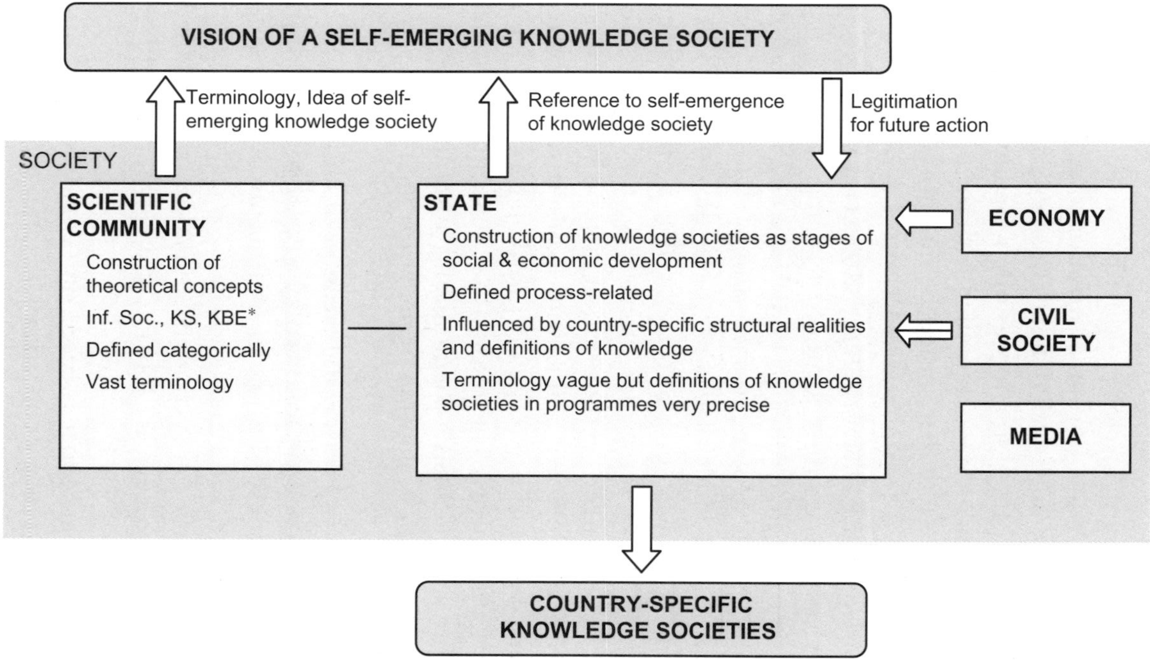

Diagram 1: The construction of knowledge society by the scientific community and state politics

*Inf. Soc. = Information Society; KS = Knowledge Society; KBE = Knowledge-based Economy.
Source: Hornidge (2007:4).

of reality and understand 'knowledge society' as it is defined by the social actors creating it (1984:16).

In the later half of the 20th century, multiple theoretical concepts of knowledge society were developed primarily by the scientific communities of Japan, USA and Europe as mentioned above. While the academic concepts were quite well defined, this new, manifold terminology lacked a clear distinction and was often used interchangeably. The terminological vagueness — combined with the picture of a self-emerging knowledge society that should be monitored, assessed and analysed — contributed to the construction of a vision of a self-emerging knowledge society (Hornidge, 2007). This vision describes a stage of development as future form of social and economic reality based on the increasing relevance of knowledge and information to social and economic development. Accordingly, knowledge society was often seen as a product of technological developments in the information and communication sector as well as economic developments in the service and knowledge intensive sectors. Governments of many countries embarked on the creation of knowledge societies as stages of national development and legitimised their actions by referring to the perceived necessity to guide, guard and monitor ongoing technological developments. The vision of a self-emerging knowledge society therefore acted as basis for legitimising government programs and activities towards the realisation of the envisioned future stage of development.

In the following, I will review the creation of knowledge society as theoretical idea and concept, followed by an outline of its adoption by US-American, Japanese, European and eventually Singaporean politics. It is the aim to show that knowledge society as theoretical concept and political vision was constructed and instrumentalised by actors worldwide. As such we are looking at a global hype with local consequences.

Methodologically this paper is based on (a) a review and discussion of primary and secondary conceptual literature on the notion of 'knowledge society' and (b) a review and qualitative assessment of US-American, Japanese, European and Singaporean state programs and initiatives towards 'knowledge society'.

2. 'Knowledge Society' — The Conceptual Development

Overall, the theoretical construction of the concepts of knowledge society can be structured into a primary[4] and a secondary phase,[5] as argued in detail by Hornidge (2007). During the primary phase and inspired by developments in the information and communication technology sector, especially the internet,[6] several scientists developed the idea of knowledge, information, as well as information and communication technologies becoming increasingly important for economic and social development and leading to a new stage of development, following the industrial society. The secondary phase of construction is characterised by the further development of the conceptual ideas, increasing their empirical base, depth and the theorising of these empirical data. While the conceptual ideas leading to the concepts 'knowledge society' and 'information society' were mainly developed in the 1960s to 1980s, the conceptual basis of the 'knowledge-based economy' was formed mainly in the 1990s.

---

[4] Here, theorists such as Umesao (1963); Nora/Minc (1979) and Castells (1989, 1996, 1997, 1998) can be named as contributors to the concept of a technology determined society, often called 'information society'. Lane (1966); Bell (1973, 1987); Touraine (1969); Kreibich (1986); Böhme/Stehr (1986); Willke (1998) and Gibbons *et al.* (1994) worked on a concept of a knowledge-driven society, generally labeled 'knowledge society', while Machlup (1962); Porat (1976) and Drucker (1969, 1993a, 1993b) can be listed together with international organisations such as OECD (1996a, b) and APEC (1998, 2000) as theorists constructing the concept of a 'knowledge-based economy'.

[5] Contributors to this secondary phase of construction include Kumar (1978); Gershuny (1978); Collins (1981); Lyon (1988, 1996); Dordick/Wang (1993); Stehr (1994, 1999, 2001a, 2001b); Webster (1995); Willke (1998, 1999); Maasen (1999); Dunning (2000); Evers (2000, 2002a, 2002b, 2003, 2005); Evers *et al.* (2000); Hofmann (2001); Steinbicker (2001); David/Foray (2002); Lloyd/Payne (2002); Evers/Menkhoff (2003); Mattelart (2003); Evers/Gerke (2005); Knoblauch (2004, 2005); Kübler (2005); Tänzler/Knoblauch/Soeffner (2006) and Evers/Hornidge (2007).

[6] On the development of the internet, see King and Kraemer (1995). Information infrastructure, national policy, and global competitiveness. *Information Infrastructure and Policy*, March. Until today, the internet forms the technological backbone of the 'knowledge society' by enabling simultaneous knowledge sharing which facilitates accelerated knowledge production.

2.1 *Phase I: Ideas, Terms, Concepts*

The concept of knowledge and its importance to society is not new. While, for instance, the philosopher Plato (428–347 BC) rated intelligence as the most important quality of a political leader, the philosopher and economist Mill argued, in 1863, that intellectual and moral education even surpasses industry and wealth in its effects on societal development (Mill, 1974). But if one wants to identify a founder of the notion of 'knowledge society', it should be the American sociologist Robert E. Lane. In 1966, Lane developed, based on his works on the US-American society, the concept of a 'knowledgeable society', assuming that knowledge, mainly referring to scientific, philosophical and cultural knowledge, replaces industrial organisation and production as the major source of productivity. Lane stated as "a first approximation to a definition", that "the knowledgeable society is one in which, more than in other societies, its members: (a) inquire into the basis of their beliefs about man, nature, and society; (b) are guided (perhaps unconsciously) by objective standards of veridical truth, and, at the upper levels of education, follow scientific rules of evidence and inference in inquiry; (c) devote considerable resources to this inquiry and thus have a large store of knowledge; (d) collect, organise, and interpret their knowledge in a constant effort to extract further meaning from it for the purposes at hand; (e) employ this knowledge to illuminate (and perhaps modify) their values and goals as well as to advance them". Lane described further: "Just as the 'democratic society' has a foundation in governmental and interpersonal relations and the 'affluent society' a foundation in economics, so the knowledgeable society has its roots in epistemology and the logic of inquiry" (1966:650). The common criticism towards the concept of the 'knowledgeable society', that knowledge is present and always has been present in all types of human society, Lane answered by mentioning that the elements of knowledge creation, consumption and furthering are present in some degree in every society; but "in the knowledgeable society they are present to the greatest degree" (1966:650).

In 1973, the American sociologist Daniel Bell then popularised the concept with his book "The Coming of Post-Industrial Society". Bell focused on the transformation from industrial to post-industrial society in which theoretical knowledge as axial principle forms the central, economic growth enhancing power. He stated; "the post-industrial society, it is clear, is a knowledge society" (1973:212) and identified two indicators for its 'emergence': (a) "the sources of innovation are increasingly derivative from research and development (...)", and (b) "the weight of the society — measured by a larger proportion of Gross National Product and a larger share of employment — is increasingly in the knowledge field" (1973:212). Parting the society into economic sectors, he argued that the post-industrial sector is vastly developing and changing due to telecommunication and computer technology.[7] The technological determinism, supported by Bell from the 1980s onwards, can also be found in the literature on the concept 'information society'. Today, four decades later, universities and research institutions are nevertheless far from being the central, axial institutions guiding societal development that Bell envisioned. Instead, several authors state that the university and research sector have long lost their dominant position (Evers, 2000; Knorr-Cetina, 1999; Willke, 1999; Heidenreich, 2003).[8] Despite this and further criticism[9] Bell's work on the emergence and rise of post-industrial

[7] In order to define the term 'post-industrial society', Bell identifies five components (1973:14):

1. Economic sector: the change from a goods-producing to a service economy;
2. Occupational distribution: the pre-eminence of the professional and technical class;
3. Axial principle: the centrality of theoretical knowledge as the source of innovation and policy formulation for society;
4. Future orientation: the control of technology and technological assessment;
5. Decision-making: the creation of a new 'intellectual technology'.

[8] Bell's reference to the increasing research budget of the United States in the 1960s and 70s (the time of writing) does not take into account their decrease from the 1980s onwards (Steinbicker, 2001:72).

[9] For details, see Hornidge (2007:33–35).

society remains to be the first detailed study on the change-enhancing aspect of increasing knowledge production and dissemination in society. As described by Bell, "the post-industrial society (...) is primarily a change in the character of social structure — in a dimension, not the total configuration of society. It is an 'ideal type', a construct, put together by the social analyst, of diverse changes in the society which, when assembled, becomes more or less coherent when contrasted with other conceptual constructs" (1987:73).

In Europe before the 1980s, only few European scientists took part in the discussion. An exception is the French sociologist Alain Touraine (1969). He published his thoughts on an evolving 'societé postindustrielle' in opposition to Bell's thesis and discussed the possibly arising conflict between knowledge 'have' and knowledge 'have-nots' as potential carrier of societal innovation and change.

The term 'information society' goes back to the Japanese economist Tadao Umesao, who in 1963 published his analogy between evolution and the three economic sectors of society. He argued that the agricultural sector can be regarded as an organism simply digesting the production of material goods, while he regarded the production of intellectual goods as analogous to organisms using their nerve systems for planning their actions and controlling their environment. This analogy between evolution and the three economic sectors led Umesao to his argument that society, once it reaches the highest level of societal development, concentrates on the production of intellectual goods (Umesao, 1963). The development of 'joho shakai' ('information society') became the aim of industrial development and since then has shaped the economic and research politics of Japan (Dordick and Wang, 1993:37). Eleven years later in Europe, Simon Nora and Alain Minc published a report on the "informatisation of French society" (Nora and Minc, 1979). The authors took a clearly technological approach and argued that the development of ICTs would act as an economic growth enhancing factor leading to first, a new sector of production specialised on the required hard and software, second, a technology driven productivity push in all industrial sectors and third, a diffusion of ICTs in all sectors of society and materialisation as central factor in the social infrastructure of nations worldwide. Hence, an 'information society', a society based on ICTs would arise. The process leading to this new state of society was labelled 'informatisation'. The combination of telecommunications and automatic data processing that was seen as the main drivers of this development was named "télématique" (Minc, 1987:134). Similar to the thoughts of Umesao in Japan, the report heavily shaped the politics of France in the field of ICT development (Nora and Minc, 1979:7).

Critically inspired by McLuhan's idea of the global village (1962),[10] Manuel Castells in 1989 published his thoughts on the transformation of spatial orders due to ICTs as manifestations of interaction between the restructuring of capitalism as a social system and informationalism as a new form of socio-technical organisation, with the informational city in its centre. (Castells, 1989, 2004; Castells and Laserna, 1989). In his trilogy "The Information Age" (Castells, 1996, 1997, 1998), Castells drew the picture of an 'informational capitalism', first (1996) outlining the informational society as revolution of the information technology, the global informational economy, the network cooperation, the transformation from labour and occupational structure and the evolution of a culture of virtuality with its final culmination in the network society; second (1997) discussing new social movements that oppose the instrumental and universal order of the networks; and third (1998) assembling analyses of the collapse of socialism and the second world, the downfall of the fourth world (referring to development countries as well as to peripheries in the metropoles), the rise of the tiger states in the Asia-Pacific-Rim, as well as the unification process of Europe. In the first volume, Castells developed his concept of an 'informational society'. Most central to his approach is the distinction between the capitalist mode of production and the informational mode of development. While the former is a way of organising a social system, the later is presented as a means of generating a given level of production.

[10] McLuhan assumed that ICTs would decrease the role of distances and hence leading to a restructuring of spatial orders with a decreasing importance of cities.

According to Castells, different societies operate with different modes of development, such as today ICTs announce "the rise of a new technological paradigm, which heralds a new mode of development" (1989:12). Parallel to Bell, Castells views the changes in techniques of production and development — due to ICTs — as well as the increasing importance of information and knowledge as central, but analytically independent axes of societal change. Thus, Castells regards the ICT revolution as main driver to all major structural transformations (Webster, 1995:196). He reasoned that networks form the new social morphology of society and the expansion of the network logic changes the functions and results of production processes, experiences, power and culture (1996:528). As an example, Castells identified the international financial flows as densest, most flexible and efficient global network, which — based on knowledge and information that continuously flow via ICTs around the globe — basically cannot be controlled by national or democratic institutions (1996:530).

Besides these academic works, popular scientists like Toffler (1970, 1980, 1990), Naisbitt (1982), Vester (1968), Buchholz (1970) and Jungk (1973) contributed to the conceptual development of a mainly technologically-determined 'information society' as well as at the time to the conceptual blurredness and lack of clear-cut definitions. As such, the terms and concepts left the academic field and entered political and popular culture, where they nourished the hopes and dreams towards a better future as discussed in the second half of this chapter.

So while the notion 'knowledge society' underlines the increased role of different types of knowledge for overall societal development, 'information society' regards ICTs as central to a new form of societal and economic development. In addition to this, the term 'knowledge-based economy' emphasises the increased role of knowledge, information and data to economic prosperity, an aspect basically already acknowledged by Adam Smith in 1776 by stating: "man educated at the expense of much labour and time (...) may be compared to one of those expensive machines" (Smith, 1910 quoted in Machlup, 1962:5). Two-hundred years later (1962), the US-American economist Fritz Machlup took up this thought and argued that a fourth economic sector can be added to the traditional three, namely agriculture, industry, services. He labelled this fourth sector the 'knowledge industry'. For the empirical analysis of it, he combined two approaches: (a) the industry approach and (b) the occupational approach (1962:44–50). In both approaches, he ascribed an economic value to industrial sectors and occupational groups and calculated their contribution to the US-American GNP.[11] Due to a proportionate contribution of the fourth sector, Machlup argued that a 'knowledge economy' is emerging. Machlup's statistical approach to grasping the 'knowledge economy' was later developed further by Marc Porat (1976), who calculated the contribution of 'information workers' and the 'information economy' to the overall GNP of the USA. His approach was nevertheless criticised for overlaps in his defining categorisation of occupations as 'information workers'. While he for example, counted judges and rent collectors as 'information workers', doctors were not. Today, Porat's calculations are mainly conjoined with Machlup's founding work. Despite existing criticism, Machlup's and Porat's calculations contributed to an easy-to-grasp and purely economy-oriented definition of 'knowledge society'. It later contributed to the fact that the terms 'information society' and

[11] The industry approach groups information goods and services that are not produced by information workers under the fourth sector. As an example, Machlup mentions the process of paper manufacturing. Within this fourth sector he distinguishes five industry groups (split into fifty sub-branches): (a) education (e.g., schools, libraries, and universities); (b) media of communication (e.g., radio and television, advertising); (c) information machines (e.g., computer equipment, musical instruments); (d) information services (e.g., law, insurance, and medicine); as well as (e) other information activities (e.g., research and development, and non-profit activities). In the occupational approach, all occupations concerned with the production and use of knowledge and information are listed. Yet, the disadvantage of the occupational approach is according to Machlup, that *firstly*, a connection between using information and knowledge at the work place and the production of information goods does not necessarily exist. *Secondly*, the occupational approach disrespects qualitative differences in the use of information/knowledge. He therefore concluded: "We conclude that both industry analyses and occupation analyses are needed in order to find out about the past development and present role of knowledge-production" (1962:48).

'knowledge-based economy', much less 'knowledge society', heavily entered the political sphere.

Parallel to Machlup and Porat's work, the US-American economist Peter F. Drucker (1969) was convinced that knowledge "has become the foundation of the modern economy" as we have shifted from an 'economy of goods' to a 'knowledge economy' (1969). For his analysis, he distinguished the "age of continuity" between 1913 and the late 1960s and the "age of discontinuity" after 1960. He argued that the main technical inventions took place in the years from 1913 to the beginning of World War I in the industrialising countries, while in the subsequent 50 years, economic development took place, but no change in structure. This changed with the age of discontinuity after 1960, which brought about fundamental changes in the areas of technology, economy, political structure and society. He holds four factors responsible for the emergence of the age of discontinuity: (a) the development of information and data configuration technologies; (b) the internationalization of the economy; (c) an individualization that leads to a neutralization of the main social and political organisations; as well as (d) the emergence of a 'knowledge society' in which knowledge becomes the central element. Drucker (1969:60) pointed to the development of ICTs which embody a new economic reality. Similarly to Bell, Drucker addressed the knowledge-based character of these technologies as a central aspect of knowledge society. Yet, for Drucker, this new economic sector is directed to a new expansive economic phase in which the state merely creates the legal and infrastructural frame, which then is filled by the industry itself. For Bell, in opposition to Drucker, these new industries point to an increasing dependence of economic growth from state organised basic R&D. Hence, the same focus (on ICTs as new industrial sector) was interpreted by Drucker and Bell very differently (Steinbicker, 2001:23).[12]

[12] In the economic sphere, Drucker predicted the development of a world economy that is characterised by increasing global integration, disregarding national borders. As an institution, guarding the production and distribution of goods worldwide, Drucker (1969:103–107) suggested a multinational world corporation, not national governments. Looking at the micro-level of economy, he emphasised the increasing

Similarly to Machlup, Drucker pointed to the growth of the 'knowledge industries'[13] and described the development of knowledge, its character and importance in economy and society from knowledge of salvation (*Erlösungswissen*) in 1700 to the linkage of technology and science as well as the application of knowledge in industrial processes and finally the application of knowledge in knowledge production today. He concluded that the increase in formal education after World War II (an increase in 'knowledge workers') as well as the character change of knowledge are the main driving forces for the emergence of a 'knowledge economy'.[14]

The above outline illustrates that the members of the scientific community who originally developed the multiple concepts of 'knowledge society' defined those categorically, meaning by dividing the assessed changes in society and economy into certain categories. The concepts as well as the terminology labelling the differing concepts are both manifold in character, used interchangeably and with textual overlaps. While the definitions of the varying knowledge society concepts are rather distinct, the interchangeably used and manifold terminology blurs the understanding of the notion of a knowledge society. Common to most academic works outlined above is nevertheless the belief that some kind of knowledge society

---

importance of the 'knowledge worker' in these 'knowledge industries'. In the political sphere, Drucker developed a theory of organisation which stated that the modern society is increasingly structured by specialised organisations that concentrate on certain social and political aspects in society. The interweaving of organisations with autonomous orientation created a new pluralistic order in society, according to Drucker (1969:219–223), which in turn witnessed the state loosing its central role. Drucker saw the state as increasingly dysfunctional and argued that a reorganisation of the state and its roles is required. In the social sphere, he clearly saw a 'knowledge society' arising. He argued, that "the central wealth-creating activities will be neither the allocation of capital to productive uses, nor 'labour'... Value is now created by 'productivity' and 'innovation', both applications of knowledge to work" (1994:8).

[13] He defined 'knowledge industries' as industries, producing ideas and information rather than goods and services.

[14] In his later works (Drucker, 1993b, 1994), Drucker developed the above outlined hypothesis further.

is emerging due to the technological developments in the information and communication industries, the growth of the service sector and the increasing knowledge-intensity of industrial products. But before discussing how the notions of 'knowledge society', 'information society' or 'knowledge-based economy' entered national politics, I shall outline the secondary phase of developing the theoretical concepts further.

2.2 *Phase II: Concepts, Theories and Recommendations towards Their Realisation*

The primary phase of constructing the theoretical concepts 'knowledge society', 'information society' and 'knowledge-based economy' was followed by a secondary phase. Here, scholars built on the above outlined works and attempted to specify the analyses of the social and economic changes, the developed concepts and the introduced terminology in order to offer a comprehensive picture of the assessed changes. During this secondary phase, the primary theories on the concepts continued to act as main reference theories which were scrutinised and theorised further. Common to most of these concepts has been the belief that the stage of societal development which they describe emerges as a result of the rapid technological developments in the information and communication sector, the growth of the service sector and the high profit margin of knowledge intensive goods. Hence, that the stage of development here named knowledge society is self-emerging.[15] This section will briefly discuss scholars contributing to the secondary phase. A more detailed outline of their works is, for reasons of limited space, provided in Hornidge (2007) only.

In 1986, Stehr and Böhme in "The Knowledge Society" aimed to contribute to a new approach towards formulating "a theory of

[15] Only few scholars such as Lyon (1988, 1996); Webster (1995); Lloyd/Payne (2002); Mattelart (2003); Knoblauch (2004, 2005); Tänzler/Knoblauch/Soeffner (2006); Kübler (2005); (Hornidge, 2007) point to the aspect of knowledge societies being constructed by social actors.

society which captures the dynamics of science, technology and society" (Böhme and Stehr, 1986:7). They agreed with the assertion of theorists such as Bell that knowledge arises as an 'axial principle' in highly developed societies. Yet, they criticised mainly three aspects: (a) a missing sociology of knowledge in the existing theories on knowledge societies defining its core element — knowledge (1986:16); (b) the usage of the term 'post-industrial society' rather than 'knowledge society' (Stehr, 1994:12); and (c) the internalism of the new sociology of science, neglecting the impacts of scientific knowledge on societal development (Böhme and Stehr, 1986:4). They stressed the importance of scientific knowledge and concluded that "contemporary society may be described as 'knowledge society' based on the penetration of all its spheres of life by scientific knowledge" (1986:8).[16] In 1994, Stehr (1994:9) defined knowledge as the capacity for social action, which — according to him — emphasises the aspect of value added due to knowledge (1994:95).

In 1998, the sociologist Helmut Willke underlined the independent production of new knowledge in all functional areas of

---

[16] It advances as follows (1986:8):

1. penetration of most spheres of social action by scientific knowledge ('scientification');
2. replacement of forms of knowledge by scientific knowledge (e.g., professionalisation). The role of experts and consultants is further discussed by Stehr in 1992 (Stehr/Ericson, 1992);
3. emergence of science as an immediately productive force;
4. differentiation of forms of political action (e.g., science and educational policy);
5. development of a new sector of production (the production of knowledge);
6. change of power structures (technocracy debate);
7. emergence of intellectuals as a new social class.

   In 1994, Stehr completed this list by replacing point 7 with point 8 and adding point 9 and 10 (1994:10/11):

8. emergence of knowledge as the basis for social inequality and social solidarity;
9. trend to base authority and expertise;
10. shift in the nature of societal conflict from struggles about the allocation of income and divisions in property relations to claims and conflicts about generalised human needs.

society as defining aspect of 'knowledge society'. He saw this in differentiated and highly technological societies of the West, where the structures and processes of the material and symbolic reproduction of society are penetrated by knowledge-based operations up to a degree that the importance of information processing, symbolic analysis and expert systems decreases compared to other factors of reproduction (1998:162). While Stehr and Böhme heavily emphasised the importance of scientific knowledge for the emergence of a 'knowledge society', Willke, as well as Michael Gibbons *et al.* argued that science and academic centres loose its former monopoly-status. According to Willke and Gibbons *et al.*, every sector of society, including the cultural, judiciary, economic and health systems, reproduces itself by producing its own knowledge independently (Willke, 1998; Gibbons *et al.*, 1994). This varying emphasis on the production of scientific knowledge stated the main disagreement between Stehr as well as Willke and Gibbons *et al.* It determined the further development of two divergent concepts of 'knowledge society'.

Analysing the production of this knowledge, Gibbons *et al.* stated that the increasing diversification and specialisation of the localities of knowledge production results in new forms of production. The scholars developed a 2-mode-concept: Mode 1 is the traditional way of producing knowledge, characterised by its homogeneity and disciplinary focus. Research problems are solved within academic institutions that are hierarchically organised. Mode 2 is the new form of producing knowledge, characterised as reflexive, multi- and trans-disciplinary and therefore dynamic and heterogeneous. Mode 2 knowledge is produced in a multiplicity of different organisations and institutions and is carried out in a context of application. It is generally a very problem-oriented form of knowledge production. In their work, Gibbons *et al.* prognosticated that Mode 1 is slowly replaced by or integrated into Mode 2. This results in a socially distributed knowledge production system which enables most members of society to take part in knowledge production as well as in the consumption of new knowledge (Gibbons, *et al.*, 1994:1–16). Evers took up this thought and argued that a triple helix of science-industry-university has emerged producing knowledge polycentrically with a multitude of 'epistemic cultures' and 'milieus of knowledge construction' (Evers, 2000, 2005). In the words of Knorr-Cetina: "A knowledge society is not simply a society of more experts, more technological gadgets, and more specialist interpretations. It is a society permeated with knowledge cultures, the whole set of structures and mechanisms that serve knowledge and unfold with its articulation" (Knorr-Cetina, 1999:7–8).

The role of national governments in influencing the definitions of 'knowledge society' according to their perspective of future, not current society, was critically questioned by Knoblauch (2004:360–361). For him the discourse surrounding 'knowledge society' in the political sphere represents the aim to construct different types of social reality rather than assessing it.[17] Kübler in 2005 underlined that 'knowledge society' is merely a myth constructed by academics, politicians and the media.

Similarly to the notion of 'knowledge society', 'information society' was developed further.[18] In 1995, Webster assessed several theoretical perspectives on 'knowledge' or 'information society' by identifying two groups: (a) those who proclaim the emergence of a new type of society (e.g., Bell, Castells, Baudrillard etc.); and (b) those who emphasise continuities (e.g., Schiller, Harvey, Giddens, Habermas, Garnham, etc.) and aimed "to shake at least some of the presumptions

[17] In 2006, Tänzler, Knoblauch and Soeffner underlined this further and regarded 'knowledge society' as "one of the last great inventions of the social sciences" which caused some sensation also outside of the academic world (Tänzler/Knoblauch/Soeffner, 2006:7).

[18] In 1988, Lyon critically assessed how the emergence of an 'information society' "is orchestrated, by whom, to what purpose and with what methods and effects" (1988:20). The book concludes that the concept 'information society' is ideological as well as utopian in character, used "to disguise the reality of powerful interests and beliefs at work within it" (1988:19), but should not be abandoned. As reasons he stated: (a) the process of 'informatising' poses questions concerning social, economic and cultural life that have to be discussed; (b) the development of ICTs is of social as well as technical relevance; (c) it should be remembered that technological potential is not social destiny; and (d) ICT-policies should always also involve social analysis.

of those who subscribe to the notion of the arrival of a novel 'information society'" (1995:4). He criticised the strong emphasis on technological development as prime vehicle of social change and cautioned that the assumption of a new form of society arising caused scholars to merely seek phenomena that might characterise the new order. The analysis follows the idea rather than the observation of social phenomena of change the analysis.

In 2001, Steinbicker critically assessed the works of Drucker, Bell and Castells. He identified a structural common ground: (a) the new means of productivity that are expressed in the organisational structure, work processes and changes in the academic system, as well as in the relationship between scientific community, state and economy; and (b) the transformation of labour and work relations. For Drucker, this is the inner contradiction of 'knowledge work'. For Bell, the central characterisation of the 'post-industrial society' typifies work/labour as the 'play between humans'. For Castells, the work conditions lever through the changes in the social structure that affect society. These aspects were not considered as crucial by the authors themselves, but, according to Steinbicker, pose a possibility for connecting all three towards an empirically based theory of the 'information society'. Concerning Castells' concept, this empirical analysis could focus on the interplay of technological development, economy, state and research as well as the institutional structures evolving. An empirical assessment of the transformation of work conditions should include (a) the thesis of structural change concerning work conditions; (b) the analysis of the gap between 'knowledge workers' and lower qualified workers as pointed out by Drucker and Castells; (c) the relation between 'knowledge work' and organisations as well as the meaning of hierarchy and control in the work process; and (d) the social relevance of structural change concerning work conditions. Yet, Steinbicker merely suggested the development of this theoretical model without developing it further.

In line with Webster's criticism of technological determinism underlying many analyses of the 'information society', Mattelart (2003) assessed the magnitude of which this technological development is the result of geopolitical interests. He argued that the idea of a global 'information society' is a construction that releases symbolic powers while at the same time legitimises political activities, for the satisfaction of geopolitical and economic interests. According to Mattelart, the term 'information society' developed due to the invention of the intelligent machines built during World War II. From the 1960s onwards, it emerged as academic, political and economic aim. The combination of (a) the belief in technology and technological process as well as (b) the idea of a sanctuary of all human knowledge built the ideological foundation of the term 'information society'. Aiming for a genealogical deconstruction of the term, Mattelart takes a geopolitical perspective and goes back in time, outlining the development of informatic machines, the academic debates on post-industrialism and its impacts on the national politics of Japan, France and the USA as well as the spread of the concept into international politics. In contrast to Webster or Steinbicker, he does not discuss the theoretical works contributing to the primary phase of conceptual construction, but rather focuses on the origin and consequences of the discourse and the creation of 'information society' in reality.[19]

In comparison to the terms 'knowledge society' and 'information society', the notion of the 'knowledge-based economy' was less constructed in the academic sphere, but mainly developed by international political organisations and think tanks, before it triggered down to national politics. Since the early 1990s, multinational organisations such as the Organisation of Economic Cooperation and Development (OECD) as well as the Association of South-East Asian Nations (ASEAN) in publications spoke of a 'knowledge-based economy'. In 1996, OECD defined the 'knowledge-based economy' by emphasising the importance of knowledge as "the driver of

[19] Besides the here mentioned secondary theorists, many more contributed to the construction of the concepts. These include authors such as Bittlingmayer (2005); Cawkell (1987); David/Forray (2002); Dizard (1982); Dordick/Wang (1993); Dunning (2000); Feather (1998); Garnham (2002); Gill (1996); Lievrouw/ Livingstone (2002); Lloyd/Payne (2002); Martin (1995); Riddle (1988) and Williams (1982, 1991).

productivity and economic growth leading to a new focus on the role of information, technology and learning in economic performance" (1996a). In another article of the same year, the origin of the term 'knowledge-based economy' is seen in the "fuller recognition of the role of knowledge and technology in economic growth" (1996b). The 'knowledge-based economy' is here regarded as naturally emerging. As indicators of this emergence were identified: (a) the strongest expansion of output and employment in high-technology industries such as computers, electronics and aerospace; and (b) the rapid growth of knowledge intensive service sectors such as education, communications and information. Based on this, OECD estimated >50 percent of GDP in the major OECD economies as knowledge-based (1996b:9). Several of the conceptual ideas of OECD were also adopted by the Asia-Pacific Economic Cooperation (APEC) for the Asia-Pacific Region. In 2000, APEC defined the 'knowledge-based economy' as "an economy in which the production, distribution, and use of knowledge is the main driver of growth, wealth creation and employment across all industries". A definition that was later also adopted by the Singaporean government (Toh *et al.*, 2002). Overall, these organisations did not question the emergence of a 'knowledge-based economy' but merely assessed them along mainly economic indicators.

Despite the conceptual and terminological overlaps of the notions 'knowledge society', 'information society' and 'knowledge-based economy', the terms 'information society' and 'knowledge-based economy' entered the political sphere more rapidly than the more academic term 'knowledge society'. Reason might be the technological and economic connotations, emphasising the importance of ICTs and their infrastructure for economic prosperity, which emerged as a political focal point in many countries from the 1980s onwards.

3. Entering the Political Sphere — USA, Japan and the European Union

As pointed out by Kubicek *et al.* (1997:9), the notion of 'knowledge society' was a technology-focused idea that — until the 1990s — failed to capture the public's imagination. While in the 1970s, many industrial countries promoted microelectronics as well as the "new media" cable TV and view data, the promoted technology in the 1980s was ISDN (Integrated Services Digital Network). In the 1990s, multimedia and 'the information superhighway' emerged as new catchwords, which are currently replaced by WLAN (Wireless Local Area Network), UMTS (Universal Mobile Telecommunications System) and digital signalling (in opposition to analogue handheld two-way radio). Yet eventually the governments of many countries, with the USA, Japan and the European Union belonging to the foremost, embarked on political programmes aiming at the construction of 'knowledge societies'. In the following paragraph, I will highlight some activities of these early players.

3.1 *United States of America*

In the late 1960s, the internet, which until today can be regarded as the technological backbone of the 'knowledge society', was under the term ARPANET developed in several US-American defence-sponsored university research projects. It was designed as communication network which could withstand a nuclear attack. From 1971 to 1981, the ARPANET grew from two dozen sites to 200. Besides the ARPANET, several other private companies built their own networks. IBM for example built the VNET. In order to enable communication between these networks, a set of conventions was drafted and published in the network community. The publication of these conventions, combining the varying networks that existed until then, basically marks the beginning of the internet. In 1993, 160 countries were connected and the internet as a communication infrastructure with mailing services, file transfer and news groups was created (King and Kraemer, 1995:5). As technological foundation for increasingly faster, by now simultaneous communication and knowledge exchange around the globe, the internet fundamentally inspired the creation of the theoretical concepts of knowledge society. It highly pushed the idea of an 'arising' information or knowledge society in the public, political and academic spheres worldwide.

In the USA in 1992, Governor Bill Clinton and Senator Al Gore used the development of the enormously fast growing internet for a successful presidential campaign (Kubicek, 1999:70–71; Read and Youtie, 1995:101; Schneider, 1997:345). The 'information infrastructure', as named by Clinton and Gore, was quickly given the name 'information superhighway' by the public. A few months after taking power, the Information Infrastructure Task Force (IITF) composed of high–level representatives from various ministries was formed under the lead of the then Minister of Trade, Ronald H. Brown, as well as accompanied by an advisory council constituted by high level representatives from business, scientific community and non-government organisations. In September 1993, the IITF published the Agenda for Action, a mix between declaration and action plan. Here, the National Information Infrastructure (NII) was defined as a "seamless web of communication networks, computers, databases, and consumer electronics that will put vast amounts of information at users' fingertips" (IITF 1993). The access to information should lead to an information revolution which was supposed to introduce sustainable changes to human lives, work and interaction.

One year after founding IITF, a progress report was published (IITF, 1994a). Furthermore, the working group on applications published the broad spectrum of applications of that time (IITF, 1994b). The outline involved topics such as e-commerce, industry applications, disaster management, schools, libraries and art. The Advisory Council, founded in 1993, published three reports during its three year activity (US-Advisory Council on the National Information Infrastructure, 1995, 1996a, 1996b), which all focused on establishing universal access to NII. Consequently, the question of universal access was increasingly also recognised by IITF itself. Following from this, the government appointed the National Telecommunications and Information Administration (NTIA) with the task to suggest ways of assuring universal access to the government in preparation for a telecommunication reform. This reform of the telecommunication law was finally executed in 1996, aiming at the deregulation of the telecommunication sector (Kubicek, 1999:70). Further areas of legal reform were copyright, data security and coding, control of illegal discriminating contents as well as the equipment of schools. In the second term of the Clinton-/Gore-administration, the focus shifted to numerous projects aiming at the final user. Hence, NII was continued to be built, but at a more user-oriented level than before.

By successfully making use of the envisioning character of the Agenda for Action and the booming growth of the internet technology the Clinton-/Gore-administration clearly managed to introduce the topic of ICTs to the public (Read and Youtie, 1995:101). Here the use of the term 'information superhighway' which drew an analogy to the construction of the interstates, the US-American motorways that link the different states (Kubicek, 1999:70–71) was a cleverly chosen image for fostering public acceptance. The technological focus underlined the tangibles of 'knowledge society' and found expression in the terminological choice of 'information society'.

3.2 *Japan*

In Japan, the term 'information society' has been frequently used in government reports and publications from the late 1960s onwards. Special focus laid on the impact of technological development, specifically in the field of microelectronics, and on social and economic processes of transformation (Steinbicker, 2001:18). Nevertheless, it is important to note, that at the end of the 1960s in Japan, the stage of development labelled 'knowledge society' was regarded as a revolution inside the system of industrial society. The idea, that 'knowledge society' might replace industrial society only emerged with the beginning of the 1970s.

In 1971, the Japan Computer Usage Development Institute produced a governmental action plan with the title "The Plan for an Information Society: A National Goal towards the Year 2000" (Vogel, 2000:286–288). Similarly to the IITF in the USA, the plan identifies the private sector as main actor in the process of creating k-society. Besides this, it paints the following image of a future society: a central state con trolled database; linked up telesystems; programmed school lessons, which foster an ICT-embracing attitude; a central information system for small and medium sized enterprises; and a centre for retraining parts of the work force. Mattelart (2003) describes it as "Computepolis", a city, completely linked via personal computers, with automatic traffic planning, mega-supermarkets, computer guided transport vehicles and fully automated air-conditioning systems (2003:91–92). Interestingly the Federal Ministry of Education and Science of Germany published a German translation of this report (entitled "Japans Technologische Strategie") merely one year after its publication in Japan (BMBW, 1972). This suggests that while the German government was not yet speaking of a German knowledge or information society, the activities of other players, i.e., Japan, were monitored.[20]

By the late 1970s, attention turned towards knowledge production for development (Morris-Suzuki, 1996:212; Tuomi, 2001:4). In 1976 and with the aim to stimulate corporate creativity, the Very Large Scale Integration (VLSI) project was set up under the auspices of the National Research and Development Programme of the Ministry for Industry and Trade (MITI), focusing on the development of microchips. With the beginning of the 1980s, MITI defined new materials, biotechnology and new forms of microelectronic technology as main areas of innovation and research supported by the government and several highly government-financed research projects followed (Morris-Suzuki, 1996:214). In 1985, the Key Technology Promotion Centre was set up jointly by MITI and the Ministry of Posts and Telecommunications. By the early 1990s, the Key Technology Promotion Centre had supported several hundred research projects, mainly in the area of microelectronics. According to MITI, its activities in the information sector proved to be successful, when Japanese companies slowly took over a major market share in the hard drive and personal computer producing industry (Mattelart, 2003:93).

The hopes of a better future, fostered by the development of the microchip, video and audio systems industry were, in Japan, popu-

[20] The first appearance of the terms ‚information society' and 'information economy' in a German federal government document can be found in the final report of the enquete-commission "Future of the Media in the Economy and Society — Germany's Road into the Information Society" (DBt, 1998).

larised by the futurist Yoneji Masuda. In his book, "The Information Society as Post-industrial Society", published in 1980, Masuda describes a future society, in which intellectual creativity wins over the consumer society, self discipline is socially integrated and humans live in harmony with nature (Masuda, 1980). Nevertheless, in the 1990s, the Japanese government turned away from its former strong focus on the computer industry and increasingly emphasised the reformation of the educational system as well as basic research, in order to provide for potential long-term development (Vogel, 2000:323). Interestingly, this change in focus shows a clear parallel to the developments in Singapore, discussed below.

In 1994, MITI published a "Programme for Advanced Information Infrastructure" focusing on the expansion of the information technology network, connecting businesses, research institutions, offices and corporate production sites (MITI 1994). In 1999, the Ministry of Posts and Telecommunications in Japan published a White Paper, entitled "Communications in Japan 1999" (MPT, 1999). It assessed the impact of the internet on the status of the information and communication industry and policies of Japan and resulted in the formulation of the IT policy package of the Japanese Government, entitled "E-Japan Strategy" in January 2001. It overall focused on the further development of communication facilities[21] and therefore continued Japan's technological focus towards 'knowledge society'.

3.3 *European Union*

On the level of the European Union, 'knowledge society' as a topic only gained relevance in the early 1980s. In November 1983, the Council of Europe established the Senior Officials Group on

[21] Its main objectives were (Noguchi, 2003:71):

1. to increase the volume of e-commerce in 2003 to ten times the level of 1998;
2. to make available the world's most advanced communication networks by 2005;
3. to increase the number of MA and PhD holders in IT-related fields to a level comparable to the USA;
4. invitation of thirty thousand highly skilled foreign workers.

Telecommunications (SOGT) as an advisory group to the European Commission. Additionally, a subgroup of SOGT was formed with the name GAP (Group d'Analyse et de Prévision) in order to support the European Commission in the long-term development of the telecommunication networks (Campbell and Konert, 1998:73–74). Aiming for economic growth and employment, ICTs were identified as key technologies and therefore research in this field identified for being especially pushed (Vogel, 2000:324–333). Two examples are ESPRIT (European Strategic Programme on Research in Information Technology — since 1984) as well as RACE (Research and Development in Advanced Communications Technologies in Europe — since 1988). Providers of telematic applications were supported by programmes, such as DELTA (Developing European Learning through Technology Advance) and DRIVE (Dedicated Road Infrastructure for Vehicle Safety in Europe). These research programmes were accompanied by the Commissions Action Plan on Telecommunications (EC, 1984).

The from there following telecommunication politics of the European Union were mainly structured by the "White Paper on the Completion of the Community-wide Market for Goods and Services", published in 1985 (EC 1985) and the "Green Paper on the Completion of the Common Market for Telecommunication Services," published in 1987 (EC, 1987). With the beginning of the 1990s, the European Union widened its focus from explicitly ICT-development to ICT-applications. In 1993, the Commission published a White Paper entitled "Growth, Competitiveness, Employment — The Challenges and Ways forward into the 21st Century" (EC, 1993). It emphasizes the importance of trans-European networks as stimulation for the European economy and a decrease in unemployment. The construction of information networks and European networks in the transport and energy sectors were to be of prime importance. The increased focus on the effects of ICTs on work processes resulted in a decrease of the terms 'electronic highways' and 'information economy'. Instead the European Commission adopted the term 'information society' and reasoned that Europe focuses its activities, differently to the USA, on social and educational aspects of the assessed stage of development (Kubicek, 1999:73).

This White Paper formed the basis for the foundation of a high-level expert group in cooperation with representatives from the industry. In 1994, the expert group, headed by Martin Bangemann, presents its report "Europe and the Global Information Society — Recommendations to the European Council" at the EU-summit in Corfu (Bangemann, 1994). The group argued that the best support for information networks and services would be open and competitive markets.[22] The Bangemann-Report can be regarded as EU's master-document and key reference point for policy initiatives related to the electronic communication sector (Preston, 1997:282). On its basis and after being requested to do so by the European Heads of State and Government, the Commission of the European Union published the action plan "Europe's Way to the Information Society" in the same year (EC, 1994).[23] This was followed in July 1996 by the "Green Paper on Living and Working in the Information Society: People first" focusing on social aspects of 'knowledge society' (EC, 1996a). In November 1996, the Commission adopted the action plan "Europe at the Forefront of the global Information Society" building on completed, pending and ongoing activities (EC, 1996b). Early 1998, the liberalisation of the European telecommunication sector was completed.

In March 2000, Europe's political leaders met on an EU-summit in Lisbon. Here, the target of developing Europe into "the most dynamic, competitive and knowledge-based economy in the world by 2010" was explicitly formulated (EC, 2000a). Consequently, the

[22] The report outlines four steps to shape Europe's way into a knowledge society: (a) the liberalisation of Europe's telecommunication markets; (b) the creation of a common regulatory framework regarding standardisation; (c) the protection of intellectual property rights; and (d) respect of privacy and the security of data transmission.

[23] It focused on four areas: (a) the regulatory and legal framework; (b) the networks, services, applications, and content; (c) the social and cultural aspects; and (d) the promotion of k-society.

European Council published the eEurope 2002 action plan in June 2000 (EC, 2000b), designed to speed up and extend the use of the internet to all sectors of European society. In June 2002, at the Seville European Council, the eEurope 2005 Action Plan was launched and endorsed by the Council of Ministers in the eEurope Resolution of January 2003 (EC, 2002). It states the aim to develop public online services (eHealth, eLearning and eGovernment) and a dynamic environment for e-business through widespread availability of broadband access at competitive prices and a secure information infrastructure. In June 2005, the European Commission set out a new strategic framework, entitled i2010 — A European Information Society for Growth and Employment and the progress made by eEurope 2005 as well as by i2010 was assessed in a benchmarking report in December 2005 (EC, 2005:2).[24] While the technological focus is here further maintained, the inclusion of all members of society into the usage of ICTs and decentralised structures of knowledge production becomes an increasingly discussed topic.

3.4 *Singapore*

Similarly to the USA, Japan and the European Union, Singapore, a small island-state with limited land and labour available, adopted the image of a 'knowledge society' for long-term economic growth and social stability. From the early 1980s onwards (around the same time as the European Union), the Singaporean government has taken enormous action towards its realisation. Table 1 lists the government programs that clearly pursued this vision.

In Singapore of the 1980s to early 90s the focus of government activities towards 'knowledge society' clearly laid on the building of an

[24] Main trends identified are: (a) broadband roll-out is a clear success; (b) disparities between the member states have not yet been reduced; (c) connectivity of enterprises is high throughout EU25; (d) availability of online public services has continued to grow; (e) all member states are confronted with the challenge of extending knowledge society to people with little or no formal education, those not in employment and older people (EC, 2005:2–3).

Table 1: State activities for a Singaporean knowledge society

Year	Name of Initiative	Implementing Authority
1981	National Computerisation Plan	Ministry of Trade and Industry & Committee on National Computerisation
1981	Civil Service Computerisation Program	Ministry of Education & Civil Service Computerisation Group
1985	National IT Plan	National IT Plan Working Committee
1991	Founding NSTB/A*STAR	Minister Cabinet
1992	A Vision of an Intelligent Island — The IT2000 Report	National Computer Board
1994	Library 2000	Ministry for Information and the Arts & Library 2000 Review Committee
1996	Singapore ONE	National Computer Board
1997	Thinking Schools, Learning Nation	Ministry of Education
1997	1st Masterplan for IT in Education	Ministry of Education
2000	Infocomm 21	Infocomm Development Authority
2000	1st eGovernment Action Plan	Infocomm Development Authority
2002	2nd Masterplan for IT in Education	Ministry of Education
2002	Creative Industries Development Strategy	Economic Review Committee, Workgroup on Creative Industries
2003	Connected Singapore	Infocomm Development Authority
2003	Innovation & Enterprise	Ministry of Education
2003	2nd eGovernment Action Plan	Infocomm Development Authority
2004	Teach Less, Learn More	Ministry of Education
2005	Library 2010	National Library Board
2006	Intelligent Nation 2015	Infocomm Development Authority

Source: Hornidge (2010).

ICT infrastructure in conjunction with a supportive legal infrastructure. As such can be mentioned the "National Computerisation Plan" (1980–1985; Committee on National Computerization, 1980), the "National IT Plan" (1986–1991; National IT Plan Working

Committee, 1985), "A Vision of an Intelligent Island — The IT2000 Report" (1992–1999; NCB, 1992), "Infocomm21" (2000–2003; IDA, 2000), "Connected Singapore" (2003–2006; IDA, 2003) and "Intelligent Nation 2015" (2006–2015; IDA, 2007). It therefore was a very technologically and economically determined version of knowledge society pursued by these activities. This primary focus on ICT and legal infrastructure was complemented by the application of ICTs in the public administration, with the first program, the "Civil Service Computerisation Plan", launched in 1981. Further examples are the "eGovernment Action Plans" (NCB, 1982; IDA and Chua, 2006) and the "Masterplans for IT in Education" (MOE, 1997, 2002). The actual shift towards the wide application of ICTs, nevertheless, only took place in the late 1990s, with the above mentioned ICT focused programs increasingly emphasizing the application of ICTs in private and professional life, educational facilities and the public service. Here "Innovation and Enterprise", launched in 2003, can also be mentioned.

In 1991, these ICT focused definitions of knowledge society were challenged by activities addressing the production of knowledge, meaning basic and applied research and development. Examples are the founding of the Agency for Science, Technology and Research (A*STAR) in 1991[25] as well as the construction of a biomedical research hub, similar to an industrial park, called Biopolis in 2001 (Hornidge, 2008).[26] It derived from an increased

[25] A*STAR is a statutory board of the government under the Ministry of Trade and Industry which oversees altogether 12 research institutes.

[26] Biopolis is part of Fusionpolis, which is split into Vista X-Change (centre for private-public-partnership and industry development, financial and business services), Central X-Change (centre for ICTs, media and education industries) and Life X-Change (Biopolis). Together these three form Fusionpolis, which is stated in a newspaper article from 2003 to be "Singapore's icon of the knowledge economy where talents gravitate naturally and where diverse ideas thrive. With a focus on knowledge intensive activities in critical growth sectors, one-north would provide an intellectually stimulating and creative physical environment for entrepreneurs, scientists and researchers to congregate, interact and exchange ideas" (JTC Corporation, 20.02.2003).

awareness of the importance of local knowledge production. In the mid 1990s, this was further complemented by the realisation, that a knowledge society requires the potential of every citizen. With the building of a vast library scene, the Singaporean government aimed to allow everyone to participate in the usage of ICTs as well as in knowledge creation and transmission. As main action plans pursuing this inclusive definition of 'knowledge society' attempting to close the digital divide between social groups, "Library 2000" and "Library 2010" can be identified (Library 2000 Review Committee, 1994; NLB, 2005).

In the 2000s, the fostering of creativity among Singapore's citizens and the development of creative industries (e.g., design and arts), which in application of ICTs shall contribute to Singapore's economic development, moved into the center of government attention. This belief in creativity subscribes to an image of knowledge society as a creative economy and resulted into an increased push for a national cultural policy. Singapore's first cultural policy, the "Ong Teng Cheong Report" was launched in 1989 (ACCA, 1989). This was then followed by the foundation of the Ministry of Information and the Arts (later renamed into Ministry of Information, Communications and the Arts — MICA) and the development of "The Esplanade" into a major performance arts venue in 1990. In 1991 and 1993 respectively, the National Arts Council (NAC) and the National Heritage Board (NHB) were established as further coordinating and planning bodies. In 1995, the planning document "Singapore: Global City of the Arts" was published (MITA and STPB, 1995), closely followed by the "Renaissance City Report: Culture and the Arts in Renaissance" in 2000 (MITA, 2000). The turn towards the marketability of arts nevertheless came with the publishing of the "Creative Industries Development Strategy" in 2002 (Workgroup on Creative Industries, 2002). Here three approaches to defining the scope of the creative cluster in Singapore were identified and their focused development decided: (a) the cultural industries; (b) the creative industries; and (c) the copyright industries. Together these three pillars are hoped to develop arts and culture as economic sector, as key element for Singapore as 'global city' attracting 'foreign talents' and tourists and as creative basis for innovative R&D-work in all knowledge producing sectors and disciplines.

Taking the examples of the USA, Japan, the European Union and Singapore and their government activities towards the further development of information and communication technologies as well as into 'knowledge societies' indicates three main points. First, the internet, developed by defence-financed research projects, originates from military interests. Second, the technology then, as argued in section 2 of this chapter, inspired academic and visionary thinking that resulted in the creation of the theoretical concepts and visionary terms such as 'knowledge society', 'information society' and 'knowledge-based economy'. Third, these concepts and visions were then again adopted by governments worldwide to push the actual creation of these envisioned societies further by releasing programmes and initiatives fostering the building of information and communication networks, the application of the technology, as well as knowledge creation and transmission. Yet with regard to the USA, Japan, the EU and Singapore, the above illustrates that each of the four, legitimised their activities towards 'knowledge society' with their current situation. While Japan hoped for a solution to the hollowing out of its manufacturing industry, the European initiative has to be understood as a reaction to the pressures of global competition and as a transnational reaction to increasing unemployment. In the USA, the information superhighway was regarded as a solution to the U.S. infrastructural crisis and advanced as a presidential campaign topic. In Singapore, the drive towards 'knowledge society' has been identified as pathway to economic survival and socio-political stability.

4. Concluding Remarks

The above review of the conceptual and actual construction of 'knowledge society' and therewith the redrawing of a global hype aimed to shed light on the notion 'knowledge society' by assessing how it has been defined by academics in their writings and by governments in their actions around the world. It rests on the belief that social reality is shaped by its actors; basically that, to say it with the words of Franz Kafka, "paths are made by walking". The rapid developments in the information and communication technologies inspired scientific writers around the world, aiming to conceptually grasp the ongoing and future developments. During a primary and secondary phase a wide range of conceptual approaches to 'knowledge society', 'information society' and 'knowledge-based economy' were developed. From there and despite (possibly even fostered by) terminological overlaps, lacking clear-cut definitions, these notions entered the public sphere and were taken up by governments around the globe for either legitimising ongoing programmes or pushing future economy and technology oriented activities. Of the outlined countries above, Singapore, while nevertheless being purely economically motivated, heavily included the fostering of the arts into its later understanding of 'knowledge society'. The USA, Japan and the EU at large stayed with a technology-driven definition of the concept, focusing on the traditional sectors of economic growth.

Consequently, once the notions had entered into the public discourse, they were discussed, defined and redefined by a multitude of actors in multiple different ways, each outlining a unique path to and from 'knowledge society' as vision and stage of development. This multitude was then taken up again by academics, who developed the concepts further, while in parts the social actors creating 'knowledge society' as stage of development were continuously in the process of readjusting their original definition of it. This was portrayed above by the case of Singapore, which moved within a very short time frame from a purely technological definition to a definition integrating technological, economic aspects with areas such as culture, the arts and creativity being integrated into its economic focus.

This process of continuous further development of the notions and concepts as such and the multitude of definitions attached to them as stages of development continues. Yet, at the same time, the paths taken by governments around the world outline in detail different conceptualisations of 'knowledge society' which all are neither exactly that nor anything else than just that: 'knowledge society' as we, the social actors fostering its notion, define it.

References

ACCA (Advisory Council on Culture and the Arts) (1989). Report of the advisory council on culture and the arts. Singapore: Singapore National Printers.

APEC Economic Committee (1998). Towards an information society: Developments in apec. Singapore: APEC Secretariat.

——— (2000). Towards knowledge-based economies in APEC — Framework and summary. Singapore: APEC Secretariat.

Bangemann, M. (1994). Bangemann-Bericht: Europa und Die Globale Informationsgesellschaft. Empfehlungen Für den Europäischen Rat. Brüssel: Kommission der Europäischen Union.

Bell, D. (1973). *The Coming of Post-Industrial Society*. New York: Basic Books Inc.

——— (1987). The Post-industrial society: A conceptual schema. In *Evolution of An Information Society*. A. E. Cawkell (ed.). London: Aslib.

Berger, P. and T. Luckmann (1984). *Die Gesellschaftliche Konstruktion der Wirklichkeit — Eine Theorie der Wissenssoziologie*. Frankfurt a.M.: Fischer Taschenbuch Verlag.

Bittlingmayer, U. (2005). *Wissensgesellschaft als Wille und Vorstellung*. Konstanz: Universitätsverlag Konstanz.

Bmbw (Bundesministerium für Bildung und Wissenschaft) (1972). Japans Technologische Strategie — Wiedergabe Und Kommentar Einer Studie Des Ausschusses Für Technologischen Fortschritt Des Japanischen Wirtschaftsrates. Bonn: BMBW.

Böhme, G. and N. Stehr (eds.) (1986). *The Knowledge Society. The Growing Impact of Scientific Knowledge on Social Relations*. Boston: D. Reidel Publishing Company.

Buchholz, A. (1970). Die grosse transformation. Gesellschaftliche Zukunftserwartungen und Naturwissenschaftlich-Technischer Fortschritt. Ort 2: Reinbeck.

Campbell, P. and B. Konert (1998). Bausteine Der Informationsgesellschaft in Europa — Die Entwicklung der Europäischen Politik Zur Förderung der Informationsgesellschaft. Media Monographie 22. Ed. Bernd-Peter Lange. Düsseldorf: Europäisches Medieninstitut e.V.

Castells, M. (1998). *The Information Age: Economy, Society, and Culture, Volume 3: End of Millenium*. Oxford, MA: Blackwell Publishers.

——— (1997). *The Information Age: Economy, Society, and Culture, Volume 2: The Power of Identity*. Oxford, MA: Blackwell Publishers.

——— (1996). *The Information Age: Economy, Society, and Culture, Volume 1: The Rise of the Network Society*. Oxford, MA: Blackwell Publishers.

——— (2004). *The Network Society: A Cross-Cultural Perspective*. Cheltenham, UK; Northampton, MA: Edward Elgar.

——— (1989). *The Informational City — Information Technology, Economic Restructuring, and The Urban-Regional Process*. Oxford: Basil Blackwell.

Castells, M. and R. Laserna (1989). The new dependency: Technological change and socioeconomic restructuring in Latin America. *Sociological Forum*, 4(4): 535–60.
Cawkell, A. E. (1987). *Evolution of an Information Society*. London: Aslib.
Collins, R. (1981). *Postindustrialism and Technocracy. Sociology since Midcentury*. Randall C. (ed.). New York: Academic Press.
Commission of the european communities (EC) (1984). Action plan on telecommunications, Com(84) 277, 17.05.1984.Brüssel: EC.
——— (2000b). Europe 2002: An information society for all, 14.6.2000. Brüssel: EC.
——— (2002). Europe 2005: An information society for all, Com(2002) 263, 28.5.2002. Brüssel: EC.
——— (1994). Europas Weg in Die Informationsgesellschaft. Brüssel: EC.
——— (1996b). Europe at the forefront of the global information society. Rolling action plan. Communication from the commission Com(96) 607 Final. Brüssel: EC.
——— (1996a). Green paper on living and working in the information society: People first, Com(96) 389, 24.07.1996. Brüssel: EC.
——— (1987). Green paper on the completion of the common market for telecommunications services, Com(87)290, 11.06.1987. Brüssel: EC.
——— (2005). Information society benchmarking report. Brüssel: EC.
——— (2000a). Lisbon european council presidency conclusions. Lisbon: Kommission der Europäischen Gemeinschaften.
——— (1994). Weißbuch Wachstum, Wettbewerbsfähigkeit Und Beschäftigung. EGBulletin 6.
——— (1985). White paper on the completion of the community-wide market for goods and services by 1992, Com(85)310. Brüssel: EC.
Committee on National Computerisation (1980). Report on national computerisation. Ed. Tony Tan Keng Yam. Singapore.
David, P. A. and D. Foray (2002). *An Introduction to the Economy of the Knowledge Society*. Oxford: Blackwell Publishers.
Deutscher Bundestag (DBt) (1998). Dritter Zwischenbericht der Enquete-Kommission Zukunft der Medien in Wirtschaft und Gesellschaft — Deutschlands Weg in Die Informationsgesellschaft. Bonn: Deutscher Bundestag, 13. Wahlperiode, 04.05.1998.
Dizard, W. P. (1982). *The Coming Information Age — An Overview of Technology, Economics, and Politics*. New York: Longman Inc.
Dordick, H. S. and G. Wang (1993). *The Information Society: A Retrospective View. Newbury Park*. California: SAGE Publications.
Drucker, P. F. (1969). *The Age of Discontinuity. Guidelines to Our Changing Society*. London: Heinemann.
——— (1993a). Die Postkapitalisitsche Gesellschaft. Düsseldorf: Econ Verlag.
——— (1993b). The rise of the knowledge society. *The Wilson Quarterly*, 17(2): 52–72.

——— (1994). Knowledge work and knowledge society — The social transformations of this century. The 1994 Edwin L. Godkin Lecture: John F. Kennedy University, 04.05.1994.

Dunning, J. H. (2000). *Regions, Globalisation, and the Knowledge-Based Economy.* Oxford: Oxford University Press.

Evers, H.-D. (2000). Die Globalisierung der Epistemischen Kulturen: Entwicklungstheorie und Wissensgesellschaft. In *Vom Ewigen Frieden und dem Wohlstand der Nationen*, U. Menzel (ed.). Frankfurt: Suhrkamp.

——— (2002a). Southeast Asian knowledge societies. Bonn: University of Bonn.

——— (2002b). Malaysian knowledge society and the global knowledge gap. Bonn: University of Bonn, Department of Southeast Asian Studies.

——— (2003). Transition towards a knowledge society: Malaysia and Indonesia in comparative perspective. *Comparative Sociology*, 2(2):355–373.

——— (2005). Global knowledge: The epistemic culture of development. In *Local and Global: Social Transformation in Southeast Asia*. Hassan Riaz (ed.), pp. 3–17. Leiden and Boston: Brill.

Evers, H.-D. (2000). Knowledge societies: An overview of issues and theories. Singapore: National University of Singapore (unpublished manuscript).

Evers, H.-D. and S. Gerke (2005). Closing the digital divide: Southeast Asia's path towards a knowledge society. Bonn: Center for Development Research, University of Bonn, 2005. ZEF Working Paper No. 1.

Evers, H.-D. and A.-K. Hornidge (2007). Knowledge hubs along the straits of Malacca. *Asia Europe Journal*, 5(3):417–422.

Feather, J. (1998). *The Information Society. A Study of Continuity and Change.* London: Library Association Publishing.

Garnham, N. (2002). The information society: Myth or reality. 2001 Bugs. Globalism and pluralism. Montreal.

Gershuny, J. (1978). Die Ökonomie der Nachindustriellen Gesellschaft. Produktion und Verbrauch von Dienstleistungen. Frankfurt a.M./New York: Campus.

Gibbons, M., H. Limoges, S. Nowotny, P. Scgwartzman, P. Scott and M. Trow (eds.) (1994). *The New Production of Knowledge: The Dynamics of Science and Research in Contemporary Societies.* London: Sage Publications.

Gill, K. S. (ed.) (1996). *Information Society — New Media, Ethics and Post-Modernism.* London: Springer.

Heidenreich, M. (2003). Die Debatte um die Wissensgesellschaft. In *Wissenschaft in der Wissengesellschaft*, S. Böschen and I. Schulz-Schaeffer (Eds.). Opladen: Westdeutscher Verlag.

Hofmann, J. (2001). Digitale Unterwanderungen: Der Wandel im Innern des Wissens. *Aus Politik und Zeitgeschichte B*, 36:3–6.

Hornidge, A.-K. (2010). An uncertain future — Singapore's search for a new focal point of collective identity and its drive towards 'knowledge society'. *Asian Journal of Social Sciences*, 38(5): 785–818.

——— (2008). From trading goods to trading knowledge: Singapore's development into a knowledge hub. In *The Straits of Malacca — Knowledge and Diversity*, S. Gerke, H.-D. Evers and A.-K. Hornidge (eds.), pp. 63–84. Münster/Penang: Lit Verlag.

——— (2007). *Knowledge Society. Vision and Social Construction of Reality in Germany & Singapore.* Münster: Lit-Verlag.

Infocomm Development Authority (IDA) (2003). Connected Singapore — a new blueprint for infocomm development. Singapore: Infocomm Development Authority.

———. Infocomm21 — Singapore: Where the Digital Future Is. Singapore: Infocomm Development Authority, 2000.

——— (2007). Intelligent Nation 2015—iN2015. Singapore: Infocomm Development Authority.

Infocomm Development Authority (IDA) and C. Chua (2006). The Singapore e-government experience. E-Gov Forum and Integrating Trade Facilitation: Asian Development Bank.

Information Infrastructure Task Force (IITF) (1994b). Committee on applications and technology: The information infrastructure: Reaching society's goals. Washington D.C.

——— (1993). The national information infrastructure — Agenda for action. Washington DC: U.S. Department of Commerce.

——— (1994a). The National information infrastructure committee on applications and technology: Putting the Information Infrastructure at Work. Washington D.C.

JTC Corporation (2003). Fusionpolis Launched in One-North. JTC Corporation 20.02.2003.

Jungk, R. (1973). Politik und Technokratie. Aufgabe Zukunft, Qualität des Lebens, Beiträge zur 4. Internationalen Arbeitstagung zur Industriegewerkschaft Metall, Bd. 8: Demokratisierung (ed. G. Friedrichs). Frankfurt a.M.

King, J. L. and K. L. Kraemer (1995). Information infrastructure, national policy, and global competitiveness. *Information Infrastructure and Policy*, March.

Knoblauch, H. (2004). Informationsgesellschaft, Wissensgesellschaft Und Die Neuere Wissenssoziologie. In *Neue Perspektiven der Wissenssoziologie*, H.-G. Soeffner, D. Tänzler and H. Knoblauch (Eds.). Konstanz: Universitätsverlag Konstanz.

——— (2005). Wissenssoziologie. Konstanz: UVK Verlagsgesellschaft mbH.

Knorr-Cetina, K. (1999). Chapter 1.2: The cultures of knowledge societies. In *Epistemic Cultures. How the Sciences Make Knowledge*, K. Knorr-Cetina (ed.), pp. 5–11. Cambridge, MA: Harvard University Press.

Kreibich, R. (1986). *Die Wissenschaftsgesellschaft — Von Galilei Zur High-Tech-Revolution.* Frankfurt am Main: Suhrkamp Verlag.

Kubicek, H. (1999). Möglichkeiten and Gefahren der "Informationsgesellschaft. In *Tübinger Studientexte Informatik und Gesellschaft*, S. Rizvi and H. Klaeren (Eds.). Tübingen: Universität Tübingen.

Kubicek, H., W. H. Dutton and R. Williams (eds.) (1997). *The Social Shaping of Information Superhighways. European and American Roads to the Information Society.* Frankfurt: Campus Verlag.

Kübler, H.-D. (2005). Mythos Wissensgesellschaft — Gesellschaftlicher Wandel zwischen Information, Medien Und Wissen. Eine Einführung. Wiesbaden: VS Verlag für Sozialwissenschaften.

Kumar, K. (1978). *Prophecy and Progress: The Sociology of Industrial and Post-Industrial Society.* New York: Penguin Books.

Lane, R. E. (1966). The decline of politics and ideology in a knowledgeable society. *American Sociological Review*, 5(31): 649–62.

Library 2000. Review Committee. *Library 2000: Investing in a Learning Nation.* Singapore: Ministry of Information and The Arts, 1994.

Lievrouw, L. A. and S. Livingstone (eds.) (2002). *Handbook of New Media.* London: SAGE Publications.

Lloyd, C. and J. Payne (2002). In *Search of the High Skill Society: Some Reflections on Current Visions.* Warwick: ESRC Research Centre on Skills, Knowledge and Organisational Performance (SKOPE).

Lyon, D. (1988). *The Information Society: Issues and Illusions.* Cambridge: Polity Press.

——— (1996). The roots of the information society idea. In *Information Technology and Society*, N. Heap and *et al.* (Eds.), pp. 54–73. London.

Maasen, S. (1999). Wissenssoziologie. Bielefeld: transcript Verlag.

Machlup, F. (1962). *The Production and Distribution of Knowledge in the United States.* Princeton, New Jersey: Princeton University Press.

Martin, W. J. (1995). *The Global Information Society.* Aldershot, Hampshire: Aslib Gower.

Masuda, Y. (1980). The information society as post-industrial society. Tokyo: Institute for the Information Society.

Mattelart, A. (2003). Kleine Geschichte Der Informationsgesellschaft. Berlin: Avinus Verlag.

McLuhan, H. M. (1962). The Gutenberg Galaxy. London.

Mill, J. S. (1974). Das System Der Deduktiven Und Induktiven Logik. In *Geschichte Der Soziologie*, F. Jonas (ed.), pp. 434–445. Hamburg: Rowohlt.

Minc, A. (1987). The Informatisation of Society. In *Evolution of an Information Society*, A. E. Cawkell (ed.), pp. 134–140. London: Aslib.

Ministry of Information and the Arts (MITA) (2000). Renaissance city report: culture and the arts in renaissance Singapore. Singapore: MITA.

Ministry of Information and the Arts and Singapore Tourist Promotion Board (MITA and STPB). (1995). Singapore, Global City for the Arts. Singapore: Singapore Ministry of Information and the Arts; Singapore Tourist Promotion Board.

Ministry of International Trade and Industry (MITI) (1994). Programme for advanced information infrastructure. Tokyo: Ministry of International Trade and Industry, Japan.

Ministry of Education (MOE) (1997). Masterplan for it in education. Singapore: Ministry of Education.

——— (2002). Masterplan for it in education. Singapore: Ministry of Education.

Ministry of Posts and Telecommunications (MPT) 91999). Communications in Japan 1999. Tokyo: Ministry of Posts and Telecommunications, Japan.

Morris-Suzuki, T. (1996). *The Technological Transformation of Japan.* Cambridge: Cambridge University Press.

Naisbitt, J. (1982). *Megatrends. Ten New Directions Transforming Our Lives.* New York: Warner.

National Computer Board (NCB)(1982). Vendor's Conference. Singapore: National Computer Board.

——— (1992). A vision of an intelligent island: The IT2000 Report. Singapore: National Computer Board Singapore.

National IT Plan Working Committee (1985). National it plan: A strategic framework. Singapore: National IT Plan Working Committee.

National Library Board (NLB) (2005). L2010 — Our vision for the future (Overview). Singapore: National Library Board.

Noguchi, Y. (2003). Is the it revolution possible in Japan? In *Internet, Economic Growth and Globalisation — Perspectives on the New Economy in Europe, Japan and the US*, C. E. Barfield, G. Heiduk and P. J. J. Welfens (Eds.), pp. 67–83. Berlin: Springer.

Nora, S. and A. Minc (1979). Die Informatisierung der Gesellschaft. Ed. Uwe Kalbhen. Frankfurt/New York: Campus Verlag.

Organisation for Economic Co-operation and Development (OECD) (1996a). The knowledge-based economy in 1996: Science, technology and industry outlook. Paris: OECD.

——— (1996b). The knowledge-based economy. Paris: OECD.

Porat, M. (1976). The information economy. Stanford: University of Stanford.

Preston, P. (1997). The information superhighway and the less developed regions. In *The Social Shaping of Information Superhighways. European and American Roads to the Information Society*, H. Kubicek, W. H. Dutton and R. Williams (Eds.), pp. 277–297. Frankfurt: Campus Verlag.

Read, W. H. and J. L. Youtie (1995). Policy strategies along the information superhighway. *Policy Studies Review*, 14(1/2):99–106.

Riddle, D. I. (ed.) (1988). Information economy and development. Bonn: Friedrich-Ebert-Stiftung.

Schneider, V. (1997). Different roads to the information society? Comparing U.S. and European approaches from a public policy perspective. In *The Social Shaping of Information Superhighways. European and American Roads to the Information Society*, H. Kubicek, W. H. Dutton and R. Williams (Eds.), pp. 339–358. Frankfurt: Campus Verlag.

Stehr, N. (1994). *Knowledge Societies.* London: Sage Publications.

——— (2001a). Moderne wissensgesellschaften. *Aus Politik and Zeitgeschichte*, 36:7–13.

——— (2001b). A world made of knowledge. *Society*, 39(1):89–92.

——— (1999). "Wissensgesellschaften" oder die Zerbrechlickeit Moderner Gesellschaften. In *Wissen Und Arbeit — Neue Konturen Der Wissensarbeit*, W. Konrad and W. Schumm (Eds.), pp. 13–23. Münster: Verlag Westfälisches Dampfboot.

Steinbicker, J. (2001). Zur Theorie der Informationsgesellschaft. Ein Vergleich der Ansätze von Peter Drucker, Daniel Bell und Manuel Castells. Opladen: Leske + Budrich.

Tänzler, D., H. Knoblauch, and H.-G. Soeffner (2006). *Zur Kritik der Wissensgesellschaft.* Konstanz: UVK Verlagsgesellschaft mbH.

Toffler, A. (1970). *Future Shock.* New York: Random House.

——— (1990). *Powershift — Knowledge, Wealth, and Violence at the Edge of the 21st Century.* New York: Bantam Books.

——— (1980). The third wave. New York: William Morrow and Company, INC.

Toh, M.-H., Hsiu Chin Tang and A. Choo (2002). Mapping Singapore's knowledge-based economy. Economic Survey of Singapore. Singapore: Economics Division, Ministry of Trade and Industry.

Touraine, A. (1969). *La Société Post-Industrielle.* Paris: Denoël.

Tuomi, I. (2001). *From Periphery to Centre: Emerging Research Topics on Knowledge Society.* Helsinki: TEKES.

Umesao, T. (1963). Information industry theory: Dawn of the coming era of the ectodermal industry. *Hoso Asahi*, Jan.: 4–17.

United States Advisory Council on the National Information Infrastructure (1996b). Common Grounds. A nation of opportunity: Realizing the promise of the information superhighway. Washington D.C.

——— (1995). Common grounds. Fundamental principles for the national information infrastructure. First Report. Washington D.C.

——— (1996a). Common grounds. Kickstart initiative: Connecting America's communities to the information superhighway. Washington D.C.

Vester, F. (1968). Bausteine Der Zukunft. 22 Berichte Von Neuen Forschungsergebnissen und Ihren Konsequenzen. Frankfurt a.M.

Vogel, C. (2000). Deutschland Im Internationalen Technologiewettlauf — Bedeutung der Forschungs — Und Technologiepolitik Für Die Technologische Wettbewerbsfähigkeit. Berlin: Duncker und Humblot GmbH.

Webster, F. (1995). *Theories of the Information Society.* London: Routledge.

Williams, F. (1982). *The Communications Revolution.* Beverly Hills: SAGE Publications.

——— (1991). *The New Telecommunications — Infrastructure for the Information Age.* New York: The Free Press.

Willke, H. (1998). Organisierte wissensarbeit. *Zeitschrift für Soziologie*, 27:161–77.

——— (1999). Die Wissensgesellschaft — Wissen ist der Schlüssel zur Gesellschaft. In *In welcher Gesellschaft leben wir eigentlich? Gesellschaftskonzepte Im Vergleich*, A. Pongs (ed.), pp. 261–279. München: Dilemma Verlag.

Workgroup on Creative Industries (2002). Creative industries development strategy. Singapore: ERC-Subcommittee Service Industries/Ministry of Trade and Industry.

Chapter

4

Measuring Spatial Density

Knowledge Clusters in Malaysia

Hans-Dieter Evers, Pamela Nienkemper and Benjamin Schraven

1. Introduction: Epistemic Landscapes and Knowledge Maps

Following the work of Porter on the competitive advantage of nations (Porter, 1990; 2003), there has been an upsurge of research and data collection on industrial clusters (Bathelt, Malmberg and Maskell, 2004; Huggins, 2008; Karlsson, 2008; Martin and Sunley, 2003; Motoyama, 2008; Sölvell, 2009; Tallman *et al.*, 2004). These studies usually assume that modern industrial clusters are centres of innovation and therefore desirable. Our concept of knowledge landscapes is less value laden or normative; it intends to show the actual spatial distribution and density of knowledge (Evers, 2008). As discussed in Chapter 1 of this book, epistemic landscapes are formed by knowledge clusters, knowledge hubs and the distribution of knowledge assets. Which type of knowledge landscape or infrastructure optimizes intellectual or industrial output remains a question to be decided by further empirical research.

The World Development Report of 1999 has already drawn attention to the fact that knowledge is not evenly distributed within countries, regions or urban areas (World Bank, 1999). This has been described as the existence of "knowledge gaps" or, in relation to the ICT backbone of a knowledge system, as the "digital divide". Strategies on how to close these gaps by appropriate development policies were and are widely discussed. So far most studies have tried to show the narrowing or widening of knowledge gaps by using indicators, as those provided by the KAM (knowledge assessment methodology) data base of the World Bank Institute, like number of researchers per million population, investment in R&D (research and development) as percentage of GDP and other indicators. In most cases these data have measured knowledge gaps between countries or regions, without paying closer attention to the geographical distribution of knowledge assets as well as the existence of knowledge gaps within countries, provinces or cities. We intend to close this particular "knowledge gap" by providing a methodology to show and measure the geographical distribution of knowledge related assets, people or organizations, which we are referring to as "epistemic landscapes". This chapter is thus designed to discuss the methodology of creating knowledge maps and analyzing epistemic landscapes. Data and maps from ongoing research in Malaysia are used to illustrate various methodological issues and to show examples of maps and graphs.

We define epistemic landscapes in a geographical sense, i.e., we refer to the spatial distribution of knowledge assets within a predefined region (Evers, 2008). The concept is not yet standard social science terminology. We are using the term "epistemic" in line with "epistemic culture", the culture of knowledge production, as coined by Karin Knorr-Cetina (Knorr-Cetina, 1999). The term "epistemic" has been used in different contexts. One line of argument refers back to Bacon and 18th-century 'encyclopaedism' and defines an epistemic landscape as depicting a synthesis of knowledge (Wernick, 2006). In Weisberg and Muldoon's study, a single epistemic landscape corresponds to the research topic that engages a group of scientists. Agent based modelling with NetLogo software is used to model the changing epistemic landscape according to research strategies of participating scientists (Weisberg and Muldoon, 2007). Conceptually distance rather than Euclidean distance is shown in graphs, similar to those used to illustrate social networks. In our study we follow a different path and focus on the development strategies of governments, strategic groups, firms, research institutes and their success in shaping the epistemic landscape of a region. The allocation of human and financial resources creates knowledge assets which are geographically distributed and can be measured, mapped and made to depict the contours of an epistemic landscape.

Epistemic landscapes develop over long periods of time. They are seldom shaped by individual actors, but more often by the collective action of strategic groups (Evers and Benedikter, 2009; Evers and Gerke, 2009). Firms connected by a common interest to capitalize on the competitive advantage of clustering have an impact on epistemic landscapes through their location decisions. More over government strategies to develop knowledge-based societies and economies have often been decisive in shaping epistemic landscapes. Relevant development policies have been assessed in detail elsewhere for Malaysia and Indonesia (Evers, 2003), Singapore and Germany (Hornidge, 2007). In fact, developing regions of high-tech industries, clusters or knowledge hubs are, by now, standard practice in many regional planning departments around the world.

In order to visualize and analyze epistemic landscape we have developed knowledge maps.[1] A knowledge map is the cartographic representation of captured information and relationships which enables the efficient communication and learning of knowledge by observers with differing backgrounds at multiple levels of detail (Meusburger, 2000). The individual items of knowledge included in such a map can be text, stories, graphics, models or numbers. Maps can also serve as links to more detailed knowledge sources ranging

[1] All maps in this paper have been designed and produced by the authors with the assistance of Ramli Nordin and Marissa Ayesha Mohamed Idenal during a joint research project of the Center for Development Research (ZEF), University of Bonn, Germany and the Centre for Policy Research and International Studies (CenPRIS), Universiti Sains Malaysia.

from text-based groupware documents to database schemas as well as pointers to implicit knowledge (such as experts).

The creation of knowledge maps is by now a standard tool of knowledge management in organizations. "In their simplest forms, knowledge maps serve as guides to the type of knowledge held both inside and outside the organization. They serve as locators of that knowledge, and locators of people holding particular knowledge. They give a visual overview of the knowledge available to the organization" (Foles, 1998:14). The term "mapping" is, perhaps unfortunately, often used in the sense of "listing", i.e., creating a list of items, like organisations, persons or assets. "Yellow pages" listing names of persons or firms and their knowledge assets are (often used) to make knowledge easily accessible in large organizations and bureaucracies. In this paper "mapping" is, however, used in a cartographic sense.

2. Examples of Thematic Maps and Landscapes

2.1 *Distribution of Knowledge Producing Organisations*

For the mapping and spatial analysis of the data, ESRI ArcGIS 9.2 is applied, a well-established Geographic Information System software (ESRI, 1997). The knowledge producing organisations compiled in the database[2] are geo-coded in order to visualize and analyse the data in GIS. Geo-coding describes the process of locating actual geographic coordinates based on street addresses (Nolan and Kumar, 2006). The free available program Google Earth has been used to assign the addresses to coordinates. On the basis of the compiled information, the creation of point shape files is possible. Loading these shape files in ESRI ArcMap allows illustrating the spatial distribution of knowledge producing organisations in Pensinsular

[2] The data were compiled by Ramli Nordin during field research 2009. See Evers, Hans-Dieter, Ramli Nordin, and Pamela Nienkemper (2010). The Emergence of an Epistemic Landscape: Knowledge Cluster Formation in West-Malaysia. *ZEF Working Paper Series.*

Malaysia. Each dot represents one knowledge producing organisation. Knowledge producing organisations are research and educational institutions. Companies, even those involved in R&D activities, are excluded from this analysis.

The symbol sizes gradually increases according to the number of knowledge workers at the respective organisation. By knowledge worker we mean researchers and lecturers employed by the organisations or generally staff with degree or above. Based on this graduate symbol map a preliminary estimate of geographic distribution and clustering is possible. It is easily visible that the majority of dots is located in Kuala Lumpur and to the south of Kuala Lumpur in the state of Selangor (Figure 1).

2.2 *Distribution of Knowledge Assets*

Knowledge worker may act as an indicator for knowledge assets. Therefore, the distribution of knowledge worker is further investigated in the following section. In this step, the geographic location of knowledge producing organisations is assigned to administrative boundaries in order to map the number of knowledge worker by state. Thus, a simple thematic map is created which is generally used to visualize any kind of distribution based on administrative boundaries at different levels (countries, states, districts, etc.) (Parker and Asencio 2009). Again, by visual interpretation of the map, it is obvious that Kuala Lumpur and Selangor are the territories with the highest total number of knowledge worker followed by Johor, Perak and Penang (Figure 2). On the other hand, the distribution within the state boundary is neglected as well as the knowledge worker to total population ratio (the latter could be done in a next step). Compared to the graduate symbol map information is lost due to the interpolation of data to a spatial area. For example, the big circular symbol located in the state of Penang representing an organisation with a higher-than-average number of knowledge worker is a striking feature in the graduate symbol map, whereas in the thematic map the state of Penang does not stand out against the neighbouring state of Perak.

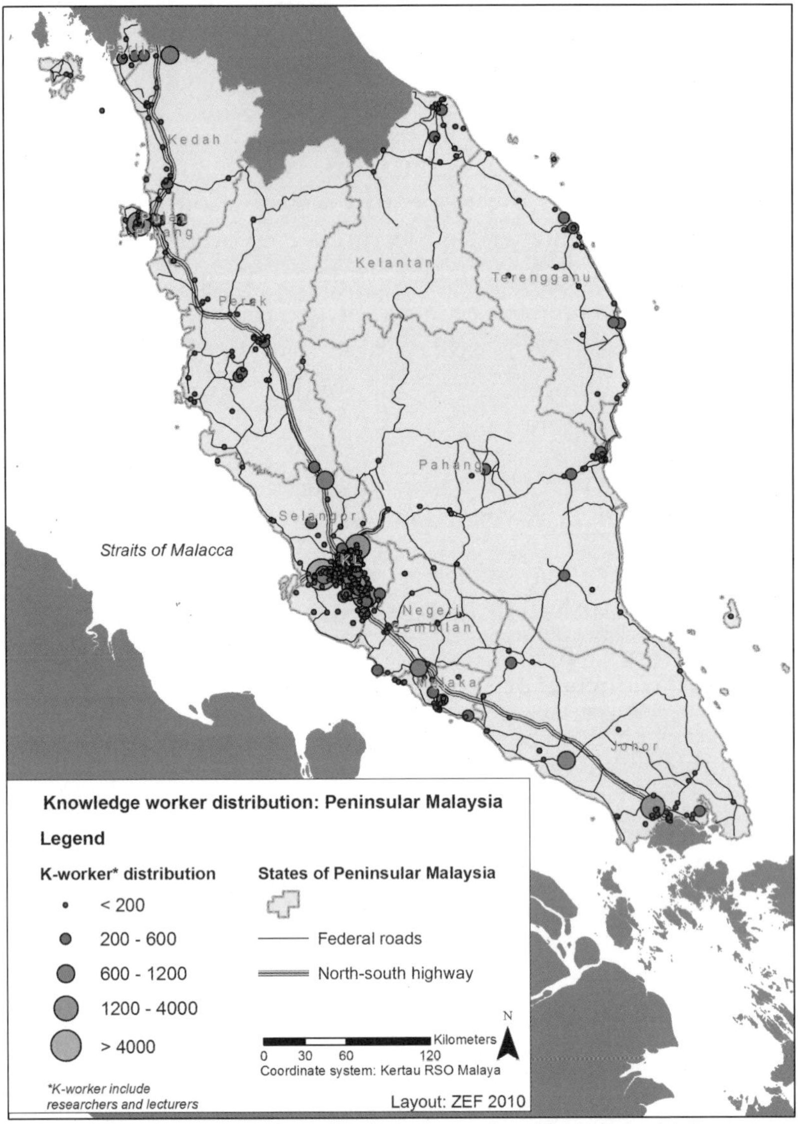

Data source: ZEF (K-worker data), Malaysia Geoportal (Street network), www.gadm.org (Administrative boundaries)

Figure 1: Graduate symbol map: Distribution of knowledge workers, West Malaysia 2009

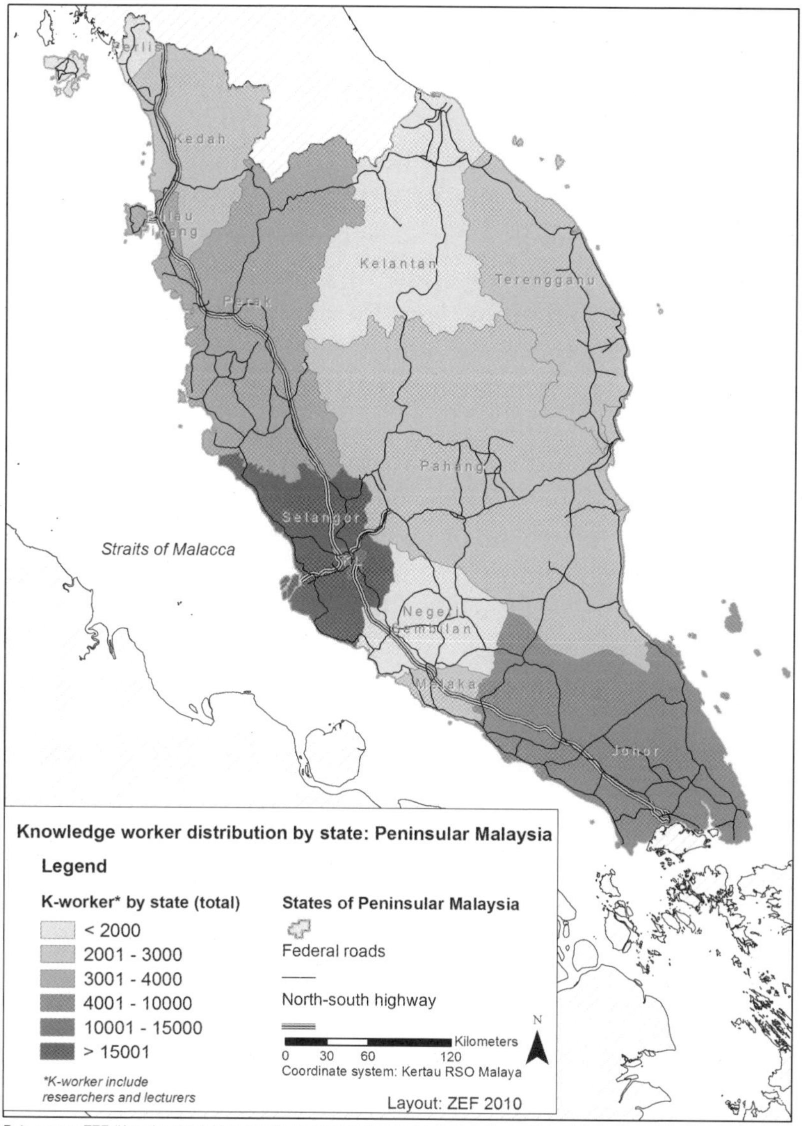

Data source: ZEF (K-worker data), Malaysia Geoportal (Street network), www.gadm.org (Administrative boundaries)

Figure 2: Thematic map: Distribution of knowledge workers, West Malaysia 2009

Following the assumption that spatial proximity is beneficial for knowledge producing organisations in order to share knowledge, utilize knowledge spillover effects and be more competitive, the analysis of agglomeration tendencies provides deeper insight. Agglomeration tendencies or clustering of point features can be analysed in ArcGIS using the Kernel Density function provided by the Spatial Analyst toolbox. It is well established to use employment data for the identification of clusters based on regional employment agglomerations (e.g., European Cluster Observatory). In this study, the points are weighted by the number of knowledge worker resulting in some points having greater influence in the density analysis than others (European Cluster Observatory 2010). This approach is considered adequate as organizations with a great number of knowledge workers should be weighted more heavily than organizations with small numbers of knowledge worker. The Kernel Density function provided by ArcGIS Spatial Analyst is based on a quadratic kernel (a bounded kernel) which is often described as Epanechnikov kernel[3] (de Smith, Goodchild and Longley, 2009). The resulting grid values are provided as relative densities, that is in our case weighted number of points per square meter. The grid resolution should be set to be reasonable within the context of the analyzed dataset, even though the grid resolution does not affect the results to any great degree (de Smith, Goodchild and Longley, 2009).

Two Kernel Density maps are created (Figures 3 and 4). One shows the clustering of knowledge workers irrespective of administrative boundaries for the whole of Peninsular Malaysia (Figure 4). Thus, this map is comparable to the graduate symbol map. The second Kernel Density map is based on a subset of the data to illustrate

---

[3] Detailed information on the formula can be found by Silverman (1986). *Density Estimation for Statistics and Data Analysis*. London: Chapman and Hall as well as de Smith, Michael J., Michael F. Goodchild, and Paul A. Longley (2009). *Geospatial Analysis: A Comprehensive Guide to Principles, Techniques and Software Tools*. Leicester: Matador.

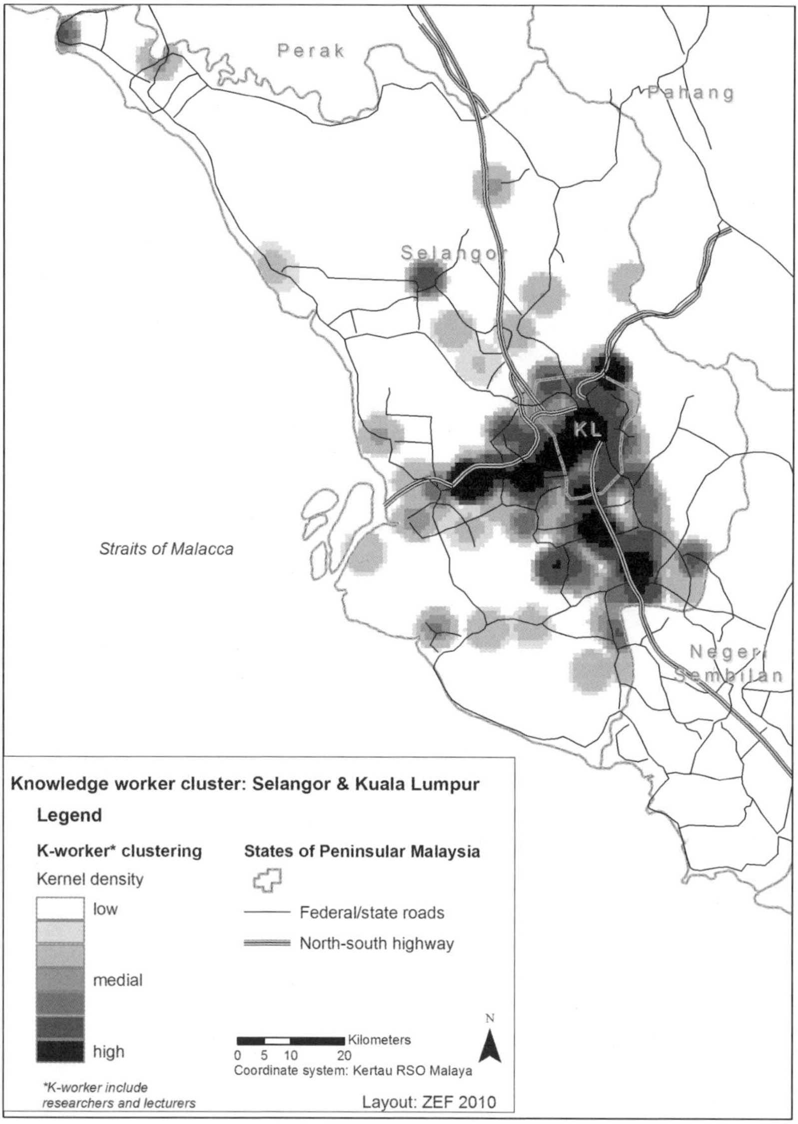

Data source: ZEF (K-worker data), Malaysia Geoportal (Street network), www.gadm.org (Administrative boundaries)

Figure 3: Kernel density map of Selangor and Kuala Lumpur, Malaysia 2009

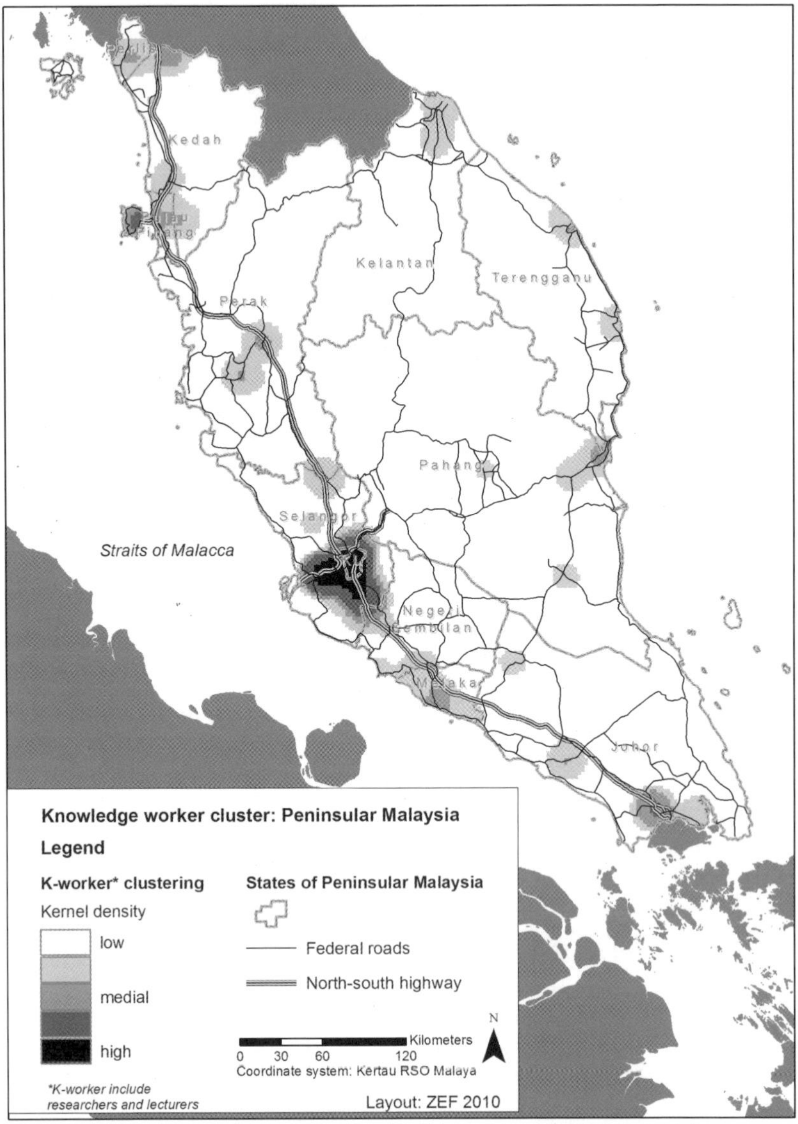

Data source: ZEF (K-worker data), Malaysia Geoportal (Street network), www.gadm.org (Administrative boundaries)

Figure 4: Kernel density map of Peninsular Malaysia, 2009

the clustering of knowledge workers in Selangor and Kuala Lumpur in greater detail (Figure 3). It is obvious that these maps provide additional information with regard to agglomeration tendencies compared to thematic or graduate symbol maps.

2.3 *Distribution of Knowledge Workers in 3D*

In the terminology of knowledge management we often talk about "landscapes", "knowledge gaps" and "knowledge hubs". All these terms are metaphorically touched. Via the illustration of data by maps, we are able to display those metaphoric terms as graphical representations. Speaking in geographical terms, an artificial landscape is created in which the surface characteristics are determined by certain (e.g., economical or social) attributes of the respective dataset. A visual alternative to flat maps is the construction of three-dimensional maps. These constructions are based on the idea that the altitude can be determined by any quantitative attribute of spatial data.

In our example, we choose to represent the altitude of the landscape by the number of knowledge worker. As mentioned above, knowledge workers holding a degree and working as lecturers or researchers are interpreted as knowledge assets in this study. By doing so, data peaks (high number of knowledge worker) are displayed as (knowledge-) hubs and data downs (low number of knowledge worker) are displayed as (knowledge-) gaps (Figure 5). Even though there is no additional information displayed in 3D maps like these, using height or elevation as a metaphor for attributes is a powerful tool for visualising relationships. It may also provide new insight into certain processes and factors that influence policy practices (Parker and Asencio, 2009). Three dimensional representations are therefore increasingly used in advertising and in science centres like the recently opened knowledge cluster Fusionopolis in Singapore. In a later section, we shall discuss some open issues of creating three-dimensional (artificial) landscapes in greater detail.

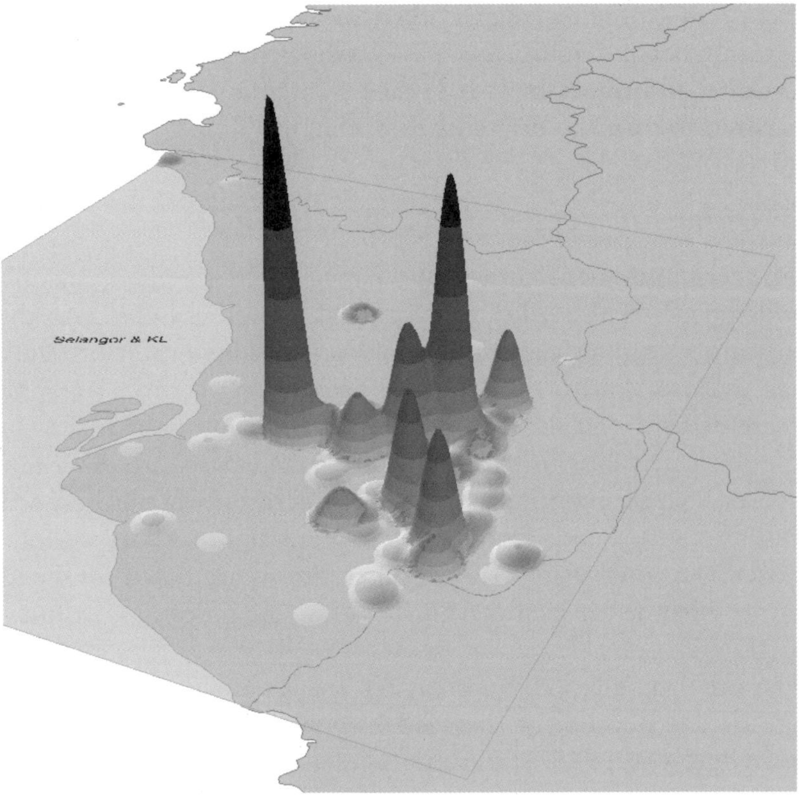

Figure 5: Three-dimensional map: Distribution of knowledge workers, Selangor D.E. and Kuala Lumpur, Malaysia 2009

3. The Spatial Distribution of Knowledge Attributes

3.1 *GIS: Linkage of Space and Content*

A Geographic Information System, GIS, is a system used for the storage, analyses and presentation of data that is somehow geographically based. The linkage between a spatial reference and attributes of any kind is a prerequisite to display data in maps. Attributes of spatial datasets can be quantitative or qualitative, personal or institutional, local or national (etc.). The crucial point is the connection to a geographic reference system to locate the attributes in space by using geographic coordinates. Working with a GIS means first of all to start the management of geo-data. This includes the storage, the transformation, the structuring, and the creation of new geo-data and the definition of attributes, properties and ranges of values.

Before creating a map one might check whether it is possible to use an already existing map instead. Once that has been done the data gathering comes into play. For this, a determination between primary, i.e., self-generated, and secondary, i.e., available data, is of importance. Even though the process of gathering primary data usually is very time-consuming, it is more detailed and represents the attributes with much contextual depth. Secondary data on the other hand are easier to find and at lower prices to get, even though it often lacks contextual quality. Since the gathered data alone do not represent any kind of information yet, the gathered data set then will have to be transformed into a meaningful map by using knowledge and transformation rules.

Because of the vast amount of information consisting in the primary as well as in the secondary data set, one will have to fulfil a reduction of complexity, in order to create the aspired map. This can also include consolidating data records to data classes (before or after collecting data).

3.2 *Data Input Formats*

With the exception of the geo-reference, there are almost no other restrictions of data input. Geo-reference is the strongest requirement and standard for mapping any kind of information within a GIS.

Generally, there are two main data formats that can be distinguished (Parker and Asencio, 2009): Vector format and Raster format. Vector formats store features as lines, points, or polygons (geometries). These shapes represent objects on the map. An advantage is that the necessary storage capacity for this format is less than for raster formats as it does not store the value for all points in a given polygon. In a raster format data is stored as series of cells with a single value assigned to each cell. It takes up more storage space as it is able to store a value for all points in a given polygon which also means that more information is maintained.

The scale levels of the attributes can vary. Their content may contain nominal, ordinal, interval or ratio scaled information. Furthermore data can be raised on different spatial scales. More examples of different scaled geo-references are: single coordinates, a city, a district, a province, a state, a continent (etc.). The differently scaled geo-references are arranged in a hierarchical manner. An up- and downscaling of data depends on their resolution.

Another important guideline is the usage of a coherent geographic and/or reference projection. If various shape files within a map project are of different projection, their visual display could be incorrect due to distortions.

In case of the simple purpose of adding additional information to already existing geo-objects there are possibilities of importing data without a manual entry. For example, to merge a table (i.e., excel, dBase) with data based on a common unit (e.g., administrative areas, city names building blocks), one just has to set up identical table structures (keys). This functionality increases the collection of data by saving a lot of time.

Another useful application is depicted by the import of GPS-coordinates. For instance ArcMap[4] offers such a feature for an easy and automatic integration and display in shape-file[5] format. By opening and editing the shapefile´s attribute table again a quick alteration of numeric and/or textual data can be done with ease.

3.3 *Spatial Data Representation*

As stated above, the representation of data is realised by points, lines, areas or volumes. Furthermore table- and text-based data can be added. For a more detailed illustration classification of data should be undertaken. By doing so, the core statements of the data set move into focus and more precise conclusions can be drawn.

---

[4] ArcMap is a component of ESRI's ArcGIS. It is the central application for the creation of maps. For this approach ArcGIS 9.2 was used.

[5] A shapefile stores nontopological geometry and attribute information for the spatial features in a data set. It consists of a main file, an index file, and a dBASE table (ESRI 1997).

Alongside the representation of data via "simple" or "nested" queries, which are characterized as direct requests without modifying data values by showing single or logical combined attributes, the data can also be manipulated by creating new geo-data from the existing geometrics. There is a variety of tools provided by GIS to manipulate or analyse data resulting in new geo-data sets (e.g., Spatial Analyst Tools, Data Management Tools, 3D Analyst Tools). For the last mentioned approach, the kernel density analysis of the knowledge producing organisations and the corresponding knowledge worker data set may be used as an example.

3.4 *Technical Considerations of Using the Third Dimension*

The three dimensional knowledge worker map of Selangor and Kuala Lumpur presented in Section 2.3 is based on the kernel density analysis output. The kernel density analysis is performed by using the point shape file (vector format) that consists of the information on geographic location (coordinates and geo-reference system) of each knowledge-producing organisation and the number of knowledge worker employed by each organisation (quantitative attribute of the point shape stored in the attribute table). After calculating the kernel density of knowledge worker, the newly generated data is stored in a raster format which means that each cell of the area of interest is assigned to a value. If the three-dimensional map is derived by using vector formats (the original shape files), the problem of missing data or interpolation occurs that is discussed by Evers, Genschick and Schraven (2009:10). In the next step, the raster format needs to be converted to a TIN (Triangular Irregular Network) using the 3D Analyst provided by ArcGIS. After importing the TIN to ArcScene it is possible to display the data in three dimensions, to work on the layout (e.g., coloured classification of value groups), and to exaggerate or smooth the slopes and elevation profile.

There are advantages and disadvantages of using TINs to display data in three dimensions. Even though a TIN is stored like a coverage in a directory of files, it is not a coverage and does not have an associated INFO file. The visualisation of differences in 3D based on specific spatial units (e.g., number of employees by states or districts) is more or less a well-established procedure in the GIS environment. But to sufficiently illustrate the data described here in three dimensions remains challenging and more different approaches will be tested in our ongoing research on the distribution and agglomeration of knowledge.

4. Results: Knowledge Maps or Epistemic Landscapes

4.1 *Knowledge Maps as Analytical Tools*

We have shown how knowledge related data sets can be visualized in thematic maps. They have one thing in common: they all use space or distance as an additional variable or, to put it differently, to geo-reference data on knowledge production, knowledge assets and knowledge flows. Data sets and maps all represent aspects of what we have termed "epistemic landscapes". The term "map" is used in the strict literal geographical sense. Our knowledge maps represent real geographical knowledge landscapes. Knowledge maps offer broad information about the spatial distribution of knowledge assets and knowledge production.

So far we have used knowledge maps mainly to study the process of clustering. The underlying hypothesis is, as explained elsewhere (Evers, 2008), that a cluster of firms and organisations producing, transmitting or using knowledge enhances innovation and productivity and thus provides a competitive advantage over other regions that show a lower degree of knowledge density or clustering. Knowledge maps visualise the degree of spatial proximity. We have tried to enhance the "visibility" of knowledge by using three-dimensional maps or even animations. The didactic value appears to be evident but the analysis has to go beyond visual impressions. We have therefore tried to use various statistical tools to measure the degree of clustering, or in other words to test the quality of epistemic landscapes and asses how they may contribute to social and economic development.

4.2 *Statistical Measures of Knowledge Clusters*

An adequate indicator for the spatial density of a cluster will have to be developed to describe the "quality" of a knowledge cluster. In the following some selected approaches shall be introduced. The density measures often used in the natural sciences (e.g., the stand density index SDI used often in forestry (Reineke, 1933), which basically work according to the function "elements per unit of area/space" are not really applicable for the measurement of spatial density of knowledge clusters. In the case of knowledge clusters, it obviously makes much more sense to use the internal distances within a cluster for the construction of a density measure.

One imaginable and very precise approach of doing this is based on the Euclidean or linear distance of each knowledge-producing organization to any other knowledge producing organizations within a knowledge cluster. Therefore, a data matrix with all coordinates of the involved organizations on both the x-axis and the y-axis has to be generated. In the matrix the Euclidean distances have to be calculated whereas the distances of the organizations to themselves have to be excluded.

Table 1 shows the data matrix that needs to be constructed for the determination of all Euclidean distances between the organizations of a knowledge cluster. The total number of distance values, which is generated by the matrix, is always determined by the equation n^2-n, whereas n is the total number of organizations. With the generated values a new variable can be constructed whereby all in all one variable per knowledge producing unit are produced. These variables contain all distances of the respective unit to all other units of the knowledge cluster. With the mean value (and additionally the standard deviation, standard deviation of the mean, variance, range etc.) a coefficients can be calculated to measure the spatial density of a knowledge cluster. The disadvantage of this approach is that there is a high vulnerability towards outliers and extreme cases. Practically, this means that if a knowledge cluster has one organization, which is located in a great distance to the rest of the cluster, this cluster will have a very high value, especially for its standard deviation even if the rest of the cluster may be located very close together.

If we take the example of research institutions in Selangor (n = 292) and the whole of peninsular Malaysia (n = 617), the average distance value for the first mentioned is 11.2 kilometres and for the last mentioned 192.6 kilometres with standard deviations of the mean of 5.6 kilometres and 76.9 kilometres. Principally problematic with this

Table 1: Example of a data matrix for the calculation of the euclidian distances (in meters) between selected knowledge producing organizations in selangor

		Name of Institute	Kolej Profesional dan Pengurusan	Politeknik Nilai, Negeri Sembilan	Kolej Universiti Nilai	Kolej MCS	Politeknik Banting, Selangor	…
		Latitude	412201	418800	418026	393136	393189	…
Name of Institute	**Latitude**	**Longitude**	304805	306292	311350	312436	312449	…
Kolej Profesional dan Pengurusan	412201	304805		6764	8762	20535	20491	…
Politeknik Nilai, Negeri Sembilan	418800	306292	6764		5117	26389	26340	…
Kolej Universiti Nilai	418026	311350	8762	5117		24914	24862	…
Kolej MCS	393136	312436	20535	26389	24914		55	…
Politeknik Banting, Selangor	393189	312449	20491	26340	24862	55		…
…	…	…	…	…	…			…

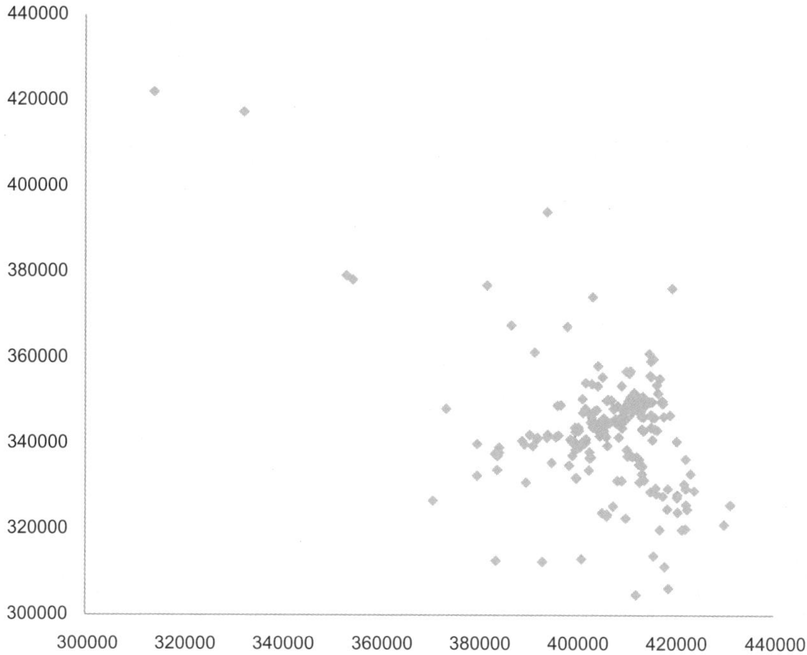

Figure 6: Spatial distribution of knowledge producing organisations in Selangor as scatter plot

approach is the relative sensitivity for outliers and extreme cases. This is also somehow the case for the example: the Selangor cluster has two institutions, which are located in some north-western distance to the rest (see Figure 6). These outliers have some impact on the density score for this cluster since without the two institutions the cluster would have a density score of 17.9 kilometres in average.

To totally overcome this potential bias caused by outliers another approach, which can be called the "neighbourhood per radius approach", can be applied. For this method, the same data matrix for the calculation of the Euclidean distances between the organizations of a knowledge cluster has to be generated. The next step is to (virtually) check all the distances whether they are bigger or smaller than a predefined distance value (e.g., 1 km, 2 km, 5 km, etc.). If they are smaller, the distance value has to be transformed into the value 1, if not it has to be transformed into the value 0. If the sum per column or per row is counted, the outcome is the number of neighbouring organizations within the pre-defined radius for every organization of the cluster. The predefined distance is not totally arbitrary, as an ideal distance for face-face communication can be assumed or calculated. With the statistical measures mentioned above coefficients measuring the spatial density of the respective knowledge cluster can be constructed (for instance the average number of neighbours per institute or the standard deviation of neighbours per institute). Contrary to the first described approach, the higher these values are, the "denser" is the cluster spatially. The disadvantage of this approach is that the value for the respective radius has to be chosen purposely in order to get sound results. The radius size is severely determining the results for the density score what can easily be demonstrated by taking the example of the Selangor cluster: the score value (= the mean number of neighbours) is 5.9 for a radius of $r = 1$ kilometre and is 13.6 for $r = 2$ kilometres (for the whole West Malaysian cluster the value changes from 3.4 for $r = 1$ kilometre to 8.2 for $r = 2$ kilometres). A solution for that could be — if existing — the analysis of respective social network data, which for instance indicates how often a member of one knowledge producing unit communicates face to face with a member of another unit. Based on that, a critical maximum distance for direct face to face communication within a knowledge cluster can be estimated. Another disadvantage of this approach is that distances are only indirectly integrated into this measure. What in the first introduced approach has a very high impact on the score values has a very low impact on the accordant measure in this approach.

A "compromise" position between these two approaches can be a third method, which can be called "nearest neighbour"-approach — based on the related cluster analysis method. For this approach, an Euclidean distance matrix (as exemplified in Table 1) is used and only the lowest or minimal value for each knowledge-producing unit is taken into consideration. Accordingly, the resulting values show the distance to the nearest neighbouring unit for every institution being part of the cluster and the matrix respectively. The so established distribution with its statistical attributes (mean value, standard deviation, median, etc.) can be used to describe the spatial density of the knowledge cluster. For the Selangor cluster the mean minimal distance has a value of 1.2 kilometres whereas the value for the West Malaysia is

3.0 kilometres. Generally, the sensitivity towards outliers and extreme cases cannot be reduced with this approach completely.

Another fourth method, which strongly mitigates the impact of outliers, is based on the natural logarithm. The values of the Euclidean distance matrix simply have to be transformed with the natural logarithm function, whereby the impact of the outliers and extreme values will be strictly mitigated. The mean for all average values per knowledge producing unit can be taken to construct a density measure for this approach. The respective density score value for the West Malaysian cluster would be 11.6 whereas the value for the Selangor cluster would have a value of 9.4. A disadvantage of this approach is that the distances between two clusters with very different measures of spread are not adequately reflected — as the example of Selangor and West Malaysia, where the accordant score values only differ by the factor of 1.2, is showing.

A last and fifth method, which takes all possible distances within a cluster into consideration and at the same time mitigates the impact of outliers, is the normal distribution probability index. In this approach the latitude and longitude variables are used. With the means and the standard deviations of both variables as well as the correlation coefficient between the latitude and longitude variables the probability that the accordant point is part of a normal distribution in the two-dimensional vector space is calculated. The probability values are then multiplied with the distance between the latitude and the longitude values to the accordant mean values (see below for the exact equations). The mean value of the so created values is then used as a density score, which is sensitive only towards very extreme outliers. The accordant value for Selangor is 2740.6 and for West Malaysia it is 16836.5.

All in all, we have thus proposed the following five measures of spatial or distance density:

1. The "Mean Euclidean Distance Index" (EDI)
2. The "Neighbourhood per Radius Index" (NRI)
3. The "Nearest Neighbour Index" (NNI)
4. The "Mean Logarithmically Transformed Euclidean Distance Index" (LTEDI)
5. The "Normal Distribution Probability Index" (NDPI)

Measure of Density/ Clustering index	Value for West Malaysia	Value for Selangor	Equation (For d_{ij} = Euclidean Distance Between x_i and x_j)
1. EDI	192551.6	18631.7	$A = \frac{1}{n}\sum_i a_i$ whereas $a_i = \frac{1}{n-1}\sum_{j \neq i} d_{ij}$
2. NRI (for $r = 1$ km)	3.4	5.9	$B^r = \frac{1}{n}\sum_i b_i^r$ whereas $b_i^r = \frac{1}{k_i - 1}\sum_{\{j \mid j \neq i \wedge d_{ij} \leq r\}} d_{ij}$
3. NNI	3006.0	1227.6	$C = \frac{1}{n}\sum_i c_i$ whereas $c_1 = \min_{j \neq i} d_{ij}$

(*Continued*)

(*Continued*)

Measure of Density/ Clustering index	Value for West Malaysia	Value for Selangor	Equation (For d_{ij} = Euclidean Distance Between x_i and x_j)
4. LTEDI	11.6	9.4	$D = \frac{1}{n}\sum_i ld_i$ whereas $ld_i = \frac{1}{n-1}\sum_{j \neq i} \ln(d_{ij})$
5. NDPI	16836.5	2740.6	$E = \frac{1}{n}\sum_i d_{i\bar{X}} * \Pr ob(N† \geq (X_i - \bar{X}))$ whereas N^2 = *bivariate normal distribution* and $d_{i\bar{x}}$ = *Euclidean distance of* x_i *to the cluster mean*

5. Conclusion: The Use of Epistemic Landscapes

In this chapter, we have demonstrated how knowledge attributes of various kinds can be made visible in form of a knowledge map. Knowledge is not evenly distributed in space but shows discontinuities, gaps and heights. We use the concept of "epistemic landscape" to allude to the complex tapestry of knowledge assets, institutions, organisations, hubs and clusters. Since antiquity maps have been used to provide orientation and direction. Moreover they "can bridge the gap between language and culture in terms of communicating" (Hatfield, 2006). Maps are thus powerful tools in aid of decision making, proper planning and resource allocation. They are put into practice by indicating "how to get from here to there". They are in themselves repositories of knowledge, which can be retrieved and put into action. Epistemic landscapes and knowledge maps are therefore important parts of any information system. By additionally computing statistical measures of the density of knowledge clusters, comparative data can be interpreted. By correlating indicators, i.e., on knowledge output (like research papers or patents) with our knowledge density measure, the effectiveness of knowledge clustering can be studied. So far a similar approach has been used to study the competitive advantage of regions, following the earlier studies of Porter and others (Porter, 1990; Sölvell, 2009).

Last not least, we refer to the didactic value of maps. Information and knowledge, in our case the potential use of knowledge for development, have to be distributed to those that either benefit or suffer from the application of knowledge for development. The concept of epistemic landscapes and maps showing the complex and complicated business of producing and using knowledge enhance, we believe, the chances of government agencies, companies and civic organisations to understand and use knowledge for development.

References

Bathelt, H., A. Malmberg, and P. Maskell (2004). Clusters and knowledge: Local buzz, global pipelines and the process of knowledge creation. *Progress in Human Geography*, 28:31–56.

de Smith, M. J., M. F. Goodchild and P. A. Longley (2009). *Geospatial Analysis: A Comprehensive Guide to Principles, Techniques and Software Tools.* 3rd Edition. Leicester: Matador.

ESRI (1997). ESRI Shapefile Technical Description. An ESRI White Paper. (http://www.esri.com/library/whitepapers/pdfs/shapefile.pdf).

European Cluster Observatory (2010). Official Homepage. (http://www.clusterobservatory.eu/).

Evers, H.-D. (2003). Transition towards a knowledge society: Malaysia and Indonesia in comparative perspective. *Comparative Sociology*, 2:355–373.

———— (2008). Knowledge hubs and knowledge clusters: Designing a knowledge architecture for development. *ZEF Working Paper Series No: 27.* Bonn: Center for Development Research (ZEF), University of Bonn.

Evers, H.-D. and S. Benedikter (2009). Strategic group formation in the Mekong Delta — The social transformation of a modern hydraulic society. *ZEF Working Paper Series*, 35.

Evers, H.-D. and S. Gerke (2009). Strategic group analysis. *ZEF Working Paper Series* 34.

Evers, H.-D., R. Nordin, and P. Nienkemper (2010). The emergence of an epistemic landscape: Knowledge cluster formation in West-Malaysia. *ZEF Working Paper Series.*

Hornidge, A.-K. (2007). *Knowledge Society: Vision and Social Construction of Reality in Germany and Singapore.* Berlin and London: LIT Verlag.

Huggins, R. (2008). The evolution of knowledge clusters: Progress and policy. *Economic Development Quarterly*, 22:277–289.

Karlsson, C. (ed.) (2008). *Handbook of Research on Cluster Theory.* Cheltenham: Edgar Elgar Publishing.

Knorr-Cetina, K. (1999). *Epistemic Cultures: How the Sciences Make Knowledge.* Cambridge: Harvard University Press.

Martin, R. and P. Sunley (2003). Deconstructing clusters: Chaotic concept or policy panacea? *Journal of Economic Geography*, 3:5–35.

Meusburger, P. (2000). The spatial concentration of knowledge. Some theoretical considerations. *Erdkunde*, 54:352–364.

Motoyama, Y. 2008. What was new about the cluster theory? What could it answer and what could it not answer? *Economic Development Quarterly*, 22:353–363.

Nolan, I. and C. Kumar (2006). Geographic information systems (GIS) in business and industry cluster analysis: A case study of Indiana. *ESRI International User Conference, Paper #2044* (http://www.pcrd.purdue.edu/documents/publications/ESRI_2006_Conference.pdf).

Parker, R. and E.K. Asencio (2009). *GIS and Spatial Analsis for the Social Sciences: Coding, Mapping, and Modeling.* London: Routledge.

Porter, M. (2003). The economic performance of regions. *Regional Studies*, 37:6–7.

Porter, M. (1990). *The Competitive Advantage of Nations.* New York: The Free Press.

Reineke, L.H. (1933). Perfecting a stand-density index for even-aged forest. *Journal of Agricultural Research*, 46:627–638.

Silverman, J. (1986). *Density Estimation for Statistics and Data Analysis.* London: Chapman and Hall.

Sölvell, Ö. (2009). *Clusters — Balancing Evolutionary and Constructive Forces.* Stockholm: Ivory Tower Publishers.

Tallman, S., M. Jenkins, N. Henry, and S. Pinch (2004). Knowledge clusters and competitive advantage. *Academy of Management Review*, 29:258–71.

Weisberg, M. and R. Muldoon (2007). Epistemic landscapes and the division of cognitive labor. *Unpublished manuscript.* Pittsburgh. Penn: University of Pennsylvania.

Wernick, A. (2006). Comte and the encyclopedia. *Theory, Culture & Society*, 23.

World Bank (1999). *World Development Report 1998–1999: Knowledge for Development.* New York: Oxford University Press.

Chapter

5

The Moment of Truth for Science

The Consequences of the 'Knowledge Society' for Society and Science

Peter Weingart

Science is probably the fastest growing enterprise in our society. Since the 17th century, every doubling of the population has seen a tripling in the number of scientists (Price, 1971). This means that, over the last 300 years, Western societies have invested a major part of their resources into the production, revision and verification of knowledge. It is a process that perpetuates itself if sufficiently nourished — the more resources go into science, the more scientists are educated, the more results they produce and the more questions arise. In this regard, the frontiers of science are indeed endless, to paraphrase Vannevar Bush's slogan. The exponential growth of science is seen as the main factor in the shift from an industrialised society to a knowledge society (Bell, 1973).

Any analysis of science and its impact on all other parts of society, such as politics, law, economics or the media, must take into account these astonishing dynamics, its causes and its consequences (Evers, 2010). And vice versa — science's expansion has repercussions not only for society but also for science. One mechanism to cope with this exponential growth has been differentiation. This strategic response is also observed in economic markets and organizations — if the competition increases, the incentive to create new 'niches' grows. The nature of these niches depends on the current alternatives and on the available resources (Whitley, 1984) — an explanation for why well-funded research areas, such as molecular biology, attract a high number of researchers and are highly competitive at the same time.

One form of differentiation is a greater abstraction towards 'a deeper understanding of the postulated structures of the object' (Stichweh, 1984). Science absorbs less knowledge from the outside but increasingly specialises in sorting, categorising and reconstructing its own knowledge in ever smaller fields. The German physicist Hermann von Helmholtz observed this more than 100 years ago: 'The obvious result is seen in the fact that every individual researcher is forced to select an ever smaller field for his research and can maintain only an incomplete knowledge of neighbouring fields' (von Helmholtz, 1896). The outstanding 'heroes' of science, people such as Newton, Leibniz or Voltaire, have made way for an army of largely anonymous workers, a sign that science has levelled any status differences and become more democratic.

The other form of differentiation is through expansion into new fields. The more abstract and the more general the scientific methods and instruments, the more effectively they can be used to investigate new objects and phenomena. This growth-driven specialisation can be regarded as the motor of science's expansion into society as more fields succumb to its scrutiny. An illustrative example is the environmental sciences that originally expanded from ecology into a wide range of specialised fields now ranging from social risk assessment to environmental psychology. In fact, they are a paradigm for a new type of science, dubbed 'post-normal science', which is characterised by its close proximity to politics, its need for social legitimisation and its dependence on outside knowledge.

This expansion has led to a tighter coupling of science with various elements of society and has consequences for both. I will investigate in particular how the increasingly tight union between science, politics and the media is changing the face of science.

On March 28, 1979, Walter Cronkite began the CBS evening news by declaring that it was the 'most confused day in the history of news media' and went on to describe the contradicting information about the accident at the Three Mile Island nuclear reactor in Harrisburg, PA. Seven years later, a similar spectacle took place in Europe following the explosion in the Chernobyl nuclear power plant. Scientists and politicians contradicted each other in the media, exaggerating or downplaying the risks of the nuclear cloud that was moving towards Western Europe. It culminated in a politician publicly consuming salad on national television in a futile attempt to prove that the fallout from Chernobyl had left vegetables untouched.

Three Mile Island and Chernobyl became symbols not only of the dangers of nuclear energy, but also of the loss of public confidence in scientific authority (Weingart, 1979). The contradictory statements of scientific experts in public hearings showed that they had only been recruited to support specific positions and interests in politics and the economy. It destroyed the image of science as an institution that is free of interests other than its own. Since then, the public has become aware of various other scientific/technical dilemmas — the safety of genetic technologies, the ethical implications of reproduction techniques, the use of biotechnology in agriculture, the political, ethical and economic implications of the sequencing of the human genome, BSE, the potential dangers of electrosmog, to name but a few. It has become common practice for the parties involved in these controversies to engage scientific experts to present evidence that supports their cause.

At the time of the nuclear accidents, scientists were already frequent visitors to the corridors of power. In fact, the last 50 years have seen the establishment of a large scientific presence within governments, ministries and parliaments. Politics either requests special scientific expertise or creates its own in the form of research institutions controlled by ministries. Few remember that in the 1950s the connection between science and politics was seen as more ambivalent — on the one hand as a guarantee that politics relied on rational and unbiased knowledge to make decisions in society's 'best interests', and on the other as a *liaison dangereuse*, where the political influence of non-elected scientific experts threatened the democratic system.

As scientific expertise became common-place in politics, these initial fears gradually disappeared, mainly because expert knowledge became equally available to all groups in the political system. This democratisation has led to an increasing competition for expertise. Surprisingly, however, this has not led to political decisions being more rational. In contrast, the controversies have become more intense and the lack of knowledge has become more apparent. But, despite the obvious absence of expert authority and rationality, the existing arrangement between science and politics continues nearly unchanged. And since there is no alternative to scientific expertise, the institutionalised coupling of science and politics is not likely to change in the future. What is changing are the borders between politics and science and how and where they are drawn.

The relationship between science and politics can be described as a recursive coupling of two interdependent developments — the scientification of politics and the politicisation of science (Weingart, 1983). One indicator of the former is that many political problems have been first recognised and described by scientists — environmental pollution, for instance, only made it on to the political agenda after scientists discovered DDT in the food chain. All debates about the risks of new technologies are, in fact, driven by research on the potential dangers and how to avoid them (Weingart, 1991; 1993). In other words, the more science becomes part of the governing process, the greater its role becomes in defining the problems that it is then asked to solve (Boehmer-Christiansen, 1995).

Vice versa, the coupling of knowledge with politics drives the politicisation of science. Knowledge, as it enters the public arena, is inevitably judged and valued by society. The expert's position in a controversy is thus seen as determined by politics and not by knowledge. 'Advisors are selected, not because members of parliament and officials from government need their advice, but because they obviously need their authoritative support for the policies they represent. By succumbing to these temptations, it is obvious that they fraudulently agree with the scientists, who exploit the prestige gained through objectivity and neutrality' (Shils, 1987). The assumption that science is always disinterested and transmits only objective knowledge is obviously a myth. Science has become one of the actors in the political arena, be it as an interested party or recruited by other actors to support their specific interests. The increasing demand for scientific expertise leads to an inflationary use of knowledge that fuels controversies. This oversupply of knowledge means that politics risks the loss of an important source of legitimisation, and science risks the loss of public confidence.

Science has also become closely interrelated with the media. In 1993, a Dutch researcher claimed to have found a vaccine against HIV. Pressed by his peers, he admitted that he had exaggerated his results because he felt it was the only way to gain public attention and support (Hagendijk and Meeus, 1993). Such spectac ular claims are a rather recent phenomenon in science and can be explained by the traditional model of popularising science. Historically, scientific knowledge has been regarded as superior and of higher value than popular knowledge or 'common sense'. Science produces 'true' knowledge, so specialised and esoteric that it needs to be translated and transmitted to the public by the media. The process is controlled by scientists — and from their point of view, popularised knowledge is simplification in the best case, contamination in the worst.

This concept has two questionable implications. One is the belief that the public is a passive and non-specialised audience, excluded from the production and verification of knowledge and incapable of judging its value (Whitley, 1985). The other is that the role of the media is limited to translating knowledge. This notion still has too many supporters to be outdated. However, a new view has taken hold more recently that emphasises the independence of the media. There is a growing consensus that they 'create their own reality', using their own selection criteria depending on the use of information, their sources and their particular readership or viewership (Dunwoody and

Peters, 1993; Marcinkowsky, 1993; Luhmann, 1996). The most important criteria for journalists in selecting their stories are actuality, sensation, personalities and local implications — in short, they cover 'what is existent, what is happening and what is important' (Dunwoody and Peters, 1993). Clearly, these criteria are different from those that the scientific community uses to communicate information. The complaints about 'wrong' or 'distorted' media coverage and 'wrong' selection of news therefore do not reflect reality. There can be no 'adequate' presentation of science in the media in a way that satisfies scientists.

Science has begun to understand this notion and started to open its doors to the media. A striking example is the recently modernised 'public relations' (PR) departments of universities, research institutes, museums and other scientific organisations that use the same techniques as professional PR companies and produce the same glossy prospectuses. Universities, with varying degrees of success, give themselves a 'corporate design' similar to large companies. The effect on the public acceptance of science when it uses the tools of commercial advertisement remains to be seen.

Another reaction of science is the increasing use of the media to mobilise public support. This is not due to unexplainable behavioural changes on the part of scientists but the increasing competition and the increasing social relevance of certain research fields. Scientists warn of global catastrophes, promise treatments for many ailments or solutions to solve mankind's problems in order to gain media attention that could increase the resources for their research. This strategy is particularly useful when the importance of the peer review system is diminished and politics influences the distribution of funds. Dorothy Nelkin showed that decisions concerning large research projects in particular are increasingly being made outside the traditional system of peer review, with the effect that external criteria gain importance (Nelkin, 1995). Consequently, funding decisions are dependent not only on potential social need but also on the potential media coverage, forcing scientists to resort to the strategy of 'selling their science to the public'.

A prime example is the field of climate research. The public debate about global warming was triggered in the USA by NASA scientist James Hansen and the pre-publication of his research in the *New York Times* before the original article appeared in *Science.* The newspaper spoke of an 'unprecedented warming' of the atmosphere in the next century and warned of rising sea levels. Hansen quickly became the central figure in the global warming debate and the most important source for journalists. The topic reappeared on the political agenda in 1998 when Hansen explained to the US Congress that carbon dioxide and other artificially produced gases would increase the temperature of the atmosphere. This time it was a politician, Senator Timothy E. Wirth, who organised the media coverage. The result was additional funds from Congress for climate research.

Similarly, in Germany in 1986, the German Physical Society warned of a climate catastrophe and called for immediate action to regulate carbon dioxide production. The unusual dramatic tone of this warning created the intended media response — the German magazine *Der Spiegel* printed a picture of Cologne cathedral sinking in the sea, which became an icon for global warming. A commission created in 1987 to investigate the threat recommended a reduction in carbon dioxide release of 25–30 percent by 2005, and the German parliament followed this with a commitment to a 25 percent decrease. The success for the scientists was the institutionalisation and continued funding of climate research.

This pattern has been repeated regularly in the establishment of new research fields. In the beginning are dire warnings in the media about a looming danger, if not catastrophe. The scientists promise to find solutions, which results in funds for research. Once established, the new research field starts its life cycle; the knowledge increases, becomes more differentiated, specialised, abstract and increasingly irrelevant to the solution of the original problem.

This 'risk of communication' is similar to the Cassandra legend in Greek mythology. Justified warnings may be overlooked because there is no way to verify their truth. Science, as society's early warning system, becomes neutralised and ignored as just one voice among many others. 'The public image of politics is nearly exclusively determined by the media. [...] Trapped in a faster and faster spinning hamster wheel, politicians run behind their own agenda and do not realise how little of their work remains visible. The more that is proposed, criticised, questioned, rejected and resurrected, the less is left in the public view. [...] Politics cannot flourish in such a climate of extreme nervousness, vain talkativeness and uncritical closeness. Disorder and loss of distance contribute to the ruin of politics; the trust in the actors is lost' (Kornelius, 1996). Stefan Kornelius described the consequences for politics, but his analysis may as well hold true for science. The strategy of using the media to generate public support for research is surely doomed to fail. Instead, it strengthens the suspicion that scientists have their own agenda and further endangers their public credibility.

It is a paradox — the more independent science and the media are, the tighter their coupling. And as the media gain importance, science is losing the monopoly of judging scientific knowledge. The abstract criterion of truth is no longer sufficient in the public debate because the media add the criterion of public acceptance. This does not mean that scientific verification is being replaced, but it is being supplemented by other measures.

The term 'information society' was originally coined in Japan at the beginning of the 1960s but has since found its way into European and US political rhetoric. But it is a rather misleading term — it is not information that will characterise future societies but the context-related production, appropriation and use of information in order to create knowledge. Consequently, the term information society is being gradually replaced by the term 'knowledge society'.

Irrespective of the description, the main problem is how science can produce and communicate reliable knowledge if the traditional means are outdated, if the distance between science and society and the resulting trust are replaced by tight coupling. As already discussed, three important tendencies in science show that this will be a major problem. First, science is increasingly influenced by politics. The use of scientific expertise by politicians and other groups in society involves science in making political decisions and installs scientists within political groups. Secondly, science is increasingly commercialised. Scientific knowledge is becoming a commodity that can be traded like any other product. Thirdly, science and the media are increasingly coupled as scientists use the media in order to raise public support for their research. All these trends point to a loss of distance between science and society, which has been the most important basis for the production of reliable knowledge.

One could conclude from these developments that science is losing its importance. But this is not quite correct. Science as an institution that produces knowledge remains irreplaceable. It rather means that the principles of science — the production and revision of knowledge — are being extended to other parts of society. Therefore, science is losing its position as the only institution that produces reliable knowledge. The point here is the validation of truths. If science fails to do so, society will, the media will demand it and politics will create the necessary trust through legal frameworks.

This loss of distance will not lead to the end of the communication of truths. Trust and confidence remain both constitutive and rare values in communication, and the more society depends on reliable knowledge the more these are required. The main characterisation of today's society is the competition for trust. Once achieved, this is invaluable and science should be keen to preserve it. Therefore, it is only the efforts needed to produce trust and confidence that have become greater.

References

Bell, D. (1973). *The Coming of Post-Industrial Society: A Venture in Social Forecasting*. New York: Basic Books.

Boehmer-Christiansen, S. (1995). Reflections on scientific advice and EC transboundary pollution policy. *Science and Public Policy*, 22:195–203.

Dunwoody, S. and H. P. Peters (1993). Massenmedien und Risikowahrnehmung. In *Risiko ist ein Konstrukt* (pp. 317–341). Munich: Bayrische Rück.

Evers, H.-D. (2010). Knowledge and the sociology of science. In *Governing and Managing Knowledge in Asia* (2nd edition), T. Menkhoff, H. D. Evers and Y. W. Chay (eds.), pp. 59–68. Singapore: World Scientific Publishing.

Hagendijk, R. and Meeus, J. (1993). Blind faith: Fact, fiction and Fraud in public controversies over science. *Public Understanding of Science*, 2:391–415.

Kornelius, S. (1996). Ein irres system von hoher geschwätzigkeit. *Süddeutsche Zeitung*, May 18.

Luhmann, N. (1996). *Die Realität der Massenmedien*. Opladen: Westdeutscher Verlag.

Marcinkowsky, F. (1993). *Publizistik Als Autopoietisches System: Politik und Massenmedien — Eine Systemtheoretische Analyse*. Opladen: Westdeutscher Verlag.

Nelkin, D. (1995). Science Controversies. In *Handbook of Science and Technology Studies*, Jasanoff *et al.* (Eds.), pp. 444–456. London: Thousand Oaks.

Price, D. J. (1971). *Little Science, Big Science*. New York: Columbia University Press.

Shils, E. (1987). Science and scientists in the public arena. *The American Scholar*, 65:185–202.

Stichweh, R. (1984). *Zur Entstehung des Modernen Systems Wissenschaftlicher Disziplinen*. Physik in Deutschland. Frankfurt am Main: Suhrkamp.

von Helmholtz, H. (1896). Über das verhältnis der naturwissenschaften zur gesamtheit der wissenschaften. In *Vorträge und Reden*, von Helmholtz, H. (ed.). Braunschweig: Vieweg und Sohn.

Weingart, P. (1979). Das 'Harrisburg Syndrom' oder die De-Professionalisierung der Experten. In *Kernenergie: Gefahr oder Notwendigkeit*, H. Nowotny (ed.), pp. 9–17. Frankfurt am Main: Suhrkamp.

Weingart, P. (1983). Verwissenschaftlichung der gesellschaft — Politisierung der wissenschaft. *Zeitschrift für Soziologie*, 12:225–241.

Weingart, P. (1991). Large technical systems, real-life experiments, and the legitimisation trap of technology assessment: The contribution of science and technology to constituting risk perception. In *Social Responses to Large Technical System, Control or Anticipation*, T.R. LaPorte (ed.), pp. 5–17. Dordrecht: Kluwer.

Weingart, P. (1993). Welche forschung ist gut? *Bild der Wissenschaft*, 5.

Whitley, R. (1984). *The Intellectual and Social Organization of the Sciences*. Oxford: Clarendon Press.

Whitley, R. (1985). Knowledge producers and knowledge acquirers: Popularisation as a relation between scientific fields and their publics. In *Expository Science: Forms and Functions of Popularization, Yearbook Sociology of the Sciences IX*, T. Shinn and R. Whitley (Eds.). Dordrecht: Reidel.

Chapter

6

The Conflicting "Norms" of Science in the Institutionalization of Scientific Knowledge

Kenneth G. Huang

1. Introduction

The norms of science as described by the prominent sociology scholar Robert K. Merton in the 1960s laid the foundation for the normative structure of scientists' behavior pertaining to scholarly enquiry into the production and accumulation of scientific knowledge (e.g., Huang and Murray, 2009). However, as the nature of science evolves and the tension between pure scientific enquiry and motivations for commercialization grows, the idealized conception of the norms or "ethos" of science as originally proposed are not necessarily adhered to in the modern science disciplines such as the life sciences. This could result in deviations and departure from the norms of science in the institutionalization of scientific knowledge.

Several contemporary scholars of sociology of science have provided similar criticisms as well to Merton's normative structure of science postulated some 50 years ago.[1] His theory had been extended, reformulated and tested empirically to a certain extent (Hagstrom, 1965; Storer, 1966; West, 1960).[2] Some even took the writings of later scholars of sociology of science such as Kuhn (1970 [1962]) as an alternative theoretical direction to Merton.[3] Lately, critics have increasingly focused on the phenomenological and "economic" perspectives, showing the growing interests among sociologists of science in these aspects of the scientists' behaviors.

Merton's initial statement held that "the institutional goal of science is the extension of certified knowledge", and knowledge comprising "empirically confirmed and logically consistent statements of regularities (...in effect, predictions)".[4] As a social institution, science is characterized by the "ethos... affectively toned complex of values and norms which are held to be binding on scientists. The norms are expressed in the form of prescriptions, proscriptions, preferences and permissions. They are legitimized in terms of institutional values. These imperatives [or norms], transmitted by percept and example and reinforced by sanctions are in varying degrees internalized by the scientist, this fashioning his scientific conscience... Although the ethos of science has not been codified, it can be

---

[1] Merton, R. K. (1973). The Normative Structure of Science. In *The Sociology of Science: Theoretical and Empirical Investigations*, Chicago, IL: The University of Chicago Press, Chapter 13.
Also see Merton, R. K. (1968). Science and Democratic Social Structure. In *Social Theory and Social Structure*, New York, NY: Free Press, Chapter 16.

[2] Both Hagstrom (1965) and Storer (1966) questioned the operation of social control in science: why scientists conform to the norms, do research and publish their results. Hagstrom proposed that scientists make gifts of new knowledge to the scientific community in exchange for recognition; Storer proposed that scientists seek "competence response" to their work, which they can only obtain if they make public contributions.

[3] For example, the work: Kuhn, T. S. (1970). *The Structure of Scientific Revolutions.* Chicago, IL: The University of Chicago Press (Original work published 1962).

[4] Merton, R. K. (1973). The Normative Structure of Science. In *The Sociology of Science: Theoretical and Empirical Investigations*, Chicago, IL: The University of Chicago Press, Chapter 13, p. 270 (Original work published 1942).

inferred from the moral consensus of scientists as expressed in use and wont, in countless writings on the scientific spirit in moral indignation directed toward contraventions of the ethos".[5, 6]

This concise paragraph describes much of the theoretical underpinnings laid out by Merton — it conveys the idea that the institution of science, specifically the life sciences in this case, contains its own norms and values. These are made rightful by its underlying set of goals, the extension of certified knowledge, and these are set in the social context and enforced by its system of rewards and punishment. In addition, the "ethos" or code of science can be drawn from the observations of the scientists' behaviors and their writings, especially from their responses to deviation from the postulated norms.

Merton further impressed that "the mores [norms] of science posses a methodologic rationale, but they are binding, not only because they are procedurally efficient but because they are believed right and good".[7] Merton expressed here that the ethos of science makes a contribution to the institutional objectives of generating knowledge and also has important communicative meanings.

This chapter focuses on the growing discussion and critiques of the idealized conceptualization of the norms (or ethos) of science in the process of institutionalizing scientific knowledge. I draw from the related themes in this contentious issue and raise anecdotal evidences to support these arguments with specific reference to the life sciences. The chapter proceeds as follows: Section 2 examines Merton's four ethos of science and to what extent (and accuracy) they can be applied to modern life sciences disciplines. Specifically, it looks at how Merton himself (in his later years) and other more contemporary sociology of science scholars have challenged (or supported) his ideal scientific

---

[5] Merton, R. K. (1968). Science and democratic social structure. In *Social Theory and Social Structure*, New York, NY: Free Press, Chapter 16, pp. 551–552 (Original work published 1957).

[6] Merton, R. K. (1973). The Normative Structure of Science. In *The Sociology of Science: Theoretical and Empirical Investigations*, Chicago, IL: The University of Chicago Press, Chapter 13, pp. 268–269 (Original work published 1942).

[7] *Ibid*. p. 270.

norms or ethos and to what extent they fit in with the life sciences. Section 3 provides some evidences of other deviant behaviors (e.g., fraud and plagiarism) and social control in the institutionalization of scientific knowledge. Section 4 describes a related phenomenon: the stratification in science. Section 5 offers some concluding thoughts.

2. The Norms of Science?

In principle, the ethos supposedly consists of "technical" (or cognitive) and "moral" (or social) norms. "The technical norm of empirical evidence, adequate and reliable, is a prerequisite for sustained true prediction; the technical norm of logical consistency, a prerequisite for systematic and valid prediction. The entire structure of technical and moral norms implements the final objective" of expanding certified knowledge.[8] The moral norms were also dealt with concisely.

The following descriptions summarize the ethos of science as proposed by Merton (and later sociology of science scholars), as well as the potential difficulties and criticisms associated with each one of them, particularly in the depiction of the life sciences.

"*Universalism*" means that knowledge claims and scientific contributions are judged (accepted or rejected) through "pre-established impersonal criteria: consonant with observations and with previously confirmed knowledge".[9] That is to say that scientific results should be analyzed objectively, held verifiable, testable and repeatable as Merton commented: "Objectivity precludes particularism. The circumstance that scientifically verified formulations refer to objective sequences and correlations mitilates against all efforts to impose particularistic criteria of validity".[10] Personal or social attributes such as race, religion, nationality, political views, class origins or gender are "irrelevant" to the validity of such judgments. Such norm also demands that scientists be rewarded in accordance with the extent of

[8] *Ibid*. p. 270.
[9] *Ibid*. p. 270–271.
[10] *Ibid*. p. 271.

their contributions in science.[11] Merton himself acknowledged that although "universalism is deviously affirmed in theory and suppressed in practice", it remains "a dominant guiding principle".[12]

Although "universalism" seems largely upheld by life scientists in their analysis, verification and judgement of scientific results, a main criticism is suggested by Barnes and Dolby (1970).[13] They argued that Merton's norms "represent attempts to describe institutional constraints upon the way the scientists think and act with respect to the published results of their peers. They are based on the view that a specifically 'scientific approach' exists common to the evaluative procedures of all scientists, and specific to them", and norms such as "universalism" is *not specific* to science and certainly not to the life sciences in particular, and therefore cannot be defined as a norm used to make a distinction between "science" from other societies or institutions.

"*Communism*", or as Barber (1952) re-named it "communalism",[14] stipulates that knowledge as a product of collective effort by the scientific community must be shared for the common good and not kept secret. Merton described "the substantive findings of science are a product of social collaboration and are assigned to the community... The scientist's claim to 'his' intellectual 'property' is limited to that of recognition and esteem..."[15] In other words, in our context, Merton would argue that the way life scientists ensure they obtain intellectual property rights to their findings is by publishing promptly

---

[11] Merton, R. K. (1973). Priorities in scientific discovery. In *The Sociology of Science: Theoretical and Empirical Investigations,* Chicago, IL: The University of Chicago Press, Chapter 14 (Original work published 1957).

[12] Merton, R. K. (1973). The normative structure of science. *The Sociology of Science: Theoretical and Empirical Investigations,* Chicago, IL: The University of Chicago Press, Chapter 13, p. 273.

[13] Barnes, S. B. and R. G. A. Dolby (1970). The scientific ethos: A deviant viewpoint. *Archives Européennes de Sociologie*, 11, 3–25 (p. 8).

[14] Barber, B. (1952). *Science and the Social Order*. Glencoe, IL: Free Press.

[15] Merton, R. K. (1973). The normative structure of science. In *The Sociology of Science: Theoretical and Empirical Investigations,* Chicago, IL: The University of Chicago Press, Chapter 13, p. 273.

and announcing the results to the scientific community. Another scholar of sociology of science, Hull (1985), described that science is structured in such a way to ensure the convergence of group and individual goals — life scientists (and scientists in general) have to make their work public or give up priorities. In essence, scientific advancement in the life sciences comes from close collaboration and cooperative sharing of results among the community biological and physical scientists and across generations of scientists.

Merton stated that "secrecy is the antithesis" of communism.[16] Although it is true that scientific collaboration takes place frequently and is important in the advancement of science and particularly the life sciences, secrecy, in modern times, plays increasingly important roles in particular scientific settings, especially with regard to life sciences projects funded by the industry. Many life scientists (and scientists in general), realizing the potential commercial rewards from their research, have become more commercially oriented. This takes place not only in industrial science laboratories but also in university settings, among academic researchers and scientists.[17] The main objective of business is to generate profit, and some level of secrecy is a necessary instrument for firms to maintain a competitive advantage. "Accordingly, the norms of science which traditionally condemn profit-making motives are beginning to change to allow for such a kind of entrepreneurship; and varying institutional structures are experimented with which fit to these new cognitive and normative patterns" (Merton, 1973 [1942]; Etzkowitz, 1994).

---

[16] *Ibid.* p. 274.

[17] This was further encouraged by the setting up of Technology Licensing Offices (and equivalent) in various academic institutions and universities around the U.S., as well as the passage of a series of Acts in the 1980s such as the Stevenson-Wydler Technology Innovation Act (1980), which explicitly set the mission of technology transfer for all federal laboratories and federal agencies, and the Bayh-Dole Act (1980), which was also designed to encourage technology transfer and commercialization and to give academic institutions, non-profit institutions, and related institutional entities the automatic rights to retain titles to patents from federally funded R&D, such as a biomedical research grant from the National Institute of Health (NIH).

Another sociology of science scholar, Ziman (1985) argued that to think scientists (and life scientists) exchange information for professional and peer recognition "is not yet out of date" but such consideration must be radically modified to account for "collectivized science", which does not only take place in industrial corporations but also in large-scale academic science demanding significant management and coordination.

In addition, as "multiples", or discoveries being made independently and often at the same time by two or more scientists that convey identical "central message", (Merton, 1973 [1961])[18] frequently occur in the history of scientific discoveries, there is little or no value in discovering a scientific, in this case, biological phenomenon or theory such as the double helix structure of DNA (Watson and Crick, 1953) second, or third and so on.[19] The scientific reward system therefore sets up an "immediate tension between cooperative compliance with the norm of full disclosure (to assist oneself and colleagues in the communal search for knowledge), and the individualistic competitive urge to win priority races" (Dasgupta and David, 1994:500). Subsequently, life scientists may depart from the norm of disclosure (and communism) by either remaining taciturn until "a result" has been obtained and can be publicly announced (e.g., results are codified into drafts of publishable papers but not circulated for sometime) or maintaining secretiveness in the form of partial disclosure, which is often a more "pervasive and persistent pattern of behavior among researchers" and "it seems potentially the more serious source of wastage of social resources" (Dasgupta and David, 1994:501). (This ties in with the discussion of the other forms of deviation from social norms in Section 3.)

---

[18] Merton, R. K. (1973). Singletons and Multiples in Science. In *The Sociology of Science: Theoretical and Empirical Investigations,* Chicago, IL: The University of Chicago Press, Chapter 16, pp. 352–356 (Original work published 1961).

[19] Watson and Crick beat Rosalind Franklin (1920–1958) and Linus Pauling (1901–1994) to the publication of the double helix structure of DNA and were rewarded copiously for their discovery — the 1962 Nobel Prize in Physiology or Medicine, alongside significant peer recognition in the scientific community.

This is to avoid being "scooped" on a particular idea.[20] As soon as the work is published, it is made public and the knowledge becomes freely available to all, with the credit going to the publishing authors.

"*Disinterestedness*" refers to the restraining of personal emotions, biases and financial attachment to their work. It involves institutional mechanisms to control the motives for scientific research to advance scientific knowledge. This does not imply altruism as social arrangements that rely entirely on altruism are unstable. Rewards constitute peer recognition of scientific achievement, not through monetary gains. Ideally, the system of reward and punishment in science should align the interests of the scientists with the scientific institution to act in a disinterested manner.[21]

However, in addition to what has been discussed previously on the shifting and growing acceptance of motives from pure academic recognition to a balance with financial incentives and entrepreneurship,[22] life scientists sometimes become emotionally attached to their work — for example, "pet theories" and projects are held onto despite opposition, as they have spent considerable amount of time and energy in developing such theories. Life scientists may vehemently defend their theories and methodological merits (viewed as territories) as people do in other social groups. However, for the true advancement of science to occur, a scientist has to be open minded, studies all valid alternative hypotheses and accept the one that is best support by experimental evidence — ideally, scientific progress should stem from the rejection of weakly supported hypotheses and acceptance of the strongly supported ones.

"*Organized skepticism*" is "both a methodologic and an institutional mandate" that requires "the suspension of judgment" about a

---

[20] Sociologist Warren Hagstrom found in a 1967 study, that 30–40 percent of American university scientists in some fields are concerned about being "scooped" in their current work, to the extent that it could restrict their discussion with their colleagues.

[21] Merton, R. K. (1973). The normative structure of science. In *The Sociology of Science: Theoretical and Empirical Investigations*, Chicago, IL: The University of Chicago Press, Chapter 13, pp. 275–277.

[22] For example, in the survey conducted by Etzkowitz (1998), although it identified half the faculty in MIT Biology department as having industrial ties in the late 1980s, the informant could identify only one of his colleagues as uninvolved.

particular scientific theory until "the facts are at hand". Merton further stated that "the scientific investigator does not preserve the cleavage between the sacred and the profane, between that which requires uncritical respect and that which can be objectively analyzed".[23] Once again, the emphasis is on the institutional arrangements of the normative structure of the institution of science. "Organized skepticism" should occur in activities like refereeing and other critical appraisals of work by competent peers.

Oftentimes, waiting for all the facts to be assembled before their scrutiny and making a judgment takes an unreasonably long time and effort — this norm needs to be reconsidered under certain circumstances. For example, to solve problems in Genetics, scientists make use of information such as the vast sequencing database derived from the recently completed Human Genome Project in 2003 (Huang and Murray, 2010). If the biologists had waited until all "the facts are at hand" before making judgments on what gene carries what hereditary diseases, and how they would be expressed, they would have to wait for several years, even decades, before the judgment can be made. Under most circumstances, theories are advanced before much is known about the actual mechanism or pathways behind a particular observation — research is often an ongoing process. Scientific publications represent and build on previous works, which would pave ways for further experimentation and research. To solve modern life science problems, complexity of research increases. Incremental research results based on what information scientists already have at hand are often as important as waiting for all the requisite evidence to be in before making a judgment call.

2.1 *Further Comments on the Ethos of Science*

The ethos of science, as with norms in general, stipulates shared expectations or ideals on how scientists (and life scientists) should behave in relation to other scientists. It is difficult, as in other fields,

[23] Merton, R. K. (1973). The normative structure of science, *The Sociology of Science: Theoretical and Empirical Investigations,* Chicago, IL: The University of Chicago Press, Chapter 13, pp. 277–278.

to expect that every life scientist will follow each and every norm all the time. Merton himself commented that there is often a "painful contrast" between normative expectations and actual behavior (Merton, 1976:40).[24] However, he still expects the norms of science to exercise patterned control over behavior of scientists.

"The normative theory as first proposed is exceedingly brief, almost elliptically so. It does not adequately anticipate criticism nor does it place the theory in the framework of social and cognitive norms in science, much less in the larger framework of cultural structures generally" (Smelser, 1988:516). Other critics of Merton have focused on problems such as the nature of the norms governing the scientists' behaviors; objectives they serve; how scientists have shared these norms; whether such norms are unique to science; and the explanation that seeming conformity to such norms implies scientists' commitment to them. For example, some assert that only cognitive (or "technical") norms are institutionalized in modern science since scientists are rewarded only for conforming to them (Mulkay, 1976:641) but not so much for social norms. Most of these criticisms are directed at scientists in general but can be observed in life scientists.

3. Other Deviant Behaviors and Social Control

Extending from the previous discussion, it is worthwhile to consider some other forms of deviant behaviors and social controls among modern scientists that Merton (1973 [1957]) and other sociology of science scholars have described (Mulkay, 1969; Hagstrom, 1974; Zuckerman, 1977). Priorities in particular scientific discoveries have often become the causes of contention among life scientists and other physical scientists alike — "on being first in contributing new knowledge and in being recognized for it" and there is "often the ironic consequences of this institutionalized pattern" (Smelser, 1988).

[24] Merton, R. K. (1976). The Sociology of Social Problems. In *Contemporary Social Problems*, R. K. Merton and R. Nisbet (eds.), pp. 3–43. New York, NY: Harcourt Brace Jovanovich.

Sometimes, the competition may get too intense in the life sciences, and fraud, plagiarism and other forms of deviant behaviors can occur when the desire for recognition surpasses the endeavor and will to advance knowledge (Smelser, 1988).

Fraud is typically concerned with *deviant behavior from the norms of "disinterestedness" and "organized skepticism"*. Fraud usually occurs when scientists knowingly attempt deceptive activities for peer recognition or monetary gains. It can come in the forms of "forging" or data fabrication, "trimming" or data manipulation, and "cooking" or data suppression (Babbage, 1976 [1830]). Data suppression is perhaps among the most difficult form of fraud to identify as no lies are told and scientists have to decide about the selection of trustworthy data. All three forms of fraud described by sociology of science scholars have occurred occasionally in the life sciences.

Plagiarism, on the other hand, typically involves *deviation from the norms of "communism" and "universalism"* as scientists illegitimately claim credits for work that was not originally performed by them. When scientists engage in deliberate acts of deception by taking credits where they are not theirs, including the copying of documents and wrongful appropriation of intellectual properties, plagiarism occurs. The latter is usually harder to detect especially under the circumstances or regions of weak intellectual property rights protection (such as copyrights and patents) (e.g., Huang, 2010), or a lack of consensus on what constitutes property rights in science.[25] This situation can be further complicated by instances of "multiples" as mentioned previously. This illegal act, although sometimes

---

[25] A particularly controversial example of intellectual property claim in genetics is a DNA patent application demanding protection for cDNAs associated with Expressed Sequence Tags (ESTs). The U.S. Patent and Trademark Office (USPTO) takes the position that an EST may be patentable if it meets the utility, novelty and unobviousness requirements (e.g., ESTs sequenced from areas of the genome that have never been sequenced before would meet the novelty and unobviousness requirements.) The issue can get even more controversial and complicated when the invention claims not only the EST itself, but also the full-length gene that encompasses the EST and other related genes that will hybridize to the EST under specified stringency conditions.

observed in the scientific community, remains an ill-defined normative area in modern science (Smelser, 1988).

Sociologists of science have considered that the social control of deviant behavior to entail procedures for deterrence, detection and punishment. For example, to instill the life scientists with the norms through education and social pressure is one way of deterrence. Another particularly effective social control is the "methodological canon of reproducibility and actual replication" of scientific results (Smelser, 1988). The former is another form of deterrence for fraud while the latter serves to detect such activity in the event that it actually happens. Knowledge by life scientists (and scientists in general) that such control mechanisms exist and fraud in all its forms may potentially be detected in the future prevents deviation from the norms. Other slightly less effective forms of social controls include planned and unplanned scrutiny and public "whistle blowing".

It is clear why life scientists (and other scientists) take these acts of deviant behaviors very seriously. To the biological and medical scientists, fraud is "heinous",[26] "scandal" with a "quality of desecration about it",[27] "shocking",[28] and "intolerable".[29] In modern scientific community, as many sociology of science writers have observed, scientists take what other scientists claim as almost always trustworthy and reliable. This widely held belief has saved much precious time and efforts to verify what other scientists claim in every case. Zuckerman (1977) sums it up succinctly, "the institution of science involves an implicit social contract between scientists so that each can depend on the trustworthiness of the rest... the entire cognitive system of science is rooted in the moral integrity of individual scientists". This

[26] Immunologist — Medawar, P. (1976). The strange case of the spotted mice. *New York Review of Books*, 23:6–11.
[27] Molecular biologist — Luria, S. (1975). "What Makes a Scientist cheat?" *Prism*, (May): 15–18, 44.
[28] Medical scientist — Petersdorf, R. G. (1986). The Pathogenesis of Fraud in Medical Science. *Annals of Internal Medicine*, 104:252–254.
[29] Biochemist and editor of Science, Koshland, D. (1987). Fraud in Science. *Science*, 235:141.

applies to the life sciences, as well as other modern science disciplines, particularly well.

4. Stratification in Science

Now we turn to consider another important characteristic of the sciences (and life sciences) that serves as an underlying motivation in explaining some of the deviant behaviors of scientists from the norms — the stratification in science. Stratification in science was first laid out as one of the research themes by Merton (1973, [1957]).[30] It stems from the fact that *the scientific community* is not created equal — *it is clearly stratified as a small number (and percentage) of scientists contribute disproportionately to scientific progress and in return, obtain a disproportionately greater portion of rewards, recognition and resources for research* (e.g., funding, grants etc.).[31] This definitely applies well to the life sciences. This stratification tends to be "self-reinforcing" over a life scientist's career although it remains largely imperceptible and unknown to people outside the scientific community.

Stratification in the life sciences is reflected in the ranking of prestige (implicit and explicit) of individual scientists, research groups, laboratories, schools, universities, journals, specialties, theories and methods. The top tier consists of elite biological scientists who have displayed consistent high social performance and superior contribution in terms of theories or findings to their specialties. These elites generally receive most financial rewards and high level of peer recognition.

---

[30] Merton, R. K. (1973). Priorities in Scientific Discovery. In *The Sociology of Science: Theoretical and Empirical Investigations,* Chicago, IL: The University of Chicago Press, Chapter 14 (Original work published 1957).

[31] There is much evidence for such marked stratification in science. For example, one crude measure of recognition is citation count, or the extent of citation of the scientists' work in research publications of their peers. Citation counts are highly skewed — of all scientists whose work was cited at all between 1961 to 1980, 62 percent were cited five times or less, 6 percent were cited up to 100 times and 1 percent were cited 500 times or more (Institute for Scientific Information, 1981). In terms of awards, university hiring of Nobel Laureates, and universities which graduate future members of the National Academy of Sciences, the statistics are just as skewed.

Collective judgment and recognition of a life scientist's work by his or her competent peers are particularly important because, aside from the joy of discovery and making original and meaningful contribution to the field, it is the only "unambiguous demonstration" of the importance of his or her findings to the advancement of science.[32] This kind of collegial recognition is also manifested in all kinds of prizes, fellowships, professorships and on rare occasions, eponymy.[33]

This kind of sharp stratification in the life sciences, as well as in other disciplines of sciences, partially explains the intense competition among life scientists and instances of deviant behaviors from the norms of science first proposed by Merton.

5. Concluding Thoughts

Merton laid the intellectual foundation for the sociology of science and described the norms of science in the 1960s. However, these norms apply only to a limited extent in describing the life sciences in the current decade. These norms (or "ethos") proposed by Merton are either ambiguous in uniquely describing science and specifically the life sciences, such as "universalism"; or do not aptly capture the range of behaviors of modern day life scientists under various settings, such as "communism" and "disinterestedness"; or are not practical for increasingly complex settings of biological and pharmaceutical research, such as "organized skepticism". Deviant behaviors from the norms can sometimes result. These norms represent an idealized process in the institutionalization of science and scientific knowledge to further our understanding of the behaviors of scientists and the scientific community. However, variations to these normalized structure provides evidence that under certain circumstances, the socio-political

---

32 Merton, R. K. (1973). Priorities in Scientific Discovery. In *The Sociology of Science: Theoretical and Empirical Investigations,* Chicago, IL: The University of Chicago Press, Chapter 14 (Original work published 1957).

33 It refers to the naming of the discovery after the scientist who first discovers it, such as Mendelian genetics (named after the Austrian monk, Gregor Mendel) and Okazaki fragment (named after Japanese biochemist, Reiji Okazaki).

context of the scientific community, and the inter- and intra- group behaviors and dynamics among the scientists need to be carefully reconsidered before more accurate and meaningful inferences can be drawn to fully depict modern science.

References

Babbage, C. (1976). *Reflections on the Decline of Science in England and Some of its Causes.* New York, NY: Scholarly (Original work published 1830).

Dasgupta, P. and P. A. David (1994). Toward a new economics of science. *Research Policy*, 23:487–521.

Etzkowitz, H. (1998). The norms of entrepreneurial science: Cognitive effects of the new university-industry linkages. *Research Policy*, 27:823–833.

Hagstrom, W. (1965). *The Scientific Community.* New York, NY: Basic Books.

Hagstrom, W. (1974). Competition in science. *American Sociological Review*, 39:1–18.

Huang, K. G. and F. E. Murray (2009). Does patent strategy shape the long-run supply of public knowledge? Evidence from human genetics. *Academy of Management Journal*, 52:1193–1221.

Huang K. G. (2010). China's innovation landscape. *Science*, 329:632–633.

Hull, D. (1985). Openness and secrecy in science: Their origins and limitations. *Science, Technology and Human Values*, 10:4–13.

Mulkay, M. J. (1969). Some aspects of cultural growth in the natural sciences. *Social Research*, 36:22–53.

Mulkay, M. J. (1976). Norms and ideology in science. *Social Science Information*, 15:627–656.

Smelser, N. J. (ed.) (1988). The sociology of science. In *Handbook of Sociology.* London, UK: Sage Publications, second printing 1989, Chapter 16.

Storer, N. (1966). *The Social System of Science.* New York, NY: Holt, Rinehart and Winston.

Watson, J. and F. Crick (1953). A structure for deoxyribose nucleic acid. *Nature*, 171:737.

West, S. S. (1960). The ideology of academic scientists. *IRE Transactions of Engineering Management*, EM-7:54–62.

Ziman, J. (1985). *An Introduction to Science Studies: The Philosophical and Social Aspects of Science and Technology.* Cambridge, England: Cambridge University Press.

Zuckerman, Harriet (1977). Deviant behavior and social control in science. In *Deviance and Social Change*, E. A. Sagarin (ed.), pp. 87–137. Beverly Hills, CA: Sage.

Part 2

Governing and Managing Knowledge as a Strategic Development Resource: Asian Experiences

Chapter 7

Korea's Public Research System as a National Knowledge Infrastructure

Kong-Rae Lee

1. Introduction

More than ten years have passed since the Korean economic crisis in 1997. Korean innovation scholars recommended strong innovation policies to strengthen the national innovation system of Korea at the time when the Korean economy fluctuated due to the lack of foreign currency in 1997 (Lee *et al.*, 1998). They regarded the weakness in the innovation capability of Korean firms as a fundamental cause of the economic crisis. In the 1990s when the Korean economy was unstable, not only private firms but also public research institutes were weak in innovation capability to develop and manage knowledge intensive products and services.

The Korean government has struggled to remedy the serious problems of her money market, and has eventually solved much of the problems during the last ten years. Private companies have also struggled to improve their financial positions, in some firms by hurting their R&D activities. The government has maintained a steady and constant increase in R&D expenditure over the last ten years in the hope that government R&D may substitute for the decrease of private R&D. The efforts of both the public sector and the private sector to overcome the crisis appeared to have been successful.

On the occasion of overcoming the economic crisis, the Korean government reformed the governance system of government research institutes (GRIs) in 1999. The People's Government (February, 1998–February, 2002) initiated the Administration Reform Committee and decided to separate all the GRIs from their subordinate ministries. It was believed that the degree of autonomous operation of GRIs was low because ministries intervene too much in their research activities, resulting in lowered levels of R&D performance and efficiency. Thus, as a policy, GRIs were separated from their subordinate ministries in order to activate and increase performance and efficiency.

As the result of the reform, GRIs were separated from their subordinated ministries and belonged to five research councils under the Prime Minister's Office in 1999. The Administration Reform Committee supposed that GRIs under the Prime Minister's Office may have more autonomy and less intervention from the government than before the reform. The first research council established in the area of science and technology was the Korea Research Council of Fundamental Sciences & Technology (KRCF). The second was the Korea Research Council for Industrial Science & Technology (KOCI) which has nine member institutes. The third research council was the Korea Research Council of Public Science & Technology (KORP) with ten member institutes. Research councils have governed their member institutes by setting evaluation criteria by which they guide the overall management of member institutes including R&D performance.

This paper examines the consequences of the reform. It briefly reviews the government research institutes, research council system and funding system for public research. In addition, it shows the performance of public research institutes by looking at relevant studies and some statistical indicators. It obviously includes subjective judgement and qualitative investigation.

2. Organizational Changes in GRIs

Government research institutes are classified into four types by their sources of financing in Korea: national labs (NLs), central government research institutes (GRIs), local government labs (LGLs), and local government research institutes (LGRIs). NLs are fully financed by the central government and employ their research staff directly. Currently 53 NLs are being operated mainly for developing agricultural technology under the Ministry of Agriculture and Fishery. GRIs are basically established and financed by the central government. They were established by law, holding an independent status in order to maintain autonomous research activity. They have conducted contract research with both governments and private firms on a competitive basis. 100 GRIs are currently being operated in 2004, of which 52 institutes are associated with humanities and social sciences.

Similarly, local governments have been operating their own labs known as LGLs and sponsored local research institutes or LGRIs. Generally, LGLs have a long history since they were established many years ago to support the agriculture and fishery industry. Some local governments recently established both LGLs and LGRIs to support the manufacturing industry or to cultivate emerging industries. The majority of LGRIs are not set up for R&D but for planning and linking local innovation actors to boost technological innovation in their regions, whose research fields are in humanities and social sciences (see Table 1). Currently, there are 118 LGLs and 38 LGRIs in Korea (2004).

The governance system of each type of public research institutes differs in terms of their subordinate structure. NLs are a part of the central government so that they are fully governed by their subordinate ministries. The status of R&D personnel at NLs is government official and their heads are usually appointed by their subordinate administrations. Likewise, LGLs are governed by local governments, and their research staff are local government officials. NLs and LGLs

Table 1: Distribution of public research institutes in Korea

	In Number, %			
Fields	Natural Science & Technology	Agriculture and Fishery	Humanities and Social Science	Total
Total numbers	79 (25.6)	131 (42.4)	99 (32.0)	309 (100.0)
— National labs (NLs)	7 (13.2)	34 (64.2)	12 (22.6)	53 (100.0)
— Central gov. sponsored (GRIs)	46 (46.0)	2 (2.0)	52 (52.0)	100 (100.0)
— Local gov. labs (LGLs)	21 (17.8)	95 (80.5)	2 (1.7)	118 (100.0)
— Local gov. sponsored (LGRIs)	5 (13.2)	0 (0.0)	33 (86.8)	38 (100.0)

Source: Lee, Kong-Rae and Choi, Ji-Sun (2004). *Strategy to Manage Public R&D Institutes for Building-up Open Regional Innovation systems* (in Korean). Seoul: STEPI.

mostly involve their research activities in such regulative areas as examination of toxic ingredients, quality test and assurances, setting standards, safety checks, health care, disease prevention, environment protection, prevention of natural disasters, etc. LGRIs are financially supported by local governments but they are autonomous organizations established by local laws. The local government that finances LGRIs appoints board members and so do their presidents.

GRIs are governed by their subordinate ministries like LGRIs. However, some GRIs have different governance systems compared with other GRIs as shown in Figure 1. They are governed by research councils which were established in 1999 by Mr. Kim, Dae-Jung's government. The elected president named his government as the People's Government and took many restructuring measures for the purpose of facilitating the recovery of the Korean economy that was seriously damaged by the shortage of foreign currency at the end of 1997. The People's Government reformed government organizations, the economic system including M&A of many bankrupted companies and the public research system in 1998.

The reform of the governance system of GRIs was the first signal to restructure the public sector. The People's Government organized

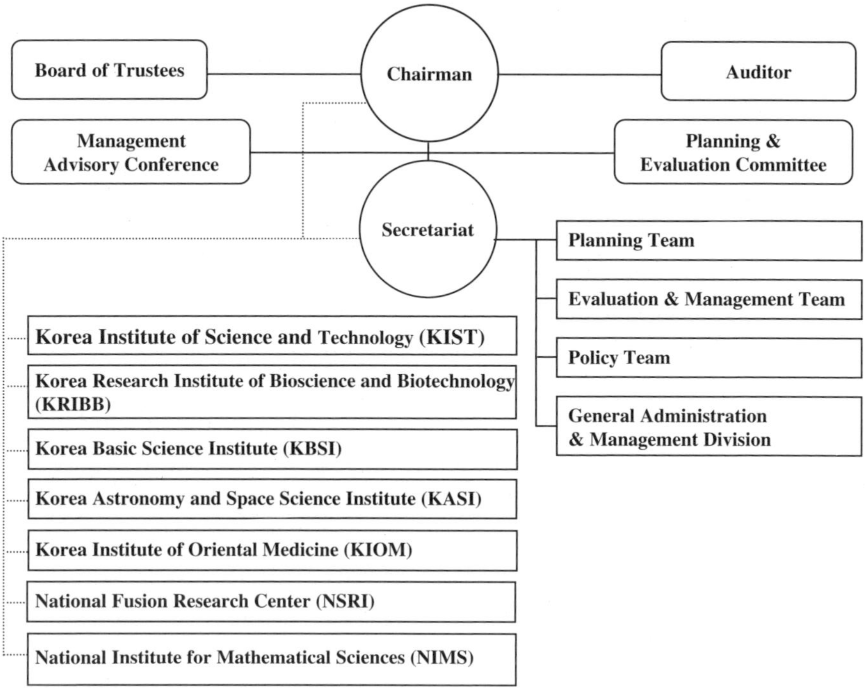

Figure 1: Korea Research Council of Fundamental Sciences & Technology (KRCF)

the Administration Reform Committee to make proposals and relevant laws for the new system in 1998. The chairman of the committee was Professor Linsu Kim who was once the president of the Science and Technology Policy Institute (STEPI). The committee decided to separate all the GRIs from their subordinate ministries. GRIs were cooperative to the restructuring measures of the People's Government.

As the result of the reform proposal, GRIs were separated from their subordinate ministries and belonged to five research councils under the Prime Minister's Office in 1999. The Administration Reform Committee supposed that GRIs under the Prime Minister's Office would be more autonomous and experience less intervention from the government than before the reform. In the field of science and technology, the Korea Research Council of Fundamental

Sciences & Technology (KRCF) had seven member institutes (see Figure 1). The University of Science and Technology (UST) was established under the KRCF in 2004. The Korea Research Council for Industrial Science & Technology (KOCI) had nine member institutes and the Korea Research Council of Public Science & Technology (KORP) ten member institutes. In 2008, Lee Myung-Bak's government reshuffled the Korea Research Council of Public Science & Technology (KORP) transferring their member institutes to other two councils and placed the KRCF under the Ministry of Education and Science and Technology (MEST) and KOCI under the Ministry of Knowledge Economy (MKE).

There were also two research councils in non-science and technology areas, the Social and Economic Research Council and the Humanity Research Council. Both these councils to which all the social science and humanity research institutes belonged, were established in 1999. When the Participatory Government initiated government reform in 2004, three research councils that specialized in science and technology were transferred to the National Science and Technology Committee (NSTC). At the same time, two research councils that specialized in social and humanity science merged into one called the National Research Council for Economics, Humanities and Social Science (NRCS), which has been under the Prime Minister's Office. 23 research institutes that conducted research on economics, social sciences and innovation policies belonged to the NRCS from 2004.

The chairmen of the three research councils are appointed by the Deputy Prime Minister for Science & Technology. The council was made up of the Board of Trustees, Auditor and Secretariat. In general, the Board of Trustees comprised 15 persons who are Vice Ministers of relevant ministries and experts invited from universities, private firms, public research institutes and mass media. It appoints presidents of member institutes whose term has been for three years. Chairman of each council generally operates planning and evaluation committee, management advisory committee, and secretariat having such functions such as policy research, planning, managing projects, evaluating and so on. Few administrative staffs work for each function.

Member institutes report their research and management plans to their research councils on a yearly basis. Research councils appoint committee members to evaluate the previous year's research and management performance of their member institutes. The results of the evaluation by the committee exert significant influence on the budget allocation of the Ministry of Planning and Budget (MPB) to each research institute as it has pursued a policy to discriminate the yearly growth rate of budget based on the outcomes of the evaluation.

Each research institute has its own organizational culture and governance system, regardless of their research councils. For instance, the Korea Institute of Science & Technology (KIST), first established as a type of GRIs in 1966, has an internal governance structure as shown in Figure 2. The KIST has its own Auditor and operates Future Fusion Technology Laboratory, KIST Europe Center, KIST Gangneng Branch, Division of Commercial Technology Development,

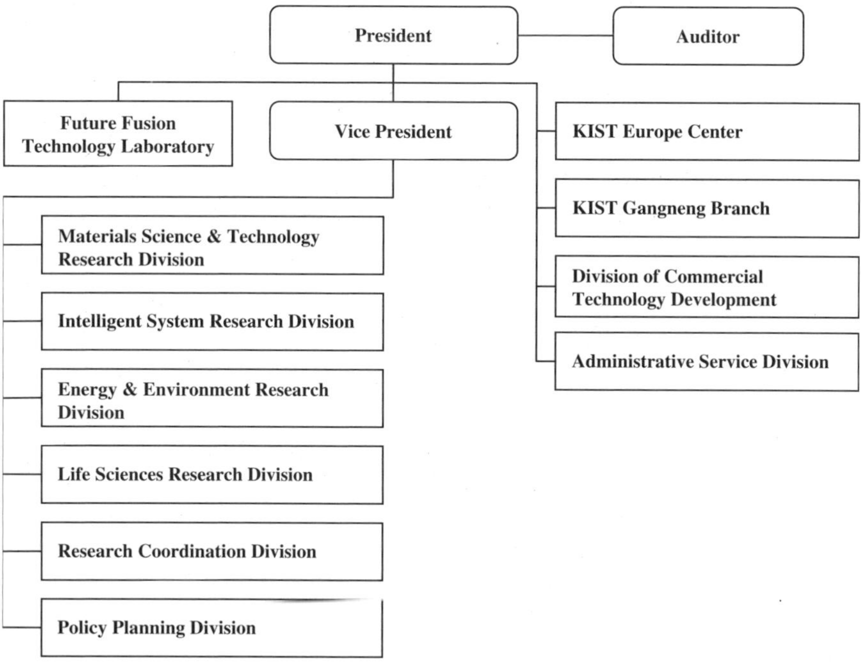

Figure 2: Organization of Korea Institute of Science & Technology (KIST)

and Administrative Service Division. There are six internal divisions for research activities: Materials Science & Technology Research Division, Intelligent System Research Division, Energy & Environment Research Division, Research Coordination Division, and Policy Planning Division. The KIST has been a leading institute in the research area of science and technology in Korea since its inception, and maintaining its excellence at the world class level.

Funding of public research institutes vary depending on the four types of institutes, (NLs, GRIs, LGLs, and LGRIs) above mentioned. NLs are completely financed by the central government, while LGLs are financed by local governments. The central government provides financial grants to GRIs and also supports LGRIs through local governments. The portion of government support out of total R&D expenditure differs by the type of research fields, institutes and their dependence on the government. It is estimated that a half of the budgets of GRIs come from the government grant, another half may come from their contract research with various organizations including the central government and local governments.

It is of worth looking at how GRIs under the three research councils are financed and how much financial grants they receive from the central government. As shown in Table 2, 24 research institutes and one university under the three research councils received about 921 billion wons of grant in total in 2007 from the central government. Amongst the three research councils, the KORP specialized in public technology received the largest amount of government grant, about 285 billion wons accounting for 41.8 percent out of the total grant to the three research councils. The KOCI which specializes in industrial technology received about 284 billion wons of government grants accounting for 30.8 percent, while the KRCF responsible for fundamental technology received about 251 billion wons of government grant that took 27.4 percent share.

The KRCF has been interestingly operating a university named as Korea University of Science and Technology (UST). The government-grant to this university was not large compared with that of other research institutes, but increased rapidly at the growth rate of 43.2 percent in 2007. The UST specialized in graduate programmes

Table 2: Government grant to GRIs under the research councils

Name of Councils and Their Member Institutes	In Million Wons, %		
	2006	2007	Growth Rate
Korea Research Council of Fundamental Science & Technology (KRCF)	8,229	13,761	67.2
Korea Institute of Science and Technology (KIST)	84,134	85,908	2.1
Korea research Institute of Bioscience and Biotechnology (KRIBB)	45,458	50,832	11.8
Korea Basic Science Institute (KBSI)	35,417	39,647	11.9
Korea Astronomy and Space Science Institute (KASI)	16,323	18,357	12.5
Korea Institute of Oriental Medicine (KIOM)	12,875	17,316	34.5
National Fusion Research Center (NFRC)	11,114	20,371	83.3
National Institute for Mathematical Sciences (NIMS)	1,000	2,105	110.5
Korea University of Science & Technology (UST)	2,059	2,949	43.2
Subtotal	216,606	251,246	16.0
Korea Research Council of Industrial Science & Technology (KOCI)	10,509	10,478	Δ0.3
Korea Institute of Industrial Technology (KITECH)	59,363	56,147	Δ5.4
Electronics and Telecommunications Research Institute (ETRI)	20,204	21,246	5.2
Korea Food Research Institute (KFRI)	15,354	16,654	8.5
Korea Institute of Machinery and Materials (KIMM)	39,830	45,780	14.9

(*Continued*)

Table 2: (*Continued*)

Name of Councils and Their Member Institutes	In Million Wons, %		
	2006	2007	Growth Rate
Korea Electrotechnology Research Institute (KERI)	32,657	35,124	7.6
Korea Research Institute of Chemical Technology (KRICT)	35,152	39,463	12.3
National Security Research Institute (NSRI)	31,788	35,182	10.7
Korea Institute of Toxicology (KITOX)	13,341	26,342	97.5
Subtotal	258,198	284,416	10.9
Korea Research Council of Public Science & Technology (KORP)	11,245	11,334	0.8
Korea Institute of Science and Technology (KISTI)	55,038	63,843	16.0
Korea Institute of Construction Technology (KICT)	24,609	23,622	Δ4.0
Korea Railroad Research Institute (KRRI)	16,238	20,053	23.5
Korea Research Institute of Standards and Science (KRISS)	53,748	56,030	4.2
Korea Ocean Research & Development Institute (KORDI)	39,929	47,119	18.0
Korea Institute of Geoscience and Mineral Resources (KIGAM)	35,557	39,056	9.8
Korea Aerospace Research Institute (KARI)	25,769	26,791	4.0
Korea Institute of Energy Research (KIER)	31,092	38,779	24.7
Korea Atomic Energy Research Institute (KAERI)	52,567	58,340	11.0
Subtotal	345,792	384,967	11.6

Source: MOST (2006), *Yearbook of Science and Technology 2006*, Seoul: MOST.

focusing only on training of research students on interdisciplinary R&D fields which other universities can not offer easily within a short time. National and private universities have showed a strong disciplinary academic culture as they prefer students who have passed one straight way of academic carrier. For instance, economics major in undergraduate course followed by economics major in graduate course. The UST found its education niche at the graduate level which is closely associated with the problems of the university society.

Among member institutes of the three research councils, the KIST took the largest grant, 86 billion wons from the central government in 2007. Many people are not surprised that the KIST received the largest support from the central government. It has been called as the mother institute of many member institutes since they started from a branch of the KIST and later split into an independent institute. The KIST has been focusing on the research of broad fundamental science and technology while other institutes tend to specialize in a specific technology area so as to gain competitive advantage.

The yearly budget size of the KIST has been far greater than the size of the government grant. In the 2004 budget and balance sheet, the total income of the KIST was 138 billion wons, of which 49.8 percent came from the government grant, and 50.2 percent from contract researches, royalty income, and others. Taking into the fact that its contract researches with the public sector, mostly the central government, accounted for 83.5 percent out of operating income, the KIST depends 91.8 percent of total income on government in 2004. The high dependency of the KIST on the government may be rationalized by the reason that it conducts research in fundamental science and technology, which is beneficial not only for Korea but also for other developing countries. In this regard, the KIST has a clear duty to share its scientific knowledge with Korean society as well as with developing countries through active training of graduate students, S&T consulting for developing countries and global collaboration in science and technology.

The second largest beneficiary of government grants was the Korea Institute of Science and Technology Information (KISTI)

Table 3: Budget size and balance sheet of KIST (2004)

		In Million Wons, %	
	Contents	Amount	Ratio
Turnover (Income)			
Government grant	Personnel remuneration	20,234	14.6
	Basic research fund	30,083	21.7
	Others	18,565	13.4
Operating income	Public-sector contracts	58,126	42.0
	Private-sector contracts	7,000	5.1
	Royalties	1,700	1.2
	Other incomes	2,758	2.0
	Total income	138,446	100.0
Expenditure			
Personnel remuneration		45,199	32.6
Direct research costs	Costs for basic research projects	30,083	21.7
	Costs for government-contract projects	35,306	25.5
	Costs for private-contract projects	4,641	3.4
Operating costs		7,529	5.4
Facility costs		9,915	7.2
Other costs		5,793	4.2
	Total expenditure	138,446	100.0

Source: Lee, Kong-Rae *et al.* (2007). *R&D Institutes of Excellency in the World* (in Korean). Seoul: Sangkake Namu.

receiving about 64 billion wons in 2007. The KISTI has evolved from the Korea Science and Technology Information Center (KOSTIC) which was established in 1962 as a part of the National Security Office. It merged with the Korea Institute for International Economics in 1982 and changed its name to the Korea Institute of Economics and Technology (KIET). The KIET again split into two institutes (KINITI and KIET) in 1991 after 10 years of experimental operation mixing economics research and science and technology information. The KINITI later merged again with the Korea R&D Information Center (KORDIC) and changed its name to the KISTI in 2001 and become a larger organization dealing with many areas of science and technology, focusing on computing solution, software, policy design and implementation rather than direct R&D of science and technology.

3. Funding System for Public Research

Ministries have allocated R&D budget usually through their funding agents, and for some programs directly allocate R&D budget to beneficiaries. Most ministries have established their own funding agents for financing, managing and evaluating R&D projects that they designed. For instance, the Ministry of Science and Technology established the Korea Science and Engineering Foundation (KOSEF) to finance science and technology related R&D projects, and the Korea Institute of Science and Technology Evaluation and Planning (KISTEP) to implement or coordinate various R&D projects and programs. The MOCIE established Korea Industrial Technology Foundation (KOTEF) and Institute for Industrial Technology Evaluation and Planning (ITEP) for similar purposes.

Likewise, the Institute for Information Technology Advancement (IITA) was born for the MIC, Agricultural R&D Promotion Center (ARPC) for the MAF, Korea Health Industry Development Institute (KHIDI) for the MOHW, Korea Institute of Environmental Science and Technology (KIEST) for the ME, Agency for Defense Development (ADD) for the MND, Korea Institute of Marine Science and Technology Promotion (KIMST) for MOMAF, Korea Institute of Construction & Transportation Technology Evaluation and Planning (KICTEP) for the MOCT as shown in Figure 3. These funding agents are legal institutes identified as public organizations so that they make yearly contracts with subordinate ministries for carrying out their assigned mission.

Among the many ministries, the MPB has been the central funding body for government R&D activities in Korea. The Office of Science, Technology and Innovation (OSTI) under the MOST began

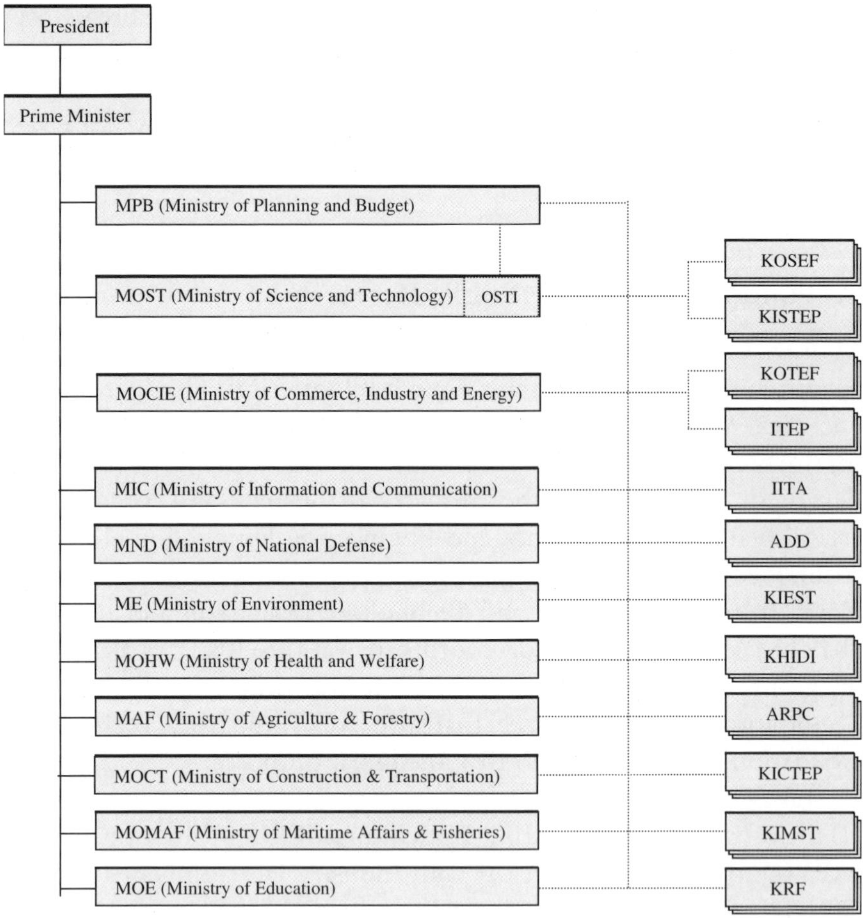

Figure 3: R&D funding agents of major ministries in 2007

to hold a pre-allocation function at the ministrial level from 2004. The funding system for public R&D has been basically a bottom-up process as many funding agents and public science and technology institutes request R&D budgets for their subordinate ministries every year. Ministries then modify their budget requests and pass them to the Ministry of Science and Technology for coordination or to get confirmation through expert review. After each ministry confirms or

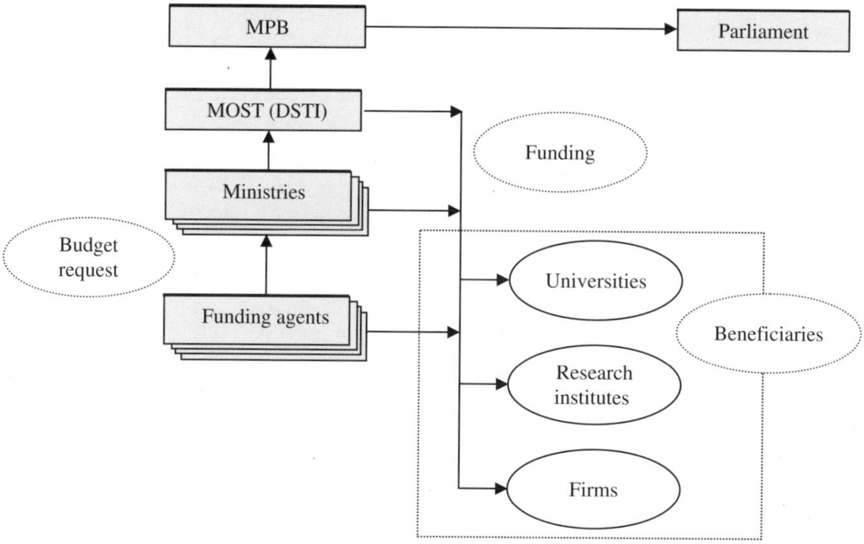

Figure 4: Budget request and funding procedure for government R&D

adjusts, each ministry sends its R&D budget proposal to the MPB, which finally decides the R&D budget size for each ministry through the examination process of Parliament (see Figure 4).

The ministry that had the largest R&D budget size was the MOST operating with about 1,955 billion wons, which accounted for 25.1 percent of total government R&D expenditure in 2005. The KOSEF has been the major agent for operating the MOST R&D budget, and the KISTEP has carried out the role of the funding. To what extent each institute deals with for the MOST is not known. Many R&D programs are also financed by the MOST directly. The MOCIE has been the second largest ministry in funding the government R&D budget operating 1,839 billion wons, taking 23.6 percent share in 2005. The KOTEF and the ITEP have been the funding agents for the MOCIE. Except for the MOST and the MOCIE, the MIC, the MND and the MOE have been operating relatively big R&D budget accounting for 8.9 percent, 11.7 percent and 10.5 percent respectively in 2005 as shown in Table 4.

Table 4: R&D budget sizes of major ministries

Ministries	Funding Agents	In 100 Million Wons, %					
		2003	%	2004	%	2005	%
MOST	KOSEF, KISTEP	12,830	19.7	16,905	23.9	19,549	25.1
MOCIE	KOTEF, ITEP	11,533	17.7	16,403	23.2	18,393	23.6
MIC	IITA	5,991	9.2	6,996	9.9	6,968	8.9
MND	ADD	117	0.2	2,931	4.1	9,112	11.7
ME	KIEST	1,074	1.6	1,301	1.8	1,365	1.8
MOHW	KHIDI	1,318	2.0	1,544	2.2	1,663	2.1
MAF	ARPC	602	0.9	674	1.0	622	0.8
MOCT	KICTEP	768	1.2	916	1.3	1,506	1.9
MOMAF	KIMST	1,081	1.7	1,086	1.5	1,405	1.8
MOE	KRF	3,340	5.1	5,278	7.5	8,209	10.5
Others	—	26,500	40.7	16,793	23.8	9,204	11.8
	Total	65,154	100.0	70,827	100.0	77,996	100.0

Source: NSTC (2006).

Beneficiaries of government R&D funding are universities, public research institutes, science and technology bridging agencies and some small and medium sized firms. The MOST has traditionally focused on funding universities and public research institutes associated with basic and applied research, while the MOE has focused on supporting R&D activities of universities associated with basic research. The MOCIE has emphasized the support of SMEs associated with developmental research at the pre-commercial stage.

However, these ministries recently recognized the importance of fundamental research so that their traditional focus, revealed in the funding of ministries, becomes blurred. The MOCIE has been increasing their support for university R&D, and so do other ministries. As a whole, universities received 32.6 percent of government R&D fund, public research institutes including hospitals and science and technology bridging agencies 52.0 percent, and small and firms 15.4 percent out of the total government R&D fund in 2006 (MOST•KISTEP, 2006).

4. Performance of Public Research System

R&D performance of GRIs has been regularly evaluated by their research councils. According to Lee Chul-Won's (2007) paper, GRIs under research councils have recently improved their R&D performance in terms of paper publications and patent applications. For example, publication of SCI papers per researcher of GRIs increased from 0.407 in 2003 to 0.465 in 2004 and 0.489 in 2005, which was higher than that (0.105) of FhG of Germany, but much lower than that (0.947) of LBNL in the U.S.A., and that (1.085) of AIST in Japan (Figure 5). The results of the comparison appeared to be similar using the index of publication of SCI papers per 100 million wons. This result implies that the government research institutes of Korea have successfully built up their research capability within a relatively short development history.

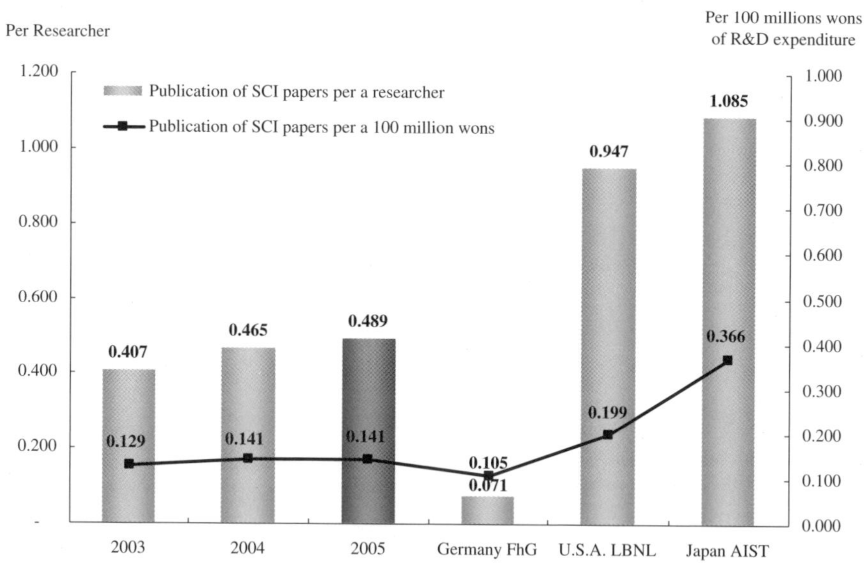

Figure 5: Paper publications of GRIs in SCI journals

Source: Lee, Chul-won (2007).

In terms of patent applications, R&D performance of GRIs turned out to be much higher than that of public research institutes in developed countries. As shown in Figure 5, patent applications per researcher of GRIs have continuously increased from 0.6754 in 2003 to 0.765 in 2005, which is much higher than that of FhG in Germany, LBNL in U.S.A. and AIST in Japan. The results of the comparison using the index of patent applications per 100 million wons show the same trends as revealed in patent applications per researcher. Also, the number of technologies developed for the period of 1998–2006 at Korean GRIs appeared to be 3,158, which is higher than that of US GRIs (1,790) and Canadian GRIs including universities (1,307).

The transfer and commercialization of R&D results from public research institutes has been also emphasized in Korea. Channels for

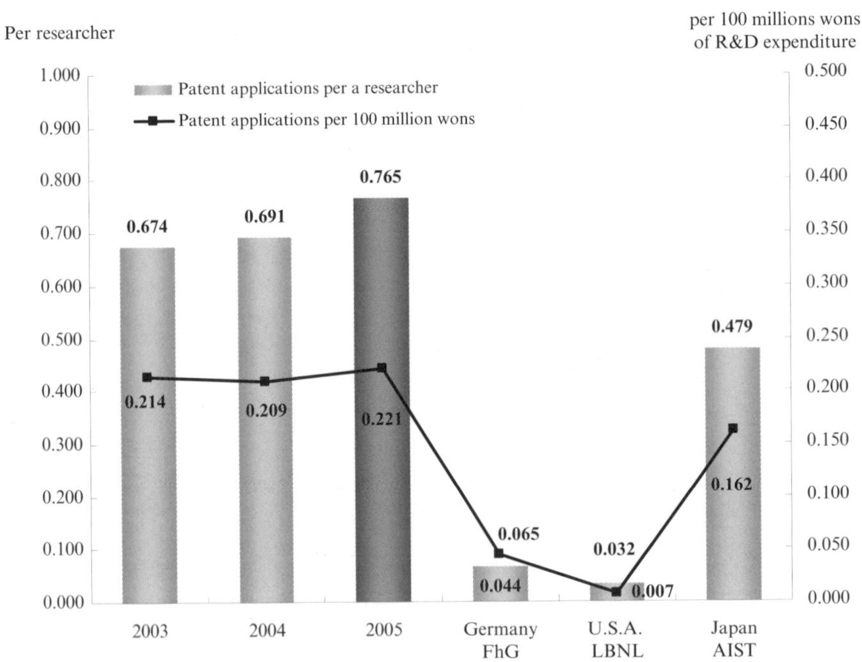

Figure 6: Patent applications of GRIs

Source: Lee, Chul-won (2007). ibid.

transferring public research results include technology transfer agreements, direct creation of venture firms, technical consulting, training of engineers and technicians, etc. GRIs usually receive royalties or fees in return for transferring their technologies. Because of income generation and promotion of technology transfer, most GRIs maintain commercialization units although their staff size is small. As shown in Table 5, their average numbers of employees are 3.6 persons, lower than that of universities (4.8). Korea's average number of employees at commercialization units of GRIs including universities are roughly half of that of the U.S.A. and Canada. This may indicate that the technological capability of Korean GRIs is not enough to employ many experts to deal with technology transfer.

Royalty income per researcher of GRIs has increased annually although slowly, as shown in Figure 7. It was about 8 million wons in 2004, 9 million in 2004 and 10 million in 2005. Royalty ratio out of R&D expenditure also showed similar trends for the period of 2003–2005. Compared with Germany, the U.S.A. and Japan, FhG of Germany revealed the highest royalty income per researcher, much higher than that of Korean GRIs. They have maintained relatively high levels of royalty income per researcher compared with the U.S.A. and Japan. These outcomes in royalty income statistics reflect that GRIs of Korea are indeed active in commercializing their R&D results.

To conclude R&D performance of GRIs, it can be said that GRIs under the research councils have strengthened their research competitiveness and capability after the reform in 1999. The People's Government separated GRIs from their subordinate ministries and put them under three research councils which were under the Prime Minister's Office, and later under the National Science and Technology Committee. This research council system seemed to have greatly influenced the management and organizational culture of GRIs.

First of all, R&D performance of GRIs has improved because competition among researchers as well as research institutes became intense as research councils announced the ranks of each member institutes after evaluation. GRIs are keen on ranking as the Ministry of Planning and Budget (MPB) has been linking the evaluation

Table 5: International comparison of R&D performance of GRIs (1998–2006)

Performance indicators	In ea, %							
	Korea			U.S.A.			Japan (Univ. + GRIs)	Canada (Univ. + GRIs)
	Univ.	GRIs	Total	Univ.	GRIs	Total		
Number of technologies developed (2006)	4,616	3,158	7,774	15,002	1,790	16,792	8,725	1,307
Number of technology transfers	629	951	1,580	4,087	671	4,758	1,171	544
Ratio of technology transfers	13.6	30.1	20.3	27.2	37.5	28.3	13.4[2)]	41.6
Yearly royalty income (million US dollars)	3.2	53.3	56.5	1,088	346	1,435	n/a	43.3
Yearly R&D expenditure (million US dollars)	2,200	2,964	5,164	3,7162	4,082	41,244	47,200	3,127
Number of employees per a commercialization unit	4.8	3.6	4.2	8.65	6.1	8.2	14.3[1)]	8.3

Source: Lee Chul-won (2007). Ibid.

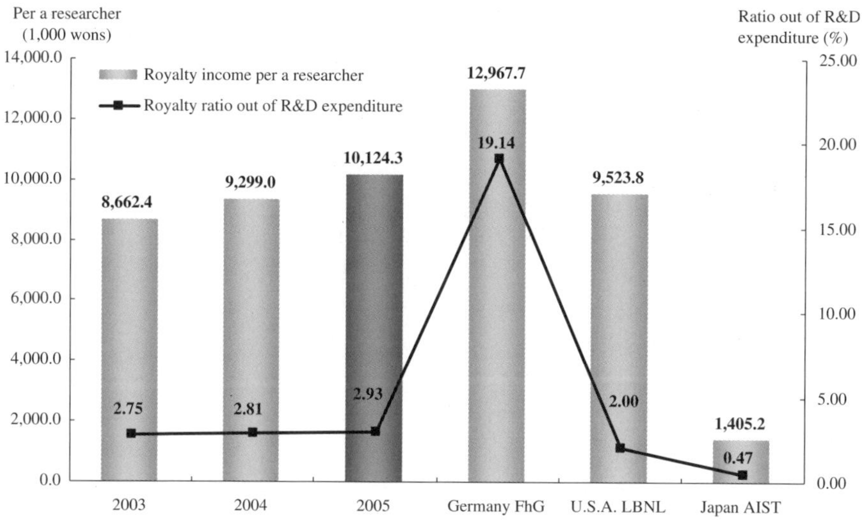

Figure 7: Royalty income of GRIs under the research councils

Source: Lee (2007).

results and budget allocation of each research institute. Research councils have governed their member institutes by setting evaluation criteria on which they can guide the overall management of member institutes including R&D performance.

Secondly, GRIs have secured autonomous research environment as intended. Many researchers at GRIs admit that government intervention on the management of GRIs has been substantially reduced since research councils have covered many relational matters with parliament and the ministries. The bureaucratic organizational culture of GRIs, although there is large variance by institute, certainly tended to be weakened under the research council system. The evaluation committee of each research council has examined the organizational structure and its operation of their member institutes every year so that each research council has been able to guide or consult the organizational change of member institutes based on the result of examination.

However, the research council system has some weaknesses. Since research councils do not have financial capacity to support member institutes, regular evaluations and request to provide management information are often regarded as interference of a higher administration body. Some executives of GRIs mention that unnecessary yearly evaluation on research institutes exerts a heavy burden of administrative work, sometimes taking one or two months to prepare the evaluation. Standardized evaluation criteria set by research councils is also problematic since the function and research area of member institutes differ so that each institute needs a suitable evaluation criteria for their mission and work characteristics.

The second weakness is not directly associated with the research council system itself but with the lack of financial support for GRIs. R&D people working in GRIs do not have a pension system as was adopted in government and education sector, and have an early retirement age of sixty-one, although secondary school teachers retire at sixty three and university professors at sixty-five. The pension system and retirement age are closely associated with long-term stability of the research environment. In addition, the government adapted the project-based research system in the 1990s in financial support for GRIs, which has been often mentioned as a factor weakening their stable research work. As a result of the poor research environment caused by the above mentioned factors, many capable researchers at GRIs have moved to university jobs. Related administration bodies including research councils should have taken complementary measures to make the system workable and stabile, i.e., institutional change as flexible operation of project accounting, adapting pension system for researchers of GRIs, equal retirement age with university professors, and so on.

Lastly, interdisciplinary research seems to be weak in the research council system. Building up an interdisciplinary research system enables the integration or combination of multiple technologies for creating new knowledge. This is perceived as being a new approach for innovation that prevails in contemporary R&D societies, particularly in advanced countries. Institutes having both research and education functions have deeply shaped their organizations to suit interdisciplinary or multidisciplinary research and education. The research council system may have certain rigidity in promoting interdisciplinary research because an institute that tries to engage interdisciplinary cooperation partner located in different research council may feel that the research council acts as an obstacle.

Along with the increasing trends of forming TLOs and royalty income, universities have created venture companies. Creation of venture companies has been encouraged by promotion policies of the government. According to statistics by the Small and Medium Business Administration (SMBA), there were 9,732 venture firms in 2005, of which roughly five percent was assumed to be created through universities.

The government has recognized the importance of creating venture companies. After the economic crisis in 1997, the government enacted the Special Law to Promote Venture Firms in 1998 in order to recover the damaged economy from the crisis by creating knowledge intensive small and medium sized firms. The law was later revised to include some articles for activating spin-offs from GRIs and universities. For example, professors and researchers can get a temporary release up to three years from their home institutes and universities for running venture business (Article 16). If they fail in running their venture business, they can return to their home institutes and universities. The law allows professors and researchers to have a position in venture companies without quitting their jobs at home institutes or universities. It also stipulates that professors and researchers can operate their factories in the laboratories of their home universities and research institute (Article 18).

The Small and Medium Business Administration (SMBA) has been responsible for pursuing policies to promote spin offs from universities and research institutes. It has pursued several policies for this purpose. One prominent policy is the Incubation Center Promotion program. The AMBA supported 15,300 million wons for 278 incubation centers mostly located in university campus in 2007. About 4,000 venture companies were in these incubators in 2007 and obtained some financial support and consulting services. Researchers and professors with more than five years experience running spin offs

Table 6: Spin offs from university and public research institutes

From Universities		From Public Research Institutes	
Region	Number	Region	Number
Seoul	161	Seoul	400
Pusan-Ulsan	35	Pusan-Ulsan	54
Taegu-Kyongbuk	56	Taegu-Kyongbuk	94
Kwangju-Chonnam	39	Kwangju-Chonnam	61
Taejon-Chungnam	52	Taejon-Chungnam	225
Kyonggi	62	Kyonggi	390
Incheon	9	Incheon	56
Kangwon	12	Kangwon	13
Chungbuk	10	Chungbuk	28
Chonbuk	26	Chonbuk	16
Kyongnam	17	Kyongnam	48
Chju	3	Chju	1
Total	482	**Total**	1,386

Source: SMBA (2007. 8).

from universities and research institutes, employing less than 10 workers are qualified to apply for this program.

Another program is the promotion of student venture clubs known as "Changup Dongari". The SMBA has selected 55 student venture clubs every year and supported operation costs of up to seven million wons. This program interestingly allows the participation of not only university students but also high school students when they have good ideas. High school students who want to apply to this program should get a recommendation letter from the president of their schools. This program intends to promote the creation of venture companies by young entrepreneurs.

Apart from the above mentioned programs, the SMBA has supported lectures to cultivate venture mind and entrepreneurship at universities. It was reported that the SMBA supported 25 lectures in 2006 granting two million wons per one credit lecture. The SMBA also supported five universities to operate graduate school programs on the creation of venture companies (Hoseo, Chungang, Hanbat,

Yewon Arts, and Jinju National University) from 2004. These five universities established graduate schools specialized in the creation of venture companies. Each university can get 400 million wons a year from the SMBA for five years.

5. Concluding Remarks

The public research system has been recognized as an important knowledge infrastructure as it plays the role of absorbing, generating, and disseminating knowledge for society. Most countries have established a number of government research institutes and have encouraged research activities of universities in order to pour knowledge into society. Since the cost to maintain the public research system is not cheap, governments face such policy issues as how to improve their efficiency and R&D performance and define desirable ways for the government to govern GRIs. Slowness and invisible nature of R&D results often make policy makers seek radical policy measures to increase R&D performance especially in developing countries which show lack of understanding about R&D activities.

Korea has experienced many reforms of her public research system over the last three decades. The Participatory Government reformed the governance system on GRIs by establishing three research councils in 1999. The Chairman of each council has managed member institutes by appointing their president and evaluating their management performance. The chairmen of research councils appoint evaluation committee members to evaluate the previous year's research and management performance of their member institutes. The results of the evaluation by the committee exert significant influence on the budget allocation of the government for each research institute.

The research council system of Korea has brought about a desirable impact on the research performance of GRIs. Their research performance has been improved because competition among researchers as well as research institutes became intense as research councils announce the ranks of each member institute after evaluation. In addition, it appeared that GRIs have secured autonomous research environment as intended. Many research people at GRIs admit that government intervention on the management of GRIs has been substantially reduced since research councils have covered many relational matters with the parliament and ministries.

Nevertheless, there still remain policy issues to further develop the public research system of Korea because the research council system also has weaknesses. Firstly, unnecessary yearly evaluation of research institutes needs to be reduced and the evaluation system of GRIs should be improved. Secondly, long-term stability of GRIs should be secured by adapting the pension system and increased retirement age for R&D personnel. Lastly, interdisciplinary research in public research institutes should be encouraged in various ways. Open and flexible operation of research councils may increase interdisciplinary and multidisciplinary research activities.

References

Ha Zun-Kyung (2004). An analysis on the effects of R&D on economic growth. *Economic Analysis* (in Korean), 11(2). Seoul: Monetary Economics Research Institute.

Hong, Yoo-Soo (2005). *Evolution of the Korean NIS and Technological Capability Building*, TAP Lecture Series.

Kim, Sung-Soo (2005). *Organizational Characteristics and Development Issues of Office, Science, Technology and Innovation (OSTI)* (in Korean). Seoul: STEPI Research Report 2006–17.

Kim, Sookon, Jaewon Kim, and Chonsun Ihm (1990). *Human Resource Policy and Economic Development: Republic of Korea in Human Resource Development — Selected Country Studies*, Asia Development Bank.

KIPO (2006). *The Patent Trends in Korea 2006*, Seoul.

KOSIME (2007). *An Evaluation of Science and Technological Innovation Policies in Korea*, (in Korean) Seoul.

KRF (2007). *2006 Korea Research Foundation*, Seoul: Korea Research Foundation.

KRF (2006). *2005 Industry-Academia Cooperation White Book of University*, Seoul: Korea Research Foundation.

Lee, Chul-won (2007). Challenges and Issues to Upgrade Government-Sponsored Research Institutes in Science and Technology in Korea. Paper presented at the *Annual Conference of the Korean Society for Innovation Management and Economics* on 20–21 July, Jeju Island.

Lee, Kong-Rae, *et al.* (1998). *National Innovation System of Korea* (in Korean), Seoul: Science and Technology Policy Institute.

Lee, Kong-Rae, *et al.* (2007). *R&D Institutes of Excellency in the World* (in Korean), Seoul: Sangkake Namu.

Lee, Kong-Rae (2007). An essay on government policies to manage public R&D institutes. *Asian Journal of Technology Innovation*, 15(1):21–34.

Meyer-Stamer, J. (1995). Micro-level innovations and competitiveness. *World Development*, 23:143–148.

Min, Chul-Ku *et al.* (2002). *Activation of University Research System* (in Korean), Seoul: STEPI.

MOST (2007). *2006 Science and Technology Yearbook* (in Korean), Seoul: MOST.

MOST (2007). *The Survey on the Current Status of Science and Engineering Manpower* (in Korean), Seoul: MOST.

National Science and Technology Committee (NSTC) (2006). *Report on the Survey of National Research and Development* (in Korean), Seoul: MOST.

National Statistical Office (1998). *Report on Mining and Manufacturing Survey*, Seoul: National Statistical Office.

OECD (2007). *Key Figure 2007 on Science, Technology and Innovation towards a European Knowledge Area*, Paris: OECD

Shin, Tae-Young, *et al.* (2006). *Effects of R&D Investment on Economic Growth and Income Distribution* (in Korean), Seoul: MOST.

Song, Jong-Kook, *et al.* (2002). *Changing Strategies for Business Innovation and the Implication for S&T Policy*, Seoul: STEPI.

STEPI (2005). *Report on the Korean Innovation Survey 2005: Manufacturing Sector*, Seoul: STEPI.

Sung, Tae-Kyung (2006). The determinants of firms' innovative activities: A comparison of manufacturing and service firms in Korea. *Journal of Business Research*, 21(4):283–304.

Sung, Tae-Kyung and Bo Carlsson (2007). Network effects, technological opportunity, and innovation. *Asian Journal of Technology Innovation*, 15(1):91–108.

World Bank (2000). *Republic of Korea: A Transition to a Knowledge Based Economy*, Report No. 20346-KO.

World Bank Institute and KDI (2005). *Korea as a Knowledge Economy: Evolutionary Process and Lessons Learned.*

Chapter

8

Practicing Knowledge Management

Thailand's Experience

Suparb Pasong

1. Introduction

Knowledge is a social creation, a product of social history and science. It is created by the interactions between people and their social and natural environments. A public good, it is also a material resource for development and a source of competitive advantage.

Knowledge management (KM) or knowledge sharing is a process by which knowledge is generated, transferred, acquired and used to create value, enhance efficiency and competitiveness. There are three types of knowledge management in international development:

1. KM for macro-economic planning of national, regional and global development as practiced by governments and intergovernmental institutions, for example, World Bank Group, ADB and OECD.

2. KM for strategic management and organizational development as practiced in the corporate and public sector.
3. KM for social development as practiced by national and international development agencies and civil society, in various fields, including poverty reduction, health, education, environment, etc.

These applications of KM vary in perspective, concept, method, process and results. This article considers the second and third type of KM applications by comparing the experiences of two organizations in Thailand which have applied knowledge management. These are the *Knowledge Management Institute (KMI) and the Office of Knowledge Management and Development (OKMD).*

The article is based on research conducted by the author in 2006–7. Research methods used included participant observation, interview and discussion with informants and KM practitioners. Documents generated by both centers were also consulted.

2. KM by Means of Practice: The Knowledge Management Institute (KMI)

The Knowledge Management Institute (KMI) was founded in 2003 by Vicharn Panich, a professor and former director of the Thailand Research Fund (TRF). Vicharn pondered about how best Thailand may benefit from the knowledge generated by TRF's research programmes. As a manager of a research fund, he felt obliged to find ways to put the research results to practical use. He came up with the idea of "downstream management", through which research results may be transferred to the general public, especially the rural community, civil society, government agency and small business. Knowledge Management is an approach that offers concepts, methods and tools that might be effective in transferring the knowledge to the broad range of knowledge users. According to Vicharn, he was not quite sure then, how KM concepts and tools could be put into practice. Encouraged by academic and policy colleagues, he decided to experiment with a homegrown KM approach, based on the conviction that to learn about KM one has to practice it. Vicharn maintains that one doesn't know about KM unless one practices it.[1]

Thai Health Promotion Fund ("*ThaiHealth*") is a public fund financed by excise tax on the country's alcohol and tobacco consumption. The fund's mission is to support civil society's social development programmes that lead to health promotion of Thai citizens. The fund was looking for ways and models to disseminate information and knowledge about practices that promote people's well-being. *ThaiHealth* provided Vicharn with funds to experiment with his idea for 5 years.

The idea is to promote the practice of knowledge management among governmental, business, educational and civil society organizations, and communities. To implement it, the *Knowledge Management Institute (KMI)*[2] was established under the aegis of the *Thailand Research Fund (TRF)*. Led by Vicharn, KMI employed 11 people including programme officers and administrative staff. The organization was advised by a policy committee of 10, mostly prominent academics and public policy officials. As part of the Thai Research Fund (TRF) which is a state agency operating outside and independently of the government structure, KMI also operates similarly.

2.1 *Goal and Strategy*

KMI's goal is to use KM as a change agent for Thai society in developing a learning and knowledge society. Like "ThaiHealth", KMI draws intellectual inspiration from a senior scholar and social activist, Professor Prawase Wasi, MD., who commands great respect from the medical, and public health profession as well as the Thai NGOs. According to Prawase, a pathway to sustainable development must be built on the synergy of 3 drivers — knowledge and wisdom, social

---

[1] Vicharn's statement to this effect is found in his many writings and lectures, for example, Vicharn Panich (2006). *KM by Means of Practice*. Bangkok: Sukaparpjai Publisher, p. 5.

[2] Knowledge Management Institute (KMI), Bangkok, http://www.kmi.or.th/ (in Thai).

movement, and public policy. Change agents must devise a strategy that synergizes these 3 drivers.[3] KMI's strategy is to mobilize knowledge and local wisdom to develop and strengthen a social movement, which uses the knowledge to influence public policy.

KMI envisages a network of KM practitioners and partners as an emerging social movement, linked by knowledge sharing and the practice of knowledge management. Its operational strategy emphasized tacit knowledge, placing the importance on people and process as the main thrust of its brand of KM.

Practice-based KM, as interpreted by KMI, is about the management of process and people, especially the knowledge accumulated in the people, i.e., tacit knowledge, so that it is effectively used to enhance work and organization. The emphasis is on the process and the people, not the knowledge as such. Allowing for simplification, KM is a cycle of learning from practice and practicing what is learned. Built on a practice of sharing and learning, knowledge grows and contributes to the development of a learning organization and society.

2.2 *Tacit Knowledge and Local Experts*

Because KMI places strong emphasis on tacit knowledge, it looks at the sources of such knowledge in the people who have accumulated knowledge over many years of practice in specific areas of livelihoods and social and cultural life. Some specific knowledge has been practiced and passed over from previous generations, for example practices in agriculture, traditional medicine and healthcare, folk arts and craft. The established knowledge and skills in such fields of practice are recognized as local wisdom. The people who master such skills and knowledge are usually much respected by the community as local experts on a particular subject. These are among the major sources of tacit knowledge which KMI is interested in with an aim of facilitating the process of its sharing.

---

[3] Prawase Wasi. Knowledge Management: An Emancipation Process for Human Potential, Freedom and Happiness, Prawase Wasi Official Website, publication not dated, p. 29 (in Thai) http://www.prawase.com (retrieved 2-3-2010).

In search for tacit knowledge, KMI scanned the civil society and rural development landscape. It found many existing local knowledge sources, i.e., experts from rural communities, for example a civic group specialized in rice farming in Kanchanaburi Province (*Kao Kwan Foundation*) and a farmer (Mr. Suthinan) who is an expert in land improvement and water management in Buriram Province. What KMI did was to build on tacit knowledge of these local masters, applying various KM tools to these practices to facilitate the learning and sharing across communities.

Kao Kwan Foundation, (KKF) a small NGO, operates a paddy farming school for farmers as a development project in Suphanburi Province. The project brought paddy farmers together to learn and share knowledge about farming methods, organic farming and farm management. KMI provided the community of farmer learners with simple techniques of knowledge management including storytelling, note taking, knowledge sharing and networking facilitation as well as after action review.

In the case of paddy farmers' school, there was a change in learners' attitudes from one influenced by commercial farm inputs, especially chemicals, to self-reliance in planting methods, fertilizers and pest control and self-sufficiency in investment, production and consumption. According to KKF, the paddy farming school by farmers via the utilization of knowledge management tools and networks helped paddy farmers rediscover sustainable agricultural practices, thereby regaining confidence in sustainable agriculture. Besides improving farming practices, members of the farmers' school have improved their livelihoods regarding income and health. The social learning and knowledge sharing of the farmers' school have received wide recognition among farmers and government agencies. A number of KM practitioners and facilitators, grown out of this practice, have assisted other organizations and communities in learning and adopting KM practices. The process has expanded KM networks among rural communities in Thailand.

The success of the paddy farmers' school was in part due to its own innovative self-organization, and assisted by the adoption of knowledge management tools and techniques in their social learning.

2.3 *Interpreting KM by Means of Practice*

After 5 years of operation, KMI has learned a lot from practicing it. Its 5th annual report (2007) recounts what has been learnt. Although it is not possible to examine all aspects of its achievements and failures, it is useful to review the general processes and the lessons learnt.

2.4 *KM Vision*

The application of KM requires a vision of what the individual practitioner, organization, or community wants to achieve. In other words, they must be clear about the purpose of the application of KM. In interpreting KM and putting it into practice, there is an underlying assumption that everyone always has something to share and something to learn. Sharing and learning, and practicing it are the basic tenets of KMI.

2.5 *KM Staff*

The application of KM requires a small core group of people, who are clear about the purpose of the organization in applying KM to its work and organizational development. At least 3 categories of people are functionally important for the process. These include a chief knowledge officer (CKO), who communicates the purpose of the organization and shows the direction to all members. For the sake of simplicity and to enhance the power of communication, KMI's version of KM has given a Thai common name to all key KM staff. The meaning of these names indicates the role and function of the particular staff in KM process. The CKO has been renamed as *Khun Ua*. He/she plans and leads the KM practice within the organizational context. A KM facilitator (KF), renamed *Khun Amnuay*, is responsible for the facilitation of the process of KM practice. The third staff is the KM practitioner (KP), renamed *Khun Kij*, who is a knowledge activist within the organization, applying KM tools and appropriate models in the context of the organization's vision. A knowledge network coordinator is given a Thai common name as *Khun Prasarn*.

2.6 *Process and Tools*

Central to the whole process of "knowledge management by means of practice" is the learning and sharing of best practices. A relationship and a forum must be established with the people or partners who are willing to share success stories. The process and activity of sharing and learning, in whatever form they take, are conducted with the attitude "appreciative inquiry". The appreciative attitude aims at building mutual trust, which, in turn, leads to positive, often innovative learning atmosphere and cooperation, further enhancing the dynamism of the sharing of experiences and enriching the results. Emphasis is given on the sharing and learning about tacit knowledge accumulated in successful individuals and organizations. In practice, the form of sharing and learning relationship among KM practitioners varies, depending on organizational context and purpose. KM staff will determine which form or pattern of sharing and learning best suits the purpose of KM application in the particular organization and its partners.

Tools used in the sharing and learning process vary. For the interpretive KM, there is no standardized KM tool. Tools are designed and practiced by KM staff and practitioners. Most tools are procedural or are sets of small, interrelated processes which are facilitated by experienced facilitators to produce desired learning results. Typical tools include success story telling, deep listening, dialogue, workshop, seminar, training, coaching, field visit, after action review, networking and knowledge market.

Note taking, for example, is a simple tool that helps document the knowledge generated or transferred in the process of sharing of best practices. Notes from learning activity may be taken and documented in any innovative form, from simple note pads to video filming and blogging on the internet.

Few hardware tools are involved in the process. Familiar hardware is the computer and the internet, which help open access to explicit knowledge, document the tacit knowledge and support networking. With broad access to the internet throughout the country, the power of ICT has been explored and used as a means of communication and

KM networking. A weblog named *GotoKnow*,[4] specifically designed to serve this purpose, has been developed and become a popular learning and networking tool among KM practitioners.

2.7 *KM Models*

The experiences gained from successfully practicing KM in different contexts, using a set of different tools have led to the development of various KM models. Within 5 years of KMI's operation, about 20 major viable models have been developed,[5] all practiced-based, largely original and appropriate for local conditions and specific KM visions of the practicing organizations, including community, civil society, government and private sector organizations.

Initially, KMI aimed at establishing KM practice among the various sectors of Thai society. Its aim was to develop an original KM model that suits the Thai context. At the end of the project, it was realized that no single model best suits all contexts. Instead, many powerful, original models have emerged from the practice.

An example of such original model is the *Pla-tu* Model,[6] also known as Tuna Model, developed by KMI to communicate KM practice to KM practitioners, and the audience new to KM. A simple image of a fish, *Pla-tu*, perceived in 3 main parts, symbolizes the structure of KM practice. The "fish head" represents the KM vision of a practicing organization. Before initiating KM, leaders or KM staff of the organization must be clear about the application purpose of KM. The CKO — *Khun Ua*, the KF — *Khun Amnuay*, have to communicate it to the KPs — *Khun Kij* and all members of the organization to generate understanding and ownership of the KM practice among the members. In effect, the "fish head" sets goal and

---

[4] http://www.gotoknow.org

[5] KMI Annual Report (2007). *Tracking the Success of KM in Thailand, Knowledge Management Institute*. Bangkok, KMI, pp. 62–99 (in Thai).

[6] *Pla-Tu* is a Thai common name of short-bodied mackerel (*Rastrelliger brachysoma)* a fish served as seafood, popular in Thailand and SE Asian countries. The description of *Pla-tu* model here is taken from Vicharn Panich (2006). *KM by Means of Practice*. Bangkok: Sukaparjai Publisher, pp. 129–136 (in Thai).

direction of the KM exercise, providing the leadership with a clear sense of purpose and ownership.

The "fish body" represents the process of knowledge sharing and learning, carried out by KP — *Khun Kij*, facilitated by KF (*Khun Amnuay*). Many of the learning tools and activities happen at this stage, including networking. The "fish tail" represents a storage of knowledge assets (KA), gained from the actual sharing and learning. *Khun Kij*, a knowledge practitioner (KP), extracts the knowledge from the sharing and learning exercises. *Khun Likhit*, the note taker, documents the relevant information and the contents of the knowledge which emerged during the sharing sessions.

Beside the Tuna Model, many other KM models emerged from actual practice, for example, Carp School Model, River of Wisdom Model, Xerox Model, etc. These models are used widely among the broad networks of KM practicing organizations, including community and civil society organizations, governmental agencies and a few private sector organizations. Corporate KM practitioners in KMI's network include Thai-German Specialty Glass Co. Ltd., Siam Cement Group and NOK Precision Co., Ltd.

The rich collection of KM models indicates that KM applicability is not restricted to the knowledge classes or educated sections of the society. Instead an interpretive KM by means of practice is open to all, regardless of age, gender, educational, social and economic status, agriculture, manufacturing, rural and urban communities, civil society, government and corporate organization.

KMI wanted to change an agriculture-based developing society into a knowledge and learning society. It realizes that change in the learning paradigm and developing learning culture in a society cannot happen overnight, or even in a few years, because human society is complex. There are many variables and conditions involved that need to be changed. In the public sector, the emphasis is placed on the thinking and behavioral change, directed at the executives and the agency's KM staff. At the end of the day, all major changes must come from within. People concerned themselves must change.[7]

---

[7] KMI Annual Report (2007). *Tracking the Success of KM in Thailand, Knowledge Management Institute*. Bangkok, KMI, p. 19 (in Thai).

2.8 *Transformation*

After 5 years of implementation, KMI as a project has accomplished its goals and transformed itself into an independent organization under its own statue, the *KMI Foundation*. It seeks to operate as a self-supporting, non-profit organization, guided by its vision, which remains unchanged. What is changing is its role and positioning. In the past 5 years it acted as a change agent, while developing more change agents who have gone through the process of knowledge sharing. It continues to help facilitate this process in their own or partner's organization. Its new role will be consulting, coaching and mentoring KM staff and practitioners and organizations. Through these roles, it seeks to sustain itself with self generated revenues. The new challenges for them are threefold[8]:

- Inspire others and help make the learning process pleasant and enjoyable for learners.
- Strengthen the capacity of the change agents to encourage innovation and development dynamism.
- Link up the networks to forge the relationships for cross-contexts learning.

By merits and achievements, KMI may be considered as a best practice in knowledge management for development. The knowledge and lessons it has generated are of great value to the science and arts of knowledge management.

3. "Value Creation" — The Case of OKMD

Unlike KMI which originated from the idea of managing research results for the use of various sectors of the economy and society, the Office of Knowledge Management and Development (OKMD) emerged out of the pressure of economic competition in the international markets, under the Thaksin Shinawatra administration (2001–2006).

[8] KMI Annual Report (2007). *Tracking the Success of KM in Thailand, Knowledge Management Institute*. Bangkok, KMI, p. 34 (in Thai).

Pansak Vinyaratn, a prominent senior journalist and chief advisor to the former Prime Minister Thaksin Shinawatra, recalled his many discussions with the prime minister while on board the plane on more than 20 official travels to foreign countries. Much of the time on board the plane was spent discussing how best for Thailand to cope with the increasingly stiff competition. The favorite subject revolved around the country's poor quality of human resource and the low level of science and technology. Poor education was singled out as one of the root causes. There was a concern that Thai education did not produce people that were able to "think" and being creative. Education in most countries produced graduates who can analyze, but not many who can consider and scrutinize things.[9]

Of all the complex issues in the discussion, the two men agreed on one thing: that Thailand is rich in history, arts, culture and traditional knowledge. These assets can be the source of strength. What was missing was good ideas and management to create economic value from them. Pansak propagated the idea of "value creation", that is, to realize value from the cultural and intellectual assets by applying design and creativity on them. Adding creativity and design that is uniquely Thai to the products and services, according to Pansak, creates the value that justifies higher prices because it contains artistic, historical and cultural contents.

In Thaksin's view, Thailand cannot compete in markets in which it doesn't have comparative advantages. It cannot afford to rely on buying technology to enhance production capacity and produce manufactured goods that sell at a low unit price, or export cheap agricultural commodities. Instead it should look for advantages in non-R&D intellectual assets, especially the cultural assets, to create value. With distinctive creativity, Thailand can shift to a knowledge-based economy, not abandon it. With its rich cultural endowments,

[9] Pansak Vinyaratn (2005). *Thai Education Needs a Revolution not a Reform.* A Public Lecture. Bangkok, 15 March 2005 http://www.okmd.or.th/en/knowledge_detail.asp?id=70 (retrieved 07-12-2005).

Thailand can evolve into a *value creation economy,* in which design and creativity play a key role.[10]

As a result of this analysis, Thailand under the Thaksin administration (2001–2006) pursued a parallel, dual track development. One track was based on the old paradigm of knowledge-based economy, while the other focused on value creation, based on cultural capital, community capacity and creativity. The showcase of the latter known as *OTOP Programme*,[11] where every rural community creates value by developing a product or service from their unique local resource and traditional knowledge, added with design and creativity, was very successful in generating rural employment and contributed to the rapid poverty reduction. However, to be successful in the value creation economy, local communities have to be provided with entrepreneurial know-how and access to capital by the government.

The shift towards the value creation economy is as much an issue of knowledge management, especially of tacit knowledge, as an issue of economic development strategy. In this context, value creation is broadly defined as the integrated process of identifying comparative advantage, interpreting cultural uniqueness, building on existing cultural strengths, adapting to new consumption styles and innovation.

For Thaksin, value creation goes beyond simply adding value. Addressing the 15th Asian Corporate Conference in Bangkok in June 2006 he said, developing countries in Southeast Asia and elsewhere may proudly point to high-tech assembly plants where value is added to high-tech products. But at the end of the day, much of the value of such products comes not from the assembly, but from intellectual

---

[10] *A Focus on "Value Creation" in the Government's Economic Strategy.* News item. Office of Knowledge Management and Development (OKMD), Bangkok, Government of Thailand, http://www.okmd.or.th/en/knowledge_detail.asp?id=218 (retrieved 30-08-2006).

[11] The *OTOP — One Tambon* (subdistrict) *One Product Programme* draws its inspiration from the Japanese successful *One Village One Product movement* (OVOP). Its background may be found at its official site http://www.thai-otop-city.com/background.asp (retrieved 31-03-2010).

property — the patents, the copyrights, the designs — that go into each product.[12]

Thaksin also pointed out that, for value creation economy to grow, a country must first identify their comparative advantages. They must then add intellectual value and skills to create pricing power for their products and services in the world market. For him, value creation means that one plus one, equals more than two. How much more depends on how highly the market values the ideas that go into the product. As for Thailand, which is rich in artistic and aesthetic traditions, value creation may mean creating distinctive and unique products, developing new brands, and differentiating products based on the unique Thai style. Or it may be based on something else entirely. What is important is that people must be given the tools and freedom to explore, to experiment, and to create. That is what makes the free market so compelling.[13]

Thaksin was convinced that value creation represented a new economic model for Thailand and, quite possibly, for Southeast Asia. As its people become more educated and its economy moves further up the development ladder, value creation will contribute substantively to GDP growth. He believes that the creation and protection of intellectual property will become increasingly important as the value creation model takes hold.[14]

To encourage value creation, for Thaksin, education is the most important variable in the equation. For Pansak, however, design and creativity are central in the equation. He argued that educational reform missed the point. The important intellectual assets of

---

[12] *A Focus on "Value Creation" in the Government's Economic Strategy.* News item. Office of Knowledge Management and Development (OKMD), Bangkok. http://www.okmd.or.th/en/knowledge_detail.asp?id=218 (retrieved 30-08-2006).

[13] *A Focus on "Value Creation" in the Government's Economic Strategy.* News item. Office of Knowledge Management and Development (OKMD), Bangkok. http://www.okmd.or.th/en/knowledge_detail.asp?id=218 (retrieved 30-08-2006).

[14] *A Focus on "Value Creation" in the Government's Economic Strategy.* News item. Office of Knowledge Management and Development (OKMD), Bangkok. http://www.okmd.or.th/en/knowledge_detail.asp?id=218 (retrieved 30-08-2006).

the Thai society are largely generated and kept in reserve outside the formal educational system. These are the abundance of skills as well as local ingenuity and wisdom that will enable the economy to input the appropriate knowledge and technology together to achieve desirable results. In addition, Thailand has a social structure that is very accommodating and receptive to diversity and creativity. The nurturing of creativity within the society should in turn lead to its own intellectual property rights that are both "high-touch" and "high-tech", creating greater income for the nation as a whole.[15]

In short, creativity takes a front line position in the value chain of goods and services that suit the modern consumption patterns and lifestyles.

As a result of the strategic discussion between the former prime minister and his chief advisor, The Office of Knowledge Management and Development (OKMD) was set up by a government decree in 2004 as a public agency to implement the strategy. OKMD takes the value creation approach as its framework. Creativity takes centre stage. One of the key excellent centers under OKMD is the Thailand Creative and Design Center (TCDC).

Although the notion of value creation is still central in the economic strategy, by 2008 the term *value creation economy* used in the literature and policy documents has been complemented by a new one: *creative economy*. A minor shift of emphasis may be observed in that the concept of value creation economy emphasizes the importance of the uniqueness of cultural and historical content in knowledge assets. Design and creativity play a complementary role in conjunction with the creation of value. On the other hand, in the creative economy model, design and creativity take the center stage in the context of creative industries. Essentially both concepts form a single strategy that seeks to capitalize on the intellectual assets and human resource to drive GDP growth and development.

[15] Pansak Vinyaratn (2003). "Asia finds its own way: The Thai roadmap", *Asia Times Online*, 30 May, http://www.atimes.com/atimes/Southeast_Asia/EE30Ae02.html (retrieved 31-03-2010).

However, the strategy does not imply that the country should not compete on these fronts. What it proposes is that the policymakers and macroeconomic planners seize the opportunities provided by its comparative advantages in various sectors, for example, fashion and product design, spa and hospitality service.

3.1 *Organizational Structure of OKMD*

OKMD aims at developing a knowledge infrastructure comprising learning centers, creative and design centers, a biotechnology research center, and a folk arts and culture museum to facilitate and support value creation economy. OKMD was conceived as a strategic instrument for managing the knowledge to drive value creation economy. The strategy is to strengthen the country's knowledge base and cultural assets, create high value for these assets through creativity, design and other skills. Its goal is economic growth through creative industries, including multi-media, entertainment, fashion, jewelry, etc.

OKMD is designed to manage knowledge about society and economy, while the concept of value creation economy allows the country to break free from the cycle of fiasco. The government is investing in knowledge infrastructure and human resource, so that Thailand can adjust to competition from countries like China and India. It is structured in 6 specialized centers and institutes, each if which is independent from the others, but coordinated by the OKMD secretariat. These excellence centers are designed to facilitate access to learning, knowledge and creativity with different, but complementary approaches. They may be considered as operational arms of OKMD in the task of knowledge management for the creation of value from intellectual and cultural assets, and bring it to the markets. The following are profile description of its functional centers.[16]

---

[16] *About Us*. Office of Knowledge Management and Development (OKMD), Bangkok, http://www.okmd.or.th/en/about.asp (retrieved 23-03-2010).

(i) *Institute for the Gifted and Innovative Learning (IGIL)*

The institute is a new approach to learning, including brain-based learning, which is designed to suit the different stages of development of brain growth. The institute focuses on children and youth aged 0–19, which are the most important ages forming an individual's character. The Center specializes in the task of searching for and supporting individuals who are gifted and talented in different fields. IGIL provides appropriate learning environments for gifted and talented individuals, hitherto neglected by the formal educational system. It is believed that the gifted and talented are a very important human resource for a value creation economy.

(ii) *Thailand Knowledge Park (TK Park)*

The knowledge park is a learning resource center where digital technology has been extensively used for learning. The center is designed as a dynamic modern library, which provides creativity-triggering environments, serving largely the younger generation. TK Park maintains services in Yala Province, one of the southern Muslim-dominant provinces.

(iii) *National Discovery Museum Institute (NDMI)*

The Institute focuses on stimulating imagination beyond reading and listening. It is designed to gather, analyze and display discovery to trigger curiosity of learning. It places the importance of first hand experiences from the interactions between the exhibitions and its visitors to create self-learning skills.

(iv) *Thailand Creative and Design Center (TCDC)*

The Thailand Creative and Design Center can be said to be a world-class excellence center of its kind, of which there are not many in the world, especially in developing countries. Creativity and design are the high value creation resources for which the Center aims to provide an enabling environment so that they can grow. The Center's main activities include exhibition, seminar and workshop and two specialized libraries. One specializes in printed and electronic resources, and the other is a collection of materials in collaboration with "Material ConecXion", an international consulting firm on materials, which are the basic ingredients for developing new, and high value products. Among the six centers, TCDC is very popular and outstanding for knowledge generation and service to a large section of young creative and design students, professionals and entrepreneurs.

(v) *Thailand Center of Excellence for Life Sciences (TCELS)*

The Center is created to manage the knowledge in life sciences. This is a value creation that builds on the country's richness in biodiversity. Biotechnology is the main thrust of the center. Research and development are the main activities of the Center.

(vi) *Center for the promotion of national strength on moral ethics and values (The moral center)*

To balance the heavy weight of other centers, OKMD has created a centre dealing exclusively with the moral dimension of development. The center is a self-appointed manager of the country's moral and ethical values. The central idea behind the center's mission is that peaceful and sustainable social development has to be based on moral and ethical strength of the nation. The idea believes that moral and ethical values are the source of compassion, which is unrealized value in the creative goods and services. The Center is, therefore, tasked with adding compassion to the value creation economy.

3.2 *Externality of a Knowledge Management Initiative*

Two years after its establishment, the first annual report was published. A few days earlier its co-founders and leaders, Thaksin Shinawatra and Pansak Vinyaratn, were removed from power by the coup on September 19, 2006. Repeated military coups have proven to be a major impediment for development.

OKMD suffered a major setback as a consequence of the coup. The organization was accused of irrelevance, extravagance and a wasteful use of state funds. However, no corruption charge was made. The military government did not appreciate its role and contribution to the economy. Budgets were severely cut.[17]

Fortunately, its own merits have come to its rescue. Three years later, its idea of "creative economy" which is an extension and a renaming of the value creation economy has been well received among macro-economic policy makers. NESDB which has been tasked with planning a knowledge-based economy for several years, suddenly shifted towards creative economy. NESDB proposed in 2009 that the 11th Five-Year National Social and Economic Development Plan for the period of 2011–2015 should focus on building a creative economy.[18] The relevance of OKMD and its specialized center, especially TCDC, have been restored. Their contributions have again been recognized. Nonetheless a new agency, the Office of Creative Economy has been set up in 2009 to coordinate the government's creative economy policy. The government invests 1 billion Baht in the project. Korea and the UK serve as role models for developing a creative economy.[19]

4. Analysis

To describe the achievements of an organization in a few short lines cannot do justice to the organizations under study. However, it serves to

---

[17] Interview with Mr. Palakorn Buppatanakorn, OKMD Programme Official, at OKMD office in Bangkok, 18 December 2006.

[18] National Economic and Social Development Board (NESDB) (2009). *Policy and Background Papers for 11th National Social and Economic Development Plan*, Bangkok: NESDB, pp. 44–91.

[19] "Design Korea 2009" and "Creative Economy and the Cultural Strategy of Ministry of Culture, Sports and Tourism, Government of Korea". Office of Thailand Creative Economy, Bangkok, http://www.thailandce.com/ (retrieved 31-03-2010).

Table 1: Comparative characteristics

KM Practice	Knowledge Management Institute (KMI)	Office of Knowledge Management and Development (OKMD)
Status	Initiated by Prof. Vicharn Panich in 2003 as a 5 year project under the *Thailand Research Fund*, funded by *Thailand Health Promotion Fund*, with about 20 million THB a year. Non-governmental	Established by law in 2004 as a statutory body, a brainchild of Pansak Vinyaratn, chief advisor to former prime minister Thaksin Shinawatra, with more than a billion THB starting budget. Governmental agency
Vision	Thailand is a knowledge and learning society.	Economic growth through value creation economy.
Objectives	To act as a change agent. To create more change agents. To develop extensive KM networks and KM models appropriate for Thai society.	To spark off new ideas and inspire creativity
Organizational Style	Small (11 staff), independent, non-bureaucratic, low budget (99 mil THB for 5 years).	Large and complex. It consists of 6 specialized centers independent from each other, under government bureaucracy, large total budget (1 bill. THB a year).
Orientation and Emphasis	Practice oriented Tacit knowledge, people and process	Wide range of tacit and explicit knowledge and multiple emphases, including cultural assets and creativity, museum, life science, brain-based and gifted learning.
Targets and Networks	Rural communities, NGOs, schools and universities, government agencies, hospital, health and agricultural offices	Government, business and young people.
Achievements	Having established KM practices and KM networks among development organizations and communities.	Having established design and creativity infrastructure (TCDC) as the drivers of value creation and creative economy.

recapitulate some of the complex stories of the people and the processes involved. The descriptions are deduced from studying both organizations' official documents, especially their respective annual reports.

4.1 *Learning from KMI and OKMD*

In terms of development goals, the KMI and OKMD are similar. Both realize that the world development is propelled by knowledge. Knowledge has become the most important asset to ensure the competitive edge of a country in the age of globalization. Most countries realize this. Each country has its own strategies, and intellectual capital resources.

KMI is convinced that KM is the necessary tool for managing and applying knowledge in the pursuit of knowledge and learning society. It looks at development at the level of practice. It therefore operates at this level on both micro and macro scales. It started small, building on existing, successful development projects and communities. It then slowly extended its activities through the building of networks, while developing KM models, which are appropriate within the Thai development context. Through a broad-based knowledge network, it has been able to advocate KM for development among these sectors: education, health, government agencies, industry, communities and civil society organizations.

4.2 *Missing Link*

For KMI, a period of 5 years is very short for a social development project that aims at changing the learning behavior of a society and a social relationship. A medium term of 10–15 years is needed to firmly establish a learning culture in the complex, modern society. It also requires more than strategic change at the micro, organizational level. The linkage between the changes at the ground level and the macro, strategic management by the government administration is apparently lacking. The sign of this missing link is obvious early on in KMI's vision aimed at building a knowledge management practice in Thai society as a social movement. The vision and mission deserve respect, considering the way KMI carried out its mission. It has credibly succeeded in developing original KM concepts, methods and tools, and establishing them as best practices among a range of communities of practitioners.

4.3 *Institutional Framework*

To develop a creative economy, a supportive institutional framework must be put in place, especially regulatory system and policy capacity. In this regard, the culture of good governance is important. Since late 1990s, Thailand has embarked on several important reforms, especially political and educational ones. The latter is aimed at creating a knowledge workforce. By 2010, the country continues with the educational reform for the second round, by a different government who pledges to promote a creative economy. The political reform set in motion by the Constitution of 1997 resulted in broad democratic participation and a strong government.

The public service reform has started since 2003 under the Thaksin administration, aiming at making the government bureaucracy more transparent, accountable, responsive and competent. A royal decree on public sector good governance was also put in place in 2003.[20] The Office of the Public Sector Development Commission (OPDC), an agency trusted with the task of reforming the public sector was established in 2004.

To enhance the competency and the quality of public service and the transparency of the public sector, OPDC requires that all government agencies implement organizational development programme, which includes knowledge management as a tool.[21] Hardly anybody understood then what knowledge management is about, and how to

---

[20] *Royal Decree on Criteria and Procedures for Good Governance,* 2003, Office of the Public Sector Development Commission (OPDC), http://www.opdc.go.th/english/main/content_view.php?cat_id=3 (retrieved 31-03-2010).

[21] *Public Sector Reform Initiative*, Office of the Public Sector Development Commission (OPDC), Bangkok, http://www.opdc.go.th/english/main/content_view.php?cat_id=2 (retrieved 31-03-2010).

implement it in their organization. KMI has bridged the gap by working in collaboration with OPDC in facilitating a knowledge sharing for the public sector. This was an opportunity for KMI to influence policy development.

After 2 years of cooperation, KMI reflected that it saw some positive signs of change in the learning culture and work paradigm of the public sector. However, because the bureaucratic culture is essentially authoritarian in nature, KMI advised that executives and staff at all levels have to help themselves and one another in sharing best practices in learning and delivering services, which is the central process in knowledge management.[22] Obviously OPDC did not "buy in" big enough the brand of KM offered by KMI to transform the public sector into a learning and knowledge organization.

4.4 *Political Instability*

Political stability is a necessity for development. Since the previous coup in 2006, 4 years have been lost. Nobody knows how many more years will be lost due to political instability. The KM organization of the OKMD type is prone to political reaction and instability, because it is completely dependent on the state's fund and the bureaucratic, political institutions. KMI on the other hand, has not been affected very badly by the political reaction, because the organization is least dependent on the state institution. Public investment in KMI's project is not costly compared with the funding for OKMD. Moreover, KMI works much at the grass-root level, while OKMD operates on high profile mode, thus likely to be misperceived and led to bias judgments by the political leaders.

4.5 *Interpreting KM in Different Contexts*

Describing KMI's style of KM as "interpretive" indicates that it is an ingenious attempt to put ideas into practice. The main idea of

---

[22] *KMI Annual Report 2007: Tracking the Success of KM in Thailand*, Knowledge Management Institute, Bangkok, 2007, pp. 102–111 (in Thai).

knowledge sharing is received as a set of management practice. Ideas about sharing knowledge remain ideas, unless they are put into practice. KMI would not know, where to find and how to extract the tacit knowledge about certain crop and cultivation technique. The idea was analyzed and interpreted to suit specific contexts. The point is, however, that KMI does not just interpret KM. It also practices it to produce an original and innovative best practice in knowledge management.

5. Conclusion

As best practices in knowledge management, the experiences of the Knowledge Management Institute (KMI) and the Office of Knowledge Management and Development (OKMD) provide valuable lessons. More rigorous research into the stories of both organizations could illuminate the path to rapidly changing, competitive, modern development.

5.1 *KMI and Replicability*

Knowledge management can be creatively interpreted and designed to suit any development context. KMI's experiences have shown that original models of KM may be developed out of practices that are intrinsic to the everyday life and work of individuals, organizations and communities. The identification of tacit knowledge as a prime source of knowledge for development suits the Thai context. Sharing and learning from a best practice is central to the processes of KM that enhance a learning organization and a knowledge society. The success of a KM application in an organization or a community depends on various conditions, including the leadership and ownership of the KM practice among the leaders and members of the organization. Networking expands and enhances KM as a tool for development. KM by KMI finds a niche by bridging the gaps between the needs for the practical use of knowledge for achieving organizational, and community's goals and the lack of practical skills how to use it.

Replicability of KMI's approach to knowledge management is an issue of interest for development practitioners. Considering major characteristics of KMI's track records, salient features indicate that it may be possible to replicate the practice in the Thai context. However, given that the KMI approach is a context specific interpretation of KM practice, it seems difficult to replicate or upscale it elsewhere. Nevertheless many practical experiences, especially those related to the various tools, may be shared across regions and cultures. The best way to determine the extent of replicability of KMI's approach in any given context is to share KMI's practice with an organizational practitioner.

5.2 *A Recipe for Success*

A system of knowledge management that is linked with a national, macro — development strategy offers plenty of room for creativity and innovation. Such KM approach is conceived on a global scale, within the context of global competition, to be implemented at a national scale of social and economic development. Naturally, political leaders and national policy makers would have to think on this scale, in other words, to think big. However, when it comes to the implementation of such strategy, it is entrusted to a government agency. Because none of the existing agencies fits the task, a new one was created: OKMD.

What is the contribution of OKMD? The most relevant one is the strategic and visionary thinking that underlies its existence: Thailand cannot afford to compete in the world, using the old paradigm which originated from the old industrial revolution. Thailand has to shift ahead of the existing mainstream knowledge-based economy, thus, a value creation economy based on the unique social and cultural assets, which is nothing but tacit knowledge. In other words, OKMD is supposed to dig into the tacit knowledge treasure of Thailand and put it to use to create high value through creativity and design and technology. The idea sounds so simple. The total value of creative economies of France and Italy has been cited as examples of why such a simple idea makes sense.

Digging into the treasure of tacit knowledge is what KMI has practiced and advocated for. However, KMI did not go as far as attempting to transform the rediscovered tacit knowledge into an economic value beyond the subsistence and self-sufficiency economy. The idea of value creation economy was obviously not known or not of interest to KMI and its networks of practitioners. It seems that exchange and sharing between KMI and OKMD and its multiple centers can be further improved.

On the other hand, while being entrusted with the task of facilitating value creation from social and cultural assets, OKMD has not established how it extracts the great wealth of such assets. TCDC is more successful in promoting the marriage between the tacit knowledge with creativity and design, materials and technology to create value. TCDC accesses tacit knowledge, partly through research, and partly through the specific search for individual artists, artisans, local masters and through the open invitation for national creativity and design exhibitions. In creating value, TCDC goes one step further by working with entrepreneurs and SMEs to channel the created value into the domestic and export markets. This appears to be a convincing business model.

A similar process takes place at the community level within the OTOP Programme, in which a community enterprise produces selected products or services that contain unique contents derived from local tacit knowledge. The programme strives successfully based on the idea of value creation. Although it does not refer to knowledge management, the essential elements of KM such as knowledge sharing, and rediscovering tacit knowledge, are inherent in their business process.

Obviously, there is a gap between the two organizations. KMI is hardly motivated to share its brand of KM with OKMD's centers. Vice versa, OKMD's centers hardly show an interest in learning from KMI's experiences. Each operates in one's own turfs, demarcated by different visions for knowledge. Each goes one's own ways, engages their own crowd. It is difficult, however, to comment on this gap without having adequately examined the underlying conditions and the real reasons behind it. Further research is necessary to shed light on this issue.

OKMD's approach to knowledge management is replicable on a national scale. What is not easily replicable, however, is visionary, enlightened political leadership and competent policymakers who can successfully translate a clear vision into policies and practical programmes. This seems to have little to do with knowledge management. However, considering KMI's *Pla-tu Model* (Tuna Model), competent leaders are the necessary "fish head", which is very important for successful KM in any organization, or country.

In light of the KM leadership deficiency, which is temporary, KM practitioners, organizations and policy circles may look at examples and get inspiration from policies and practices of foreign countries. That the present Thai government intends to look up to Korea and the UK as best practices of the strategic management of creative economy, is a commendable intention.

Another source of best practice, in other words, examples of successful strategic management of knowledge is the corporate sector. Pansak's lectures and writings point to the success stories of many global business firms, and individual artists, designers, and entrepreneurs. Drawing lessons and inspiration from these individuals and firms, he advised combining "high-touch" with "high-tech" to create high value that meets the requirements of the changing patterns of modern consumption and lifestyle.[23] Innovative knowledge management can well serve the goal of "high-touch" and "high-tech" value creation. To translate this advice into plain academic language, he advised a knowledge management that blends the lure and imaginations of the liberal arts with the power of technology. In short, the recipe for success in a value creation economy and a creative economy is liberal arts plus technology.

The basis of a "high-touch", i.e., creativity, is to be found in the academic disciplines like sociology, anthropology, history, literature, psychology, behavioral science, and economics, all summed up as "liberal arts". The other half of the story is technology, which is built

[23] Pansak Vinyaratn (2003). "Asia finds its own way: The Thai roadmap," *Asia Times Online*, 30 May, http://www.atimes.com/atimes/Southeast_Asia/EE30Ae02.html (retrieved 31-03-2010).

on the basic scientific disciplines, biology, physics, chemistry, mathematics and the like, to result in ICT, biotechnology and the like. Among the most successful companies in the US, and perhaps in the world, are Google and Apple. Steve Jobs, Apple's CEO, recently told the world that the recipe for Apple's success is no secret: "liberal arts + technology",[24] which is another way of saying "high-touch" plus "high-tech".

In short, to drive a knowledge economy, value creation and a creative economy, and a high power, competitive economy, the strategy is simple: Switch on the power of liberal arts and technology!

---

[24] Steve Jobs (2010). *Keynote Speech*, Apple Special Event, 29 January http://www.apple.com/quicktime/qtv/specialevent0110/ (retrieved 30-01-2010).

Chapter 9

Emerging Epistemic Landscapes

Knowledge Clusters in Ho Chi Minh City and the Mekong Delta[1]

Hans-Dieter Evers and Tatjana Bauer

1. Knowledge Clusters as Centres of Innovation and Development

Knowledge has been identified as one of the major factors of production, driving economies and societies towards a post-industrial stage of development (Menkhoff, Evers and Chay, 2010). Countries

[1] This study was carried out within the WISDOM Project by the Centre for Development Research (ZEF), University of Bonn, the Southern Institute of Sustainable Development, Ho Chi Minh City and the Mekong Delta Development Research Institute (MDI) of Can Tho University, with support from the German Aerospace Centre (DLR), the Vietnamese Ministry of Science and Technology (MoST) and the German Federal Ministry of Education and Research (BMBF). Useful comments by Solvay Gerke, Eckart Ehlers and Gabi Waibel are gratefully acknowledged.

around the globe, including several ASEAN nations, have adopted policies to encourage the growth of a knowledge-based economy. Building an ICT (information and communication technology) infrastructure has usually been one of the leading policy measures, in addition to developing universities and research institutes. Vietnam embarked on these policies later than Singapore and Malaysia, but appears to be on the way to building a knowledge-based economy.

Building a knowledge infrastructure means initially creating knowledge-producing and disseminating organisations such as research institutes, universities and colleges. To be effective, these have to be located closely to make use of common types of infrastructure such as laboratories, libraries and computing facilities. The geographical clustering theory assumes that proximity increases an organisation's innovative capacity when employees — especially researchers — can share ideas, products and services (Evers, 2008).

The focus will be on the southern Vietnamese city of Ho Chi Minh City and the adjacent Mekong Delta, which are both destined to play a major role in Vietnam's effort to build a knowledge economy (see Map 1). With Vietnam's first future megacity, Ho Chi Minh City, and "Asia's corn house", the Mekong Delta (Chong, 2002:25), a region is chosen that currently plays a decisive role in the development of Vietnam. Ho Chi Minh City is not only considered to be the financial centre of Vietnam, but also as an important cultural and industrial centre for the entire Southeast Asian region (Truong, 2007:24). The Mekong Delta, directly situated to the south of Ho Chi Minh City, is one of the world's most productive areas in terms of agriculture and aquaculture and ensures food security for the whole country. Given this background, we will show throughout this paper the importance of knowledge production for the region's socio-economic development.[2]

---

[2] Even though the largest concentration of knowledge-producing organisations is located in Vietnam's capital Hanoi, this paper will focus mainly on the South of Vietnam, as our field research is an innovative investigation in this area. Certainly, knowledge production as such has led to the overall development of Vietnam contributing to different regions.

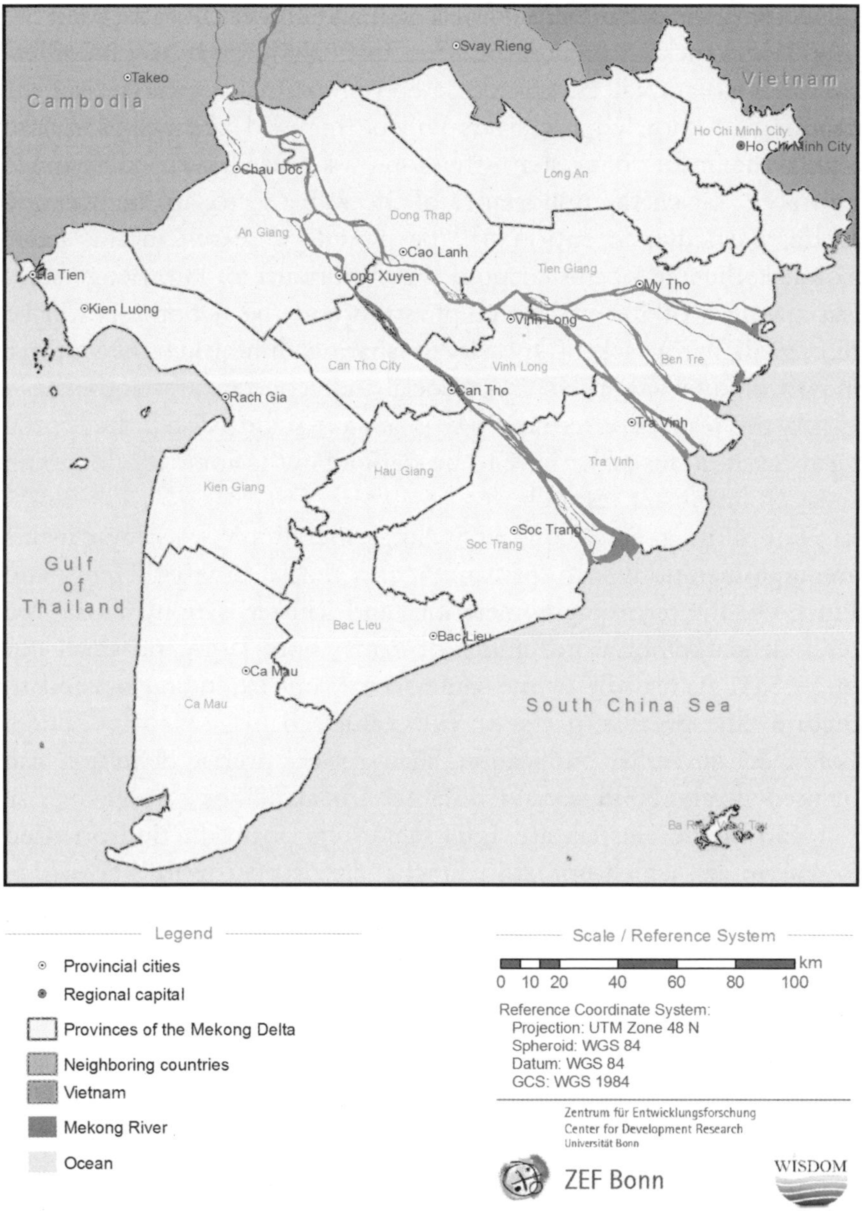

Map 1: Ho Chi Minh City and the Mekong Delta

At first, the organisation of research and higher education and the foundation of knowledge-producing institutions since the reunification of Vietnam will be traced. Based on own field research data the building of knowledge clusters will be mapped and analysed and, finally, the impact of cluster formation on knowledge production will be traced. Given the importance of the water sector in the Mekong Delta, particular attention will be paid to research in this field. Towards the end of this chapter, the importance of knowledge sharing and networking will be emphasized and the hypothesis will be discussed that a lack of knowledge sharing diminishes the positive impact of knowledge clusters on social and economic development.

As the respective terminology is not yet standardised, the central terms used in this paper have to be clarified in the context of ongoing research on cluster research.

The most general concept is 'agglomeration', whereby clusters are agglomerations with 'proximity' as a crucial variable. Henry and Pinch use the terms 'agglomeration' and 'cluster' synonymously "to refer to geographical groupings of firms (both large and small but often SMEs), broadly in the same sector, but extending beyond to incorporate greater parts of the value chain" (Henry, 2006). Following an earlier publication, shall a more precise definition will be used (Evers, 2008, see also Chapter 1 of this book).

Knowledge clusters are agglomerations of production-oriented organisations, which primarily direct their efforts toward knowledge as an output or input. Knowledge clusters have the organisational capability to drive innovations and create new industries, and are central places within an epistemic landscape, i.e., in a wider structure of knowledge production and dissemination. Common examples of organisations found in knowledge clusters are universities and colleges, research institutions, think tanks, government research agencies, and knowledge-intensive firms.

The knowledge clusters in a particular region or urban area — in our case, Ho Chi Minh City, the former Saigon of Vietnam — form what is referred to as an 'epistemic landscape', i.e., the geographical distribution of knowledge-producing organisations, their research staff and other knowledge workers and their output. The concept "epistemic landscape" is a subcategory of the more general term 'knowledge landscape'. In this usage we allude to Karin Knorr's concept of "epistemic culture, the culture of knowledge production" (Knorr-Cetina, 1999) and refer to the geographical space of knowledge production.

Epistemic landscapes develop over long periods of time. They are seldom shaped by individual actors, but more often by the collective action of strategic groups (Evers, 2009). Firms connected by a common interest to capitalise on the competitive advantage of clustering have an impact on epistemic landscapes through their location decisions. Moreover, government strategies to develop knowledge-based societies and economies have often been decisive in shaping epistemic landscapes, the relevant development policies of which have been assessed in detail elsewhere for Malaysia and Indonesia (Evers, 2003), Singapore and Germany (Hornidge, 2007). Developing industrial regions, clusters or knowledge hubs is, indeed, standard practice in many regional planning departments around the world. The allocation of human and financial resources creates knowledge-producing and disseminating organisations that can be measured, mapped and made to depict the contours of an epistemic landscape.

The assumption underlying these policies is that the clustering of knowledge-producing organisations increases knowledge output. In other words, isolated knowledge-producing institutes in knowledge-intensive industries are detrimental to innovations and economic growth. Clustering knowledge organisations is the most effective policy on the way towards a knowledge-based economy and society.

2. The Epistemic Landscape of Ho Chi Minh City and the Mekong Delta

2.1 *Knowledge-Producing Organisations in Vietnam*

This section will provide a descriptive analysis of what is referred to as an "epistemic landscape". The current literature does not provide any comprehensive elaboration on the composition and allocation of knowledge-producing organisations, particularly in Ho Chi Minh

City and the Mekong Delta. Thus, an original data set had to be compiled to acquire an overview of the status quo of Ho Chi Minh City's research environment.[3]

All data sets are based on an intensive listing of knowledge-producing organisations in Ho Chi Minh City, and were completed during a one-year field research study in Vietnam from April 2008 to March 2009.[4] This listing was finalised by means of various sources such as Ho Chi Minh City's *Yellow Pages* (Yellow Pages, 2007), directories of scientific organisations published by the city's Department of Science, Technology and Environment (DOSTE, 1998) and MoST (2004, 2008). Through an extensive internet search and telephone campaign, every knowledge-producing organisation was verified and crosschecked.

In Vietnam, scientific knowledge is produced at various levels (Figure 1). Firstly, there are large government research institutes such

Photo 1: Ho Chi Minh City Skyline

[3] A detailed analysis of the Vietnamese science and research community will be presented in the forthcoming dissertation by Tatjana Bauer.

[4] The field research was carried out by Tatjana Bauer, with occasional input by Hans-Dieter Evers. So far, there is no single source that has an overview of every organisation. Most likely, the Ministry of Science and Technology (MoST) has a list, but this not available to researchers or other users.

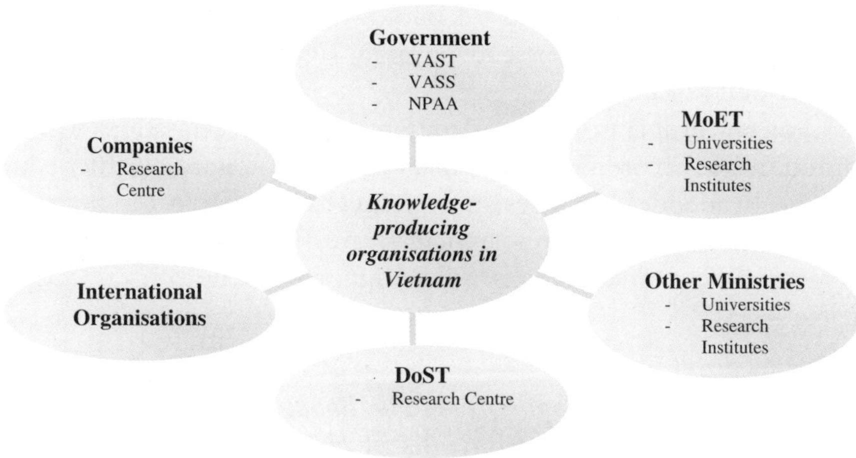

Figure 1: Overview of knowledge-producing organisations in Vietnam

as the Vietnamese Academy of Science and Technology (VAST), the Vietnamese Academy of Social Sciences (VASS) and the National Political and Administrative Academy (NPAA)[5] located in Hanoi, the capital city of Vietnam. Secondly, research is also conducted by research institutes and centres of universities under the administration of the Ministry of Education and Training (MoET). Thirdly, research institutes and ministries other than MoET are responsible for research activities related to the function of the head ministry; some of these universities are directly under the administration of the respective line ministries. Fourthly, there are a number of research centres administrated by the provincial authorities, namely the

[5] Similar to the former Soviet Union model, the term 'academy' referred exclusively to state research organisations, which led to the adoption of the system in Vietnam. Using the term 'academy' as the official translation of these organisations shows the strong connection between Vietnam and the former Soviet Union. However, in this case, the correct translation for 'viện' is 'institute'. Nevertheless, other types of academies later appeared in the form of educational organisations, e.g., Vietnam Aviation Academy, Academy of Posts and Telecommunications in Ho Chi Minh City. The term 'academy' used in the latter case is translated to 'học viện', which literally means a 'learning institute'.

Department of Science and Technology (DoST). Finally, we have international organisations and privately run companies involved in science and research.

For practical reasons, knowledge-producing organisations will be distinguished according to their main function,[6] classed as either education (academies, colleges, universities) or research (companies, centres, sub-institutes, research institutes). The affiliation to the respective head organisations will be ignored in the context of this paper.

2.2 *Historical Development — The Boom of Educational and Research Organisations in Ho Chi Minh City and the Mekong Delta after 1975*

The epistemic landscape of Ho Chi Minh City is in its infancy (Pham, 2006:238) — highlighted by the fact that 78 percent of all currently operating knowledge-producing organisations were founded after 1975, at a time when the north and the south of Vietnam were united (see Figures 2 and 3). In contrast, only 8 percent existed before 1975. For the remaining14 percent, no data was available.[7] These figures indicate that knowledge-producing organisations in Ho Chi Minh City and the Mekong Delta have developed merely within the past thirty-five years.

A second peak was reached after 1986 with the introduction of the 'Renovation policy', otherwise known as 'Đổi Mói'. During Vietnam's subsequent transition to a market economy, local or provincial authorities, ministries and universities were allowed to create R&D centres, without compulsory registration with the government, as had been the case before Đổi Mói (Annerstedt and

[6] This formal distinction according to the key task of these organisations does not have to exclude the other. As our field research has shown, in practice researchers of research organisations normally have teaching assignments at universities and university staff can also be involved in research projects.

[7] It can be assumed that a small number of organisations have been dissolved, merged and renamed, indicating that these organisations are not all newly established.

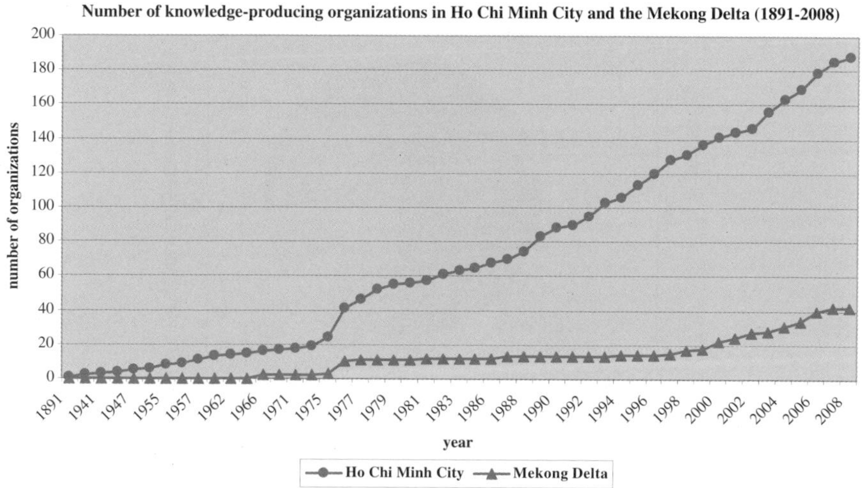

Figure 2: Number of knowledge-producing organisations in Ho Chi Minh City and the Mekong Delta (1891–2008)

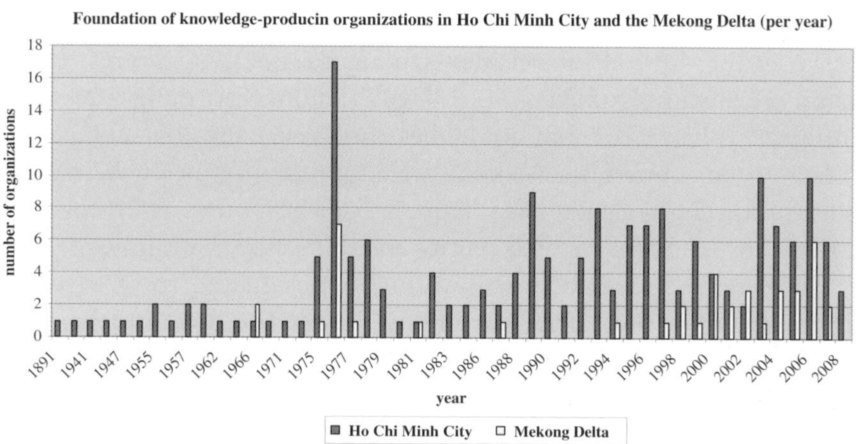

Figure 3: Foundation of knowledge-producing organisations in Ho Chi Minh City and the Mekong Delta (per year)

Nguyen, 1996:246). Although, as a consequence, the establishment of research and technology service centres has accelerated immensely, no reliable data relating to existing Vietnamese knowledge-producing organisations in Vietnam has been available until today.

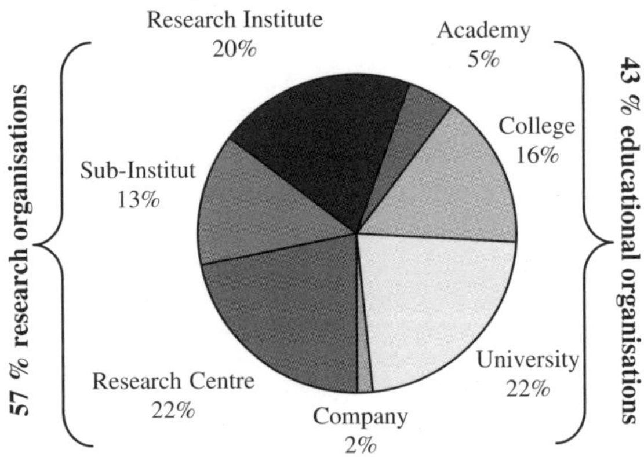

Figure 4: Distribution of knowledge-producing organisations in Ho Chi Minh City

Turning to our own compiled data set, 218 knowledge-producing organisations were identified, comprising 93 educational and 125 research organisations in Ho Chi Minh City (Figure 4), broken down further into 49 universities (truòng đại học), 48 centres (trung tâm), 44 institutes (viện), 34 colleges (truòng cao đẳng), 29 sub-institutes (phân viện), ten academies (học viện) and four companies (công ty). In contrast, the Mekong Delta accounts for merely 42 knowledge-producing organisations (Figure 5), of which there are 20 colleges, 11 universities, seven research centres and four research institutes.

The term 'research organisations' is designated to all *research institutes* under ministries or directly under the government, *sub-institutes* belonging to parent organisations in Hanoi or to a functional ministry, *research centres* under the administration of research institutes or universities, and *companies* involved in privately run research projects.

In contrast, 'educational organisations' comprise musical, military or political *academies*, *colleges* where students can get a degree after three years' study, and *universities* where students have to study for four to six years to obtain a degree. Every one of these educational organisations is under the control of the Ministry of Education and Training (MoET) or a functional ministry such as the Ministry of Agriculture and Rural Development (MARD).

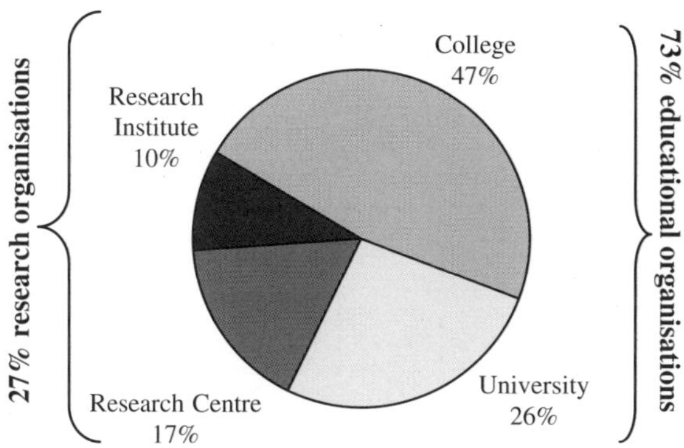

Figure 5: Distribution of knowledge-producing organisations in the Mekong Delta

According to the data collected in the field, 7.736 staff members work for research organisations and 26.970 staff members for educational organisations (Figure 4).[8] The number of staff working for knowledge-producing organisations in Ho Chi Minh City and the Mekong Delta is rather high compared to other Southeast Asian countries such as Thailand, which, for instance, has only one-third of the capacity but has developed much faster than Vietnam. Due to the constraints of this paper the question of staff qualification will not be discussed, even though it is obvious that scientific research in Vietnam results in limited scientific outcomes (Dang, 2006; Gerke, 2006:17; Dang, 2006:211).

In reality, universities have the largest number of employees in comparison with all other knowledge-producing organisations, which is of little surprise because of the size of these organisations and the additional teaching capacity undertaken by university staff, besides their research activities. Grouping organisations according to their size illustrates a tendency toward smaller-sized research organisations,

[8] The data set includes only 80 percent of all identified knowledge-producing organisations. For the remaining 20 percent of organisations, staff numbers could not be ascertained and therefore will be ignored in the following analysis. Our data shows that approximately one-sixth of the staff accounts for support staff, e.g. drivers, security guards and other service personnel.

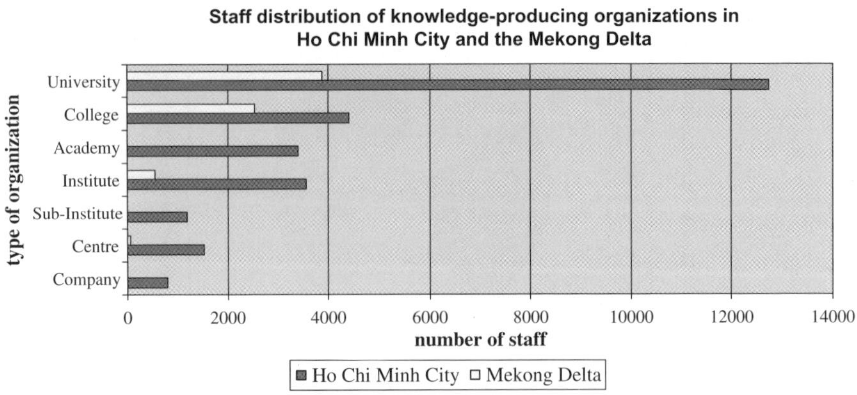

Figure 6: Staff distribution of knowledge-producing organisations in Ho Chi Minh City and the Mekong Delta

with relatively large educational organisations being the exception to the rule.

Figure 6 reveals a large gap between the staff numbers of organisations in Ho Chi Minh City and the Mekong Delta, which would be expected due to the much lower amount of knowledge-producing organisations in the Mekong Delta. Nevertheless, this difference turns out to be even bigger when considering the populations of each area — Ho Chi Minh City with 6.4 million inhabitants and the Mekong Delta with 17.5 million (GSO 2007) — revealing an immense concentration of highly qualified staff in Ho Chi Minh City.

2.3 *Knowledge Cluster Building in Southern Vietnam*

These initial considerations will assist in determining a more precise overview of knowledge clusters in Ho Chi Minh City and the Mekong Delta. Thanks to the GIS-based mapping method, described in greater detail by Evers, Genschick and Schraven (2009) two major knowledge clusters can be identified — Ho Chi Minh City and Can

Tho City — considering the proximity of these organisations as the determining factor (Map 1).

Cluster building in Vietnam has its roots in the early 1980s, when the Soviet model of 'science and production complexes' was introduced. Two of these complexes operating in the south of Vietnam were the Dyestuff complex and the Chemistry complex. In the late 1980s, this model was upgraded by the introduction of "education, research and production complexes", which incorporated academic staff into industry (Annerstedt and Nguyen, 1996:236). Even though both models were unsuccessful because of the centralised planning system that led to administrative barriers in the Vietnamese economy, these developments can be seen as a starting point for today's epistemic culture in southern Vietnam.

In the next sections, we will demonstrate how Ho Chi Minh City and Can Tho City were able to not only become knowledge clusters, but also important economic regions.

3. The Nature of Knowledge Cluster Building

Further to the fact that economic activities tend to cluster, this paper exposes similar trends in respect to research and educational activities. As indicated above, knowledge clusters are agglomerations of production-oriented organisations. Having the ability to share knowledge assets such as laboratories or libraries reduces costs and enables a knowledge-sharing environment. The reduction of transaction costs, emphasised by classical industrial agglomeration theorists, is less important for knowledge-intensive production as transaction costs are extremely low. Conversely, however, a concentration of researchers and the sharing of tacit knowledge are facilitated by proximity (Evers, 2010). By virtue of the proximity of organisations, the recruitment of highly qualified staff and knowledge exchange can be enhanced and higher productivity achieved which points to the important role of cluster building.

3.1 *Movements towards Clusters*

Clusters are attractive not only to companies and organisations due to tax incentives and enhanced infrastructure, but also because of the accumulation of highly qualified staff in these areas.

Qualified people move to organisational agglomerations as the result of a better job market and opportunities for multiple jobs, e.g., in the consultancy business, because, as stressed by interviewees, salaries for academics and scientists are exceptionally low which means that they very often rely on auxiliary income. Nevertheless, the job market is not only important with regard to job opportunities, but also in terms of physical proximity. The important roles of personal relationships and networking activities for career development were affirmed during interviews. Evidence is mounting that, through the logistical proximity of clusters, opportunities accumulate and advantageous conditions are provided.

Two Vietnamese studies demonstrate the dynamics of student movement. Can Tho City and Ho Chi Minh City are the favoured locations in southern Vietnam for those seeking a promising career and a raised standard of living. Can Tho City, home to the most important university in the Mekong Delta, attracts most of the students in the region, although it is the home province of only 20 percent of the city's university students; the remaining 80 percent come from the other twelve provinces within the Mekong Delta.[9] The fact that more than half of Can Tho's university students remain in the city after graduation points to the unbalanced development of the Mekong Delta. Can Tho City therefore functions as a hub, attracting people from the whole region and particularly graduates, who give the three main decisive factors when looking for a job as salary, secure work and promotion opportunities (Luu *et al.*, 2002:209). In this sense, these criteria are likely to be fulfilled in clusters.

Ho Chi Minh City exhibits similar characteristics. According to a study on 1,243 students from Ho Chi Minh City's largest university,

---

[9] Data refers to the time period from 1995 to 2008 and was provided by the Academic Affairs Office of Can Tho University.

the Vietnamese National University, just 30 percent stay with their families while studying. In the Vietnamese context, students normally live with their parents until they get married and find a job; however, from the perspective of the present study, the only reason for not living with their families is that they have left their home provinces to obtain higher education elsewhere. The study implies that about 70 percent of the students come from outside and thus have to rent a room, stay with acquaintances or live in a dormitory. Similar to Can Tho City, Ho Chi Minh City is a magnet point for qualified people[10] (Nguyen, 2003:46).

3.2 *Development Due to Cooperation, Exchange and Competition*

Companies and organisations are attracted by cluster advantages for practical, economic reasons. State investments are made into creating an efficient infrastructure including excellent road networks or airports such as Vietnam's largest airport, the Tan Son Nhat International Airport in Ho Chi Minh City, or the Tra Noc Airport of Can Tho City, operating since December 2008. Moreover, IT infrastructures are enhanced and the application of modern technology facilitates exchange as well as efficient management and production processes.

In addition to the supply of a transportation infrastructure, clusters have a second significant advantage by virtue of the fact that face-to-face communication is highly likely to take place, ensuring the transfer of valuable 'tacit knowledge'. Highly skilled staff is available on the spot and therein approachable for organisations in terms of consulting services, sharing experiences and elaborating new ideas with these experts.

Our survey, carried out among a selection of Vietnamese researchers, provides important data than can be utilised to verify the

[10] Additionally, Ho Chi Minh City seems to be attractive not only for students from every part of Vietnam, but also for returning overseas Vietnamese who have investment capital and management knowledge as well as contacts with foreign investors (Chong, 2002:101).

cultural and social environments that shape staff work routines and the ways in which they interact. The analysis shows that informal personal meetings and the telephone are by far the most important means of communication in Vietnam. It should be noted at this point that the effective usage of the telephone as a communication tool is only viable when the contact person is known beforehand, which also applies to email communication. The reasons for this are deeply embedded in the Vietnamese cultural suspicion of impersonal interactions, which are regarded as wholly untrustworthy. As personal relationships are inherent in professional life and the key to a project's success, they can involve high transaction costs, as explained previously. The crucial advantage of clusters, therefore, is that people can meet over short distances, which saves a lot of wasted time and related travel expenses. Another advantage is the ease and comfort of attending seminars, workshops or conferences taking place in the same city, rather than travelling many hours to the countryside to visit perhaps only one organisation or workshop. Time and costs are too high to be beneficial, but by reducing these costs and time constraints through proximity, it is possible to build networks with many organisations working in the same field. Collaborations, meetings and face-to-face interactions take place actively as a result of advantageous facilities nearby such as coffee shops and recreational after-work establishments where people can invest in valuable personal relationships that inevitably spill over into business.

From an economic perspective, the Vietnamese government shares only a small part of the knowledge-producing organisations' budget. Since the mid-1980s, technology service contracts have become a major source of funding for many Vietnamese research organisations (Annerstedt and Nguyen, 1996:230). Today, external funding through international research and development cooperation, as well as foreign investment in industry and business, constitute an important part of a research organisation's budget and therefore secure staff salaries and help to promote career opportunities.

Under these circumstances, clusters provide a platform not only for Vietnamese industrial companies, research organisations and skilled people, but also as a main access point for international companies and organisations. It is understandable that international organisations prefer to find as many convenient conditions as possible if they are to pursue business connections in other countries. Facing inconvenient or impassable roads, long distances, unclear procedures, as well as language and culture barriers will make transaction costs too high and render projects economically unviable. Furthermore, organisations located outside cluster areas are less ambitious in terms of innovation; they lack updated information about new technologies and management systems, while innovation seems to be non-profitable in a static environment that itself discourages change. Located away from clusters in the way they are isolates such companies from a network of organisations that interacts almost exclusively with the main customer base and potential collaboration partners. In this sense, regions without clusters risk being cut off from development and innovation processes and tend to lack capacity.

The survey[11] confirms a high international influence on organisations in Ho Chi Minh City and Can Tho City. Both areas create hubs in the south of Vietnam, with favourable conditions and a large pool of skilled people and advanced infrastructures.

A positive effect of international cooperation is the fact that large-scale international projects bring together different Vietnamese partners, which otherwise would never have collaborated. This intervention can be traced in the analysis of international publication output, as provided by the online academic database ISI Web of Knowledge — the majority of scientific articles with at least one Vietnamese author are joint-products. Many contributions involve Vietnamese authors from different institutes, indicating an exchange of ideas and information in terms of the joint-publication as well as a learning effect among the authors. In addition to the enhancement of collaboration activities

[11] In total, 282 questionnaires were answered by Vietnamese staff members from seven different universities, colleges, research institutes and local authorities, in which water-related activities take place. In total, 95 percent of all respondents have a BA degree, 28 percent have graduated from an MA programme, 7 percent have obtained a PhD, and 1 percent holds a professorship. It is remarkable that nearly half of the interviewees with a BA degree are currently involved in an MA programme.

among Vietnamese scientists, knowledge exchange within the international research and science arena is also encouraged. Interacting with international experts gives Vietnamese academic and research staff the opportunity to improve their skills and methods, and to work according to international standards in order to be recognised internationally. Business trips and study programmes abroad strengthen the capacity of Vietnamese organisations and likely maintain contacts for further projects. Our survey shows that 41 percent of all respondents have already participated in an international conference.

Another factor for measuring the international influence on the Vietnamese science and research community is the number of staff participating in study programmes in foreign countries (Table 1). Thirty-one per cent of the respondents in Ho Chi Minh City and Can Tho City[12] have already studied abroad, which is quite a high percentage when taking the low income of Vietnamese researchers into consideration. On average, they spent 1.6 years abroad to obtain a higher education.

The data from Table 1 produces an interesting result when taking into account Vietnam's closure to international development for a long time, indicating the recent focus on opening up the country on

Table 1: Number of respondents spending time abroad for scientific training and research[13]

Period of Time	Number of Respondents
1 year or less	47
1–2 years	11
2–3 years	6
3–4 years	6
5–8 years	5
More than 8 years	1

[12] The rate of Can Tho University staff is much higher due to the strong investment in human resources by the World Bank.
[13] 76 out of the 269 respondents have studied abroad.

the international stage. There are also many national programmes attempting to reduce over-capacity issues by sending PhD students abroad, the efforts of which enhance networking opportunities with international research institutions.

3.3 *Innovations as a Result of Cluster Building*

As the paper demonstrates, cluster building reduces transaction costs, enables the dissemination of knowledge and promotes a high mobility of highly qualified labour and other resources that can be exploited more flexibly. Consequently, productivity is higher and innovation in terms of knowledge spill-overs and cooperation are more likely to take place. Innovation can be achieved when organisations obtain updated information about research findings facilitated by a supportive economic environment so, from this point, costs can be saved and the duplication of research activities avoided. Operating in a cluster creates the necessary opportunities for exchange, cooperation and, indeed, competition (see also Sölvell, 2008).

4. Achievements of Knowledge Cluster Building

In the previous section the specific nature of clustering in southern Vietnam was discussed in order to understand the dynamics underlying this concept. To advance this argument further, the implications cluster building has in terms of achievements for regional development will be shown.

4.1 *Rise in Scientific Outcome*

The quantity and quality of scientific output is a means by which the productivity and innovation of a region can be measured. In terms of knowledge clusters, it is a common standard to use international publications by Vietnamese authors. However, most knowledge output is produced in the Vietnamese language, which is often very roughly translated at best; with international publications, very few Vietnamese authors write directly in English. In general, there is still a lack of

Figure 7: ISI journal articles published by Vietnamese authors (1977–2008)

recognition of the vast Vietnamese knowledge output; nevertheless, the list of journal articles in the ISI[14] is a measurement indicator.

The amount of Vietnamese knowledge output acknowledged internationally has increased significantly over the last twenty years (Figure 7). Compared to the Mekong Delta, Ho Chi Minh City's output production is almost ten times higher, which shows the superior role of Ho Chi Minh City in the development of southern Vietnam. Considering the Mekong Delta with only thirteen provinces, Can Tho City plays a crucial role for the region because its output accounts for 72 percent of the whole delta. Obviously, there seems to be a correlation between the number of knowledge-producing organisations and international publications — the more organisations located in an area, the higher the output.

A more complicated method is to measure Vietnam's national output, which until now has been difficult to access. Table 3 gives the example of Can Tho University, indicating that the ISI includes a bias by virtue of the fact that English language journals dominate this

[14] "ISI Web of Knowledge" is an online academic database that only considers scientific articles published in English.

Table 2: Total number of ISI journal articles published by Vietnamese authors (1977–2009)

Province/Region	Number of Published Articles
Ho Chi Minh City	2,001
Mekong Delta	265
— *Can Tho City*	*190*
— *Dong Thap*	*27*
— *An Giang*	*11*
— *Tien Giang*	*12*
— *Hau Giang*	*8*
— *Long An*	*6*
— *Kien Giang*	*5*
— *Bac Lieu*	*2*
— *Ca Mau*	*2*
— *Ben Tre*	*1*
— *Soc Trang*	*1*
— *Tra Vinh*	*0*

database. These statistics are provided on the website of Can Tho University, and show that they have more scientific articles published internationally than ISI counts — within four years, CTU published 261 articles in international scientific journals whereas the ISI only counts 190 (see Table 2). Nevertheless, the ISI gives an opportunity to compare different countries and provides a window on the development stages of particular regions.

It should not be surprising that Can Tho University publishes more articles internationally than nationally. For the past decade, the university has strongly focused on international cooperation that supports and shapes knowledge production according to international science and research standards.

Unfortunately, international researchers often neglect Vietnam's national scientific output due to language and administrative barriers. There are numerous ways for the staff of knowledge-producing organisations to publish in Vietnam. According to our survey, 99 out of the 282 respondents gave specific details about their publications, listing in total 429 national publications since 2000. Two-thirds of the 99

Table 3: Number of published articles by Can Tho University (2005–2008)

Year	No. of Published Articles — Can Tho University Scientific Journal	No. of Published Articles — National Scientific Journal	No. of Published Articles — International Scientific Journal
2005	54	33	57
2006	46	74	74
2007	46	42	77
2008	45	22	53

Source: www.ctu.edu.vn

respondents had produced only one to four national publications, while 11 percent had no national publication. Conversely, one quarter of them had published in an international journal. In total, 112 international publications were specified. On average, most of the respondents had published one or two international articles. The s show that even though the national output is much higher than on an international scale, Vietnamese scientific publications are less recognised by international scientists.

Photo 2: Can Tho University

Furthermore, most knowledge-producing organisations provide in-house publications such as annual reports, newsletters or scientific journals particular to an institute's speciality. In addition, since 2008, twenty-three national scientific journals have been uploaded onto the internet. Such an online databank[15] is a first step for national as well as international readers to gain insight into updated Vietnamese research findings.

4.2 *Successful Economic Performance*

Clustering has an effect not only on the increase of knowledge output, but also on the economic growth of these regions.[16] Bearing in mind that the strategic locations of Ho Chi Minh City and Can Tho City originate in the flourishing development of southern Vietnam, the importance of the emergence of knowledge clusters in these regions is evident.

Statistical data shows that the GDP of Ho Chi Minh City contributes one quarter of the country's GDP, even though the city holds just 7.8 percent of the total national population (see Table 4).

More significant is the comparison of the GDP per capita of Ho Chi Minh City with that of Vietnam as a whole, the former being much higher for a long time with $2,100 against $835 in 2007 (Table 5). This proves that Ho Chi Minh City has enjoyed a far higher level of

---

[15] VJOL — *Vietnamese Journal Online* is an initiative of INASP, the International Network for the Availability of Scientific Publications, based in Oxford, Great Britain. In April 2009, the management of VJOL was transferred to the Vietnamese side and is based at NACESTI, the National Centre for Scientific and Technological Information under the administration of MoST.

[16] An attempt to measure the economic growth of Vietnam's provinces is the Provincial Competitiveness Index (PCI), which was introduced four years ago. Developed by USAID and the Vietnam Chamber of Commerce and Industry (VCCI), the PCI gives the opportunity to compare the economic development of each province in Vietnam and promotes local competition in order to improve the area's economy (www.pcivietnam.org). This index is based on surveys conducted in privately run companies and allows only educated guesses.

Table 4: GDP and population of Ho Chi Minh City at the ratio of Vietnam

Year	Ratio GDP HCMC-Vietnam (%)	Ratio Population of HCMC-Vietnam (%)
2001	17.6	6.7
2002	18.0	6.8
2003	18.4	7.0
2004	18.5	7.0
2005	20.2	7.5
2006	22.1	7.6
2007	24.3	7.8

Source: www.hochiminhcity.gov.vn

Table 5: GDP per capita of Ho Chi Minh City and Vietnam

Year	GDP Per Capita of HCMC (USD)	GDP Per Capita of Vietnam (USD)
1980	384	n.a.
1985	444	251
1990	583	98
1995	937	289
2000	1365	402
2004	1720	554
2007	2100	835
2008	n.a.	1040

Sources: www.imf.org, www.pso.hochiminhcity.gov.vn

economic development than any other region in Vietnam. One of the contributing factors of this rapid development may be the positive effects of clustering. It is, however, difficult to assess whether this clustering was the outcome of a deliberate policy or merely a function of population distribution and the urban land market.

Alternatively, the distribution of industrial parks is another indicator of cluster building and economic development. In total, there are fifteen industrial parks and export processing zones alone in Ho Chi Minh City, with another fifteen located in the Mekong Delta. Compared to the other Mekong Delta provinces, Can Tho City has, with four industrial parks, the highest number of industrial parks (GSO, 2009). The advantages are evident: companies get assistance in the licensing process and local affairs, there is a reliable source of electricity, the infrastructure is well-developed, and staff recruitment is concentrated amongst a highly qualified pool of potential applicants (Chong, 2002:11). Furthermore, the parks are embedded in a region populated by knowledge-producing organisations that will guarantee the education of skilled people and scientific exchange with other industries.

5. Conclusions: Ho Chi Minh City and Can Tho City as Knowledge Clusters

Our data show two prominent cases of knowledge clusters: Ho Chi Minh City and Can Tho City. The importance of location for knowledge production and economic growth could be demonstrated; nonetheless, this development would not have been possible without the strategic location of these regions. The Mekong River not only enables ships to enter the inner city of Ho Chi Minh City, but also provides cheap and effective transportation routes into the Mekong Delta. The subsequent strategic value of the region lies in economic advantages and opportunities for traders and manufacturers alike (Chong, 2002:21), and has transformed the region into a modern hydraulic society (Evers and Benedikter, 2009). Today, Ho Chi Minh City is the focus of foreign investment in Vietnam, with half of all foreign investments flooding into the metropolis (Chong, 2002:22).

Can Tho City is following closely on the heels of Ho Chi Minh City and fast becoming the economic, political, cultural and technological centre of the Mekong Delta (Le, 2006:118). In 2002, a World Bank loan was approved to restore waterway routes and ports in the Mekong Delta. Interestingly, the 600km waterway network links Ho Chi Minh City with Can Tho City and Ca Mau, the most southerly part of Vietnam, on the one hand, and Ho Chi Minh City with Can Tho City and Ha Tien, which is close to the Cambodian border, on the other (Chong, 2002:93). Recently, construction work for the largest seaport in the region, the Cai Cui seaport, has started, and will boost Can Tho City's position as a hub in the Mekong Delta's network of waterways.[17] In addition, Can Tho City is located at the crossroads of the highway network that interlinks the Mekong Delta, and another advantage is the recently opened Tra Noc Airport in Can Tho City, which will most likely become one of the main international airports in Vietnam in the near future (Le, 2006:119). Given that this development would not be possible without qualified people, Can Tho City took the initiative of opening the first university in the Mekong Delta. Consequently, science and research are now able to develop with the assistance of international cooperation, which is vital if the region is to interact on a global level.

Vietnam is on the path towards a knowledge-based economy in which Ho Chi Minh City and the Mekong Delta are playing decisive roles. Our data shows that the number of knowledge-producing organisations is continuing to grow. Still, although there is a great deal of potential in improving the epistemic landscape of this region, it can be nevertheless assumed that the education and research sectors will develop in line with economic prosperity, since it is more likely that wealthier families — especially in urban areas with suitable facilities — will invest more in the higher education of their children.

Looking at Vietnam's southern provinces within the Mekong Delta and adjacent areas, our data shows that Ho Chi Minh City harbours one of Vietnam's major knowledge clusters, followed by the much smaller cluster of Can Tho City. The distribution of knowledge-producing organisations within Ho Chi Minh City also shows clustering, in the sense that universities and research institutes are concentrated in adjacent urban districts (Map 2).

Clustering in these two urban areas is, to a large degree, a function of population density, or 'urbanism', i.e., the availability of urban institutions and of government policy.

[17] Information taken from a newspaper article from 13 July 2009: "Work starts on Cai Cui Seaport" "http://vietnamnews.vnagency.com.vn/showarticle.php?num=03ECO130709 (accessed on 15 July 2009).

This leads to a remaining research question, namely to what extent proximity or clustering have led to inter-organisational networking and knowledge sharing. A preliminary analysis of interviews and survey data shows that the situation can be adequately described as one of hierarchical or bureaucratic sharing of knowledge and information, insofar as research results are primarily channelled into either government departments or international donor agencies. Intra-organisational knowledge sharing still seems to be in the embryonic stages of development and horizontal research cooperation and knowledge sharing between knowledge-producing organisations hardly takes place. In the terminology used in this paper, and outlined in the introductory section, the 'knowledge hubs' of networking and knowledge sharing are yet to be fully developed — the epistemic landscape has still to be completed.

Geographical clustering without knowledge sharing has greatly reduced the effectiveness of knowledge production and knowledge output. It remains to be seen whether the attempts by some Vietnamese researchers bear fruit, when they start to work toward more intensive knowledge sharing between organisations. Knowledge clustering needs to be supplemented by networking and the building of knowledge-sharing epistemic communities to produce new knowledge and economically viable innovations.

References

Annerstedt, J. and H. N. Thanh (1996). Demolishing the ivory tower — Science, technology and economic development. In *Vietnam in a Changing World*, I. Norlund, C. L. Gates and C. D. Vu (Eds.), pp. 219–250. Great Britain: Curzon Press.

Chong, L. C. (2002). *Business Environment and Opportunities in Vietnam — Ho Chi Minh City and Its Surrounding Region*. St. Gallen, Verein Business Books & Tools St. Gallen.

Dang, M. L.(2006). *Khoa học Việt Nam: Những nghịch lý và một số cách giải. Science in Vietnam: Absurdities and Some Solutions*. Khoa Học- Giáo Dục. Di tìm diện mạo mói Science — Education. Looking for new aspects). N. t. giả. TP. Hồ Chí Minh, Nhà xuất bản trẻ.

DoSTE, Department of Science, Technology and Environment of Ho Chi Minh City (1998). Giói thiệu các tổ chúc hoạt động khoa học và chuyên gia có học hàm, học vị tại TP. Hồ Chí Minh Introduction of all Scientific Organisations and Experts with an Academic Degree in Ho Chi Minh City). TP. Hồ Chí Minh, Trung Tâm Thông Tin Khoa học và Công Nghệ.

Evers, H.-D. (2003). Transition towards a knowledge society: Malaysia and Indonesia in comparative perspective. *Comparative Sociology*, 22:355–373.

Evers, H.-D. (2008). Knowledge Hubs and Knowledge Clusters: Designing a Knowledge Architecture for Development. ZEF Working Paper Series. Bonn, Centre for Development Research ZEF, University of Bonn. 27. reprinted in: S. Gopalan (ed.). *Knowledge Assets. Concepts and Measurements.* Hyderabad, The Icfai University Press, pp. 106–128.

Evers, H.-D. and S. Benedikter (2009). Hydraulic bureaucracy in a modern hydraulic society — Strategic group formation in the Mekong Delta, Vietnam. *Water Alternatives*, 2:416–439.

Evers, H.-D., S. Genschick and B. Schraven (2009). Constructing Epistemic Landscapes: Methods of GIS-Based Mapping. ZEF Working Paper Series. Bonn, Centre for Development Research ZEF), University of Bonn. 44.

Evers, H.-D. and S. Gerke (2009). Strategic Group Analysis. ZEF Working Paper Series. Bonn, Centre for Development Research ZEF), University of Bonn. 34.

Evers, H.-D., S. Gerke and T. Menkhoff (2010). Designing epistemic landscapes of knowledge clusters and knowledge hubs for development. *Journal of Knowledge Management*, 14(5):678–689.

Gerke, S. and H.-D. Evers (2006). Globalizing local knowledge: Social science research on Southeast Asia, 1970–2000. SOJOURN: *Journal of Social Issues in Southeast Asia*, 21:1–21.

GSO, General Statistics Office (2007). Industrial Parks (www.gso.gov.vn).

GSO, General Statistics Office (2009). Population Number of Ho Chi Minh City and the Mekong Delta (www.gso.gov.vn).

Henry, N. and S. Pinch (2006). *Knowledge and Clusters Clusters and Globalisation*, C. Pitelis, R. Sugden and J. R. Wilson (Eds.), pp. 114–132. Cheltenham: Edgar Elgar.

Hornidge, A.-K. (2007). *Knowledge Society: Vision and Social Construction of Reality in Germany and Singapore.* Berlin and London, LIT Verlag.

Knorr-Cetina, K. (1999). *Epistemic Cultures: How the Sciences Make Knowledge.* Cambridge, Harvard University Press.

Le, T. (2006). *Các tỉnh và thành phố Dạng Bằng Sông Cửu Long. Geography of All Provinces and Cities in the Mekong Delta.* Hà Nội, Nhà xuất bản giáo dục.

Luu, T. T. *et al.* (2002). Thủc trạng của sinh viên đối vói việc làm sau khi tốt nghiệp [Analysis of the current situation of students in finding jobs after graduation]. Tuyển tập Công trình Nghiên cửu Khoa học. Tạp chí khoa học Dại học Cần Thơ 1.

Menkhoff, T., H.-D. Evers and Y. W. Chay (eds.) (2010). *Governing and Managing Knowledge in Asia.* New Jersey, Singapore and London: World Scientific Publishing (2nd edition).

MoST, Ministry of Science and Technology (2004). *Vietnam Education and Training Directory.* Hanoi: Education Publishing House.

MoST, Ministry of Science and Technology (2008). Những Điều cần biết về tuyển sinh đại học và cao đẳng năm 2008 Everything you Need to Know about Enrolment in Universities and Colleges 2008). Hà Nội, Nhà xuất bản giáo dục.

Nguyen, V. T. (2003). Nghiên cứu sự tác động của các yếu tố kinh tế — xa hội đến hoạt động học tập và định hướng việc làm sau khi tốt nghiệp cửa sinh viên đại học quốc gia TP. HCM trong giai đoạn công nghiệp hóa hiện đại hóa. Báo cáo tổng kết đề tài. TP. HCM.

Pham, D. H. (2006). 150 triệu USD mà cỗ xe khoa học công nghệ vẫn chạy chậm 150 Mio USD But Science and Technology Still Develops Slowly). Khoa Học-Giáo Dục. Đi tìm diện mạo mói Science — Education. Looking for New Aspects). N. t. giả. TP. Hồ Chí Minh, Nhà xuất bản trẻ.

Sölvell, Ö. (2008). *Clusters Balancing Evolutionary and Constructive Forces.* Stockholm, Sweden: Ivory Tower Publishing.

Truong, T. M. S. (2007). Chuyển dịch cơ câu kinh tế khu vục dụch vi ụ thành phờ Hồ Chí Minh trong quá trình công nghiệp hóa, hiện đại hóa Transfer of Economic Service Sector Mechanisms in HCMC during the Process of Industrialization and Modernization). Hà Nội, Nhà xuất bản khoa học xã hội.

Yellow Pages (2007). Tập Đoàn bưu chính viễn thông Viêt Nam — Yellow Pages 2007 Ho Chi Minh City. Niên giám điện thoại những trắng vàng và những trang trắng. TP. Hồ Chí Minh, Nhà xuất bản bưu điện.

Chapter

10

Knowledge Sharing in the Indonesian Context

Institut Teknologi Bandung (ITB) as Potential Knowledge Hub to Create Value from Academia, Business and Government Linkages

Jann Hidajat Tjakraatmadja, Agung Wicaksono and Lenny Martini

1. Introduction

We are entering the knowledge era, where knowledge becomes the substantial capital of organization, be it in business, government, or academia. Japan's industry could grow and develop in the knowledge era due to its ability to leverage on personal, tacit knowledge of individuals who produce creative insights, and the shared, explicit knowledge which the organization needs to develop new products and innovations (Nonaka and Takeuchi, 1995).

The knowledge era requires us to better understand the role and meaning of knowledge as a form of human capital in organizations. It also demands that the individuals and society develop knowledge in an innovative and sustainable manner — be it in an organization, a region, or even a nation — as an implication of inter-organizational influence. The knowledge era has a borderless characteristic, eliminating barriers between micro-organizations in a country, or even between country states. The borderless world has intensified the influence of inter-organization relationships, be it an academia, business, or governance.

The impact of macro-level changes on the micro-level situation in Indonesia is perhaps best exemplified by the "knowledge loss phenomenon" of PT. Dirgantara Indonesia (PTDI), the national aerospace company built by former Minister of Technology, Prof. Dr. BJ Habibie, who became Vice President and later President of Indonesia. Since 1998, many technical experts from PTDI have moved overseas, to Malaysia, Brazil and Germany, among other countries, simply due to problems at the macro level. This shows that Indonesia's heavy investment in developing strategic industries has not achieved the expected outcomes in terms of creating new knowledge. A big chunk of the knowledge arising from the national investment has migrated, along with the brain drain.

It is therefore imperative to nurture a pragmatic collaboration between academic, business and goverment (ABG) organizations. This 'ABG collaboration' is also known as the triple helix model (Leydesdorff and Etzkowitz, 1998) in innovation studies. Leydesdorff and Meyer (2003) have further elaborated on the triple helix model of university-industry-government relations in the context of the knowledge economy.

In the light of Indonesia's experience with developing the knowledge economy, the key research questions in this paper are:

- How to develop a smart ABG collaboration in Indonesia?
- What are key success factors that may enable smart ABG collaboration in Indonesia?

- What would be the framework for smart ABG collaboration in Indonesia?

2. Key Concepts in Knowledge Management (KM)

2.1 *The Relevance of Knowledge Management in Organization: Macro and Micro Level*

Knowledge management (KM) has been a hot topic since the 1990s (Halal, 1998; Baker and Badamshina, 2003). Many studies have explained the importance of KM for an organization, although there are still gaps in our understanding of KM itself, both within academia and among practitioners. However, a number of companies or practitioners are confident that KM is a decisive factor for the success of the firm, mainly due to the following reasons:

- A number of academics state that the new economy will be the era of the knowledge economy, where competitiveness of the firm will be determined by the level of knowledge which can be institutionalized in the organization. In this regard, employees as a form of human capital are unique because they help to create wealth for the firm. Human capital cannot be replicated by other organizations, unlike other organizational capital such as material, financial, and even technological capital.
- KM represents a progressive logic whose value is more than what management information systems (MIS) stand for. The effectiveness of KM is determined by the quality of a supportive work environment and the depth of knowledge sharing and related knowledge outcomes. Information technology plays the role of supporting the learning process to accelerate the growth of organizational knowledge, and eventually, the firm's performance. The growth of the next generation of information technology and artificial intelligent will improve the effectiveness of KM in an organization (Sveiby, 1997).
- KM can also be manifested as the culmination of a number of existing organizational methods in an organization such as *total quality*, *reengineering*, *benchmarking*, *competitive intelligence*,

inovation, *organizational agility*, *asset management*, *supply chain*, *change management*. KM is the integrated and holistic outcome of the above concepts, focusing on the creation and implementation of knowledge in an organization (Amidon, 1998). Recently, KM has evolved into Knowledge Innovation Management.

- KM is the foundation for development in the era of the knowledge economy popularized by Drucker (1969), who explained the economic benefits from knowledge translated into tangible products or services. The maximum economic benefit of a product or service would be achieved if the organization has an education and training system which can create *knowledge workers (as the source of human capital)*, who work in an effective organizational system and technology (as the source of information and organizational capital), to yield product or services with *high value of return*, or firms with high market capitalization.

2.2 *Knowledge Innovation Management (KIM)*

With the support of an IT infrastructure, KIM can create a Knowledge Based Society (KBS) and Knowledge Based Economy (KBE) at the same time. Each dimension of KIM can be explained as follows:

- *Knowledge Based Society (KBS)* is a society consisting of people with good *human capital* capacity, able to use information efficiently, and with a vast network (*social capital*).
- *Knowledge Based Economy (KBE)* as popularized by Drucker (1969), implies that the economic benefits of knowledge are made explicit in products or services.

2.3 *The Demand for New Organizations in the Knowledge Era*

Globalization demands that people are willing and able to accept global content to be absorbed into the social environment as well as the ability to globalize local content in order to become global citizens. Nowadays, the challenge is to create value not only locally but also globally. In order to create global prosperity, we need organizations which can maintain universal principles and values.

- *The knowledge era demands learning organization*
  De Geus (1997) has described the characteristics of a long-lasting company as the *living company*. He argued that a living company has the attributes of a living creature with minds and characters. In that sense, a company can act as if it were a living entity. De Geus explained that there is a '*correlation between long-lasting companies and their ability to be a learning organization*'.
- *The knowledge era demands a learning state*
  The era of knowledge economy demands a social order and organised learning at both macro and micro levels, i.e., organizations with the ability to create, acquire and share knowledge between members of the organization, in order to develop organizational knowledge. The concept of the micro learning organisation represents the enabler for the growth of KIM at the macro level.

KIM concepts at the micro level can be useful to develop smart ABG collaboration. This smart collaboration should be focused at developing a learning society, which is an extension of the concept of a learning organization for creating, acquiring, and sharing knowledge between academia, business and government for the sake of people's welfare.

To formulate the KIM framework for collaboration among ABG organizations, we need to identify potential factors that can become enablers. Therefore, a review of the macro and micro situation as described in the next section is necessary.

3. The Indonesian Context of Knowledge Management: Socio-Technological Situation

It is important to understand the socio-technological situation in Indonesia in order to find answers for the research questions outlined above. This ought to be undertaken at the macro-level as well as the micro-level. Such an understanding serves as a foundation for building the knowledge framework for smart collaboration between ABG organizations in Indonesia.

3.1 *Contextual Experience: Social Construction of Technology*

The following section explains the contextual experience with regard to the role of technology implementation in constructing a new social order in Indonesia. The failure of implementing a new technology due to the lack of appreciation by the people is also discussed.

- *The Internet and the mobile phone: The role of technology for Indonesia's 'Reformasi' movement (1998)*

  The '*reformasi*' movement in Indonesia in 1998 was clearly the result of the widespread use of the internet and mobile phone. The internet in particular, has become the *reformasi* technology in Indonesia due to four characteristics: relatively low cost; real time information; borderless application; and creation of public spaces beyond the control of the authorities.

  When a number of mass media were banned by the Ministry of Information under Suharto's era, the internet became the alternative technology to disseminate information on actual political developments, and was later used as a means for political consolidation by pro-reform elements. The Internet has been well-appreciated by the people in Indonesia's main cities such as Jakarta, Bandung, Yogyakarta, and Surabaya. It has become instrumental in supporting social and individual change in the wake of the *reformasi* era in 1998 (Lim, 2003).

  The evolution of the internet function from being a network of computers to a medium for democratization can be framed within the theoretical framework of the social construction of technology, where technological development is not linear in nature (Bijker *et al.*, 1987). The development of a technology is socially constructed. The implementation of a technology in a society would not work if the social process between technology and the user society would not be appropriate. Appropriation is a process of valuation of a technology product by the society, based on existing values and interests in the society. Application of a new technology in a society would succeed only if the society can adapt its values to the presence of the new technology (Amir, 2002).

- *Democratization of technology*

Indonesia had a bitter experience with technology application in the area of aerospace technology during the era of Habibie. He used technology as a tool to support a political statement for the sake of Indonesia's international image and representation of vested interest groups, with the support of then president Suharto. Technology development was perceived more as an elite agenda rather than a means to improve public welfare and to achieve development progress.

The failure of implementing high technology in Indonesia was very much influenced by the authoritarian political culture of the New Order, with Habibie's reliance on Suharto and his one-man-show approach which failed to gain acceptance by the public. It was an inappropriate policy in the form of governing mentality that denied public participation in the decision making process (Campbell, 2000).

Based on that experience, it can be argued that the development of a technology would have been more effective and more appropriate had it not been dominated by certain individuals or groups. Democratization of technology, where the involvement of the public opens up space for negotiation between policy makers and the public for an appropriate national technology policy to be created, should be a public agenda. Without it, appropriation of technology application in society is difficult to achieve (Amir, 2004).

3.2 *Indonesia Today — Macro-Level Condition after* reformasi *Era*

The 1998 Asian economic crisis hit Indonesia very badly as its leaders were not open to change. External factors also affected Indonesia's economy. Declining economic growth in 1997–1998 (Figure 1) meant that thousands of firms went insolvent.

An interesting and provocative question people still pose even today — especially looking at how its neighbours have recovered from the crisis — is why it is so difficult for Indonesia to recover and learn from past mistakes? The following section explains the result of the authors' analysis based on observations at the macro and micro level in Indonesia, with reference to knowledge management concepts (Tjakraatmadja and Lantu, 2006).

- *Obsession with past glory*

  The biggest problem in facing change is the unwillingness to face new realities. This is because we are trapped in worshipping the glorious days of the past. Bureaucrats and politicians still try to tackle contemporary problems with past solutions, believing that

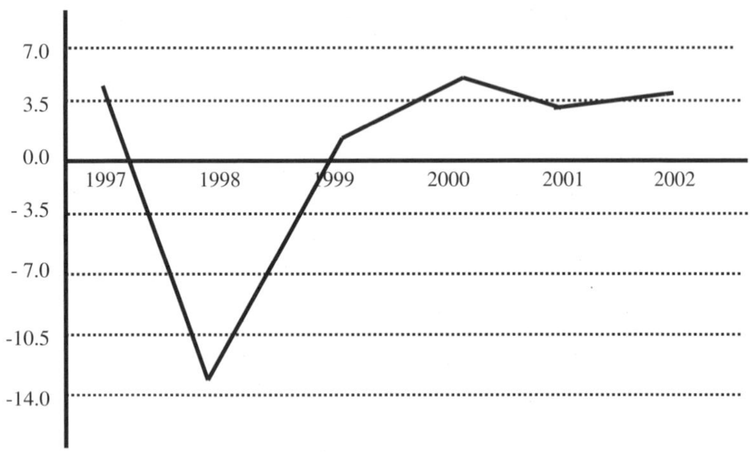

Figure 1: Indonesia's economic growth (1996–2002)

Source: Badan Pusat Statistik (BPS – Central Statistics Agency, Government of Indonesia).

if it worked before, it will work again. While this approach has to change, the fact is that it is not likely to as bureaucrats cling to well-tried, traditional methods which were successful in the past but which may not be appropriate in meeting the challenges of the future.

- *Being used to short term thinking*

  The mindset and behavior of decision makers in government and business — especially among the state-owned enterprises (SOEs) — are unfortunately short-term oriented. This is because many of the performance indicators use short-term financial targets. Managers and politicians are interested only in instant programs with gains that can be achieved during their term of office. As a result they care less about research and development as sources of innovation for future products and technology.

- *Preference to justify the habits ("membenarkan kebiasaan"), rather than creating the right habits (membiasakan kebenaran)*

  In expectation for change to happen for a better life, many do not know where to start and what to change. Therefore, people tend to keep doing the same things everyday, thus justifying that old habits are the right ones. They often do not dare to take bold right steps, let alone turning them into new habits.

- *A critical mass of human capital has not been achieved*

  For change to happen, it is necessary to have a certain amount of energy accumulated by the strength of vision, beliefs, confidence, thoughts and effort of Indonesians as a whole. Just like an atomic bomb would require a critical mass to explode, the synergy of Indonesian human capital has yet to reach a critical mass.

- *Need for more real leaders*

  Change cannot happen if leaders do not set an example as role models. Indonesia needs leaders who can inspire and move people, leaders who do not just think, but also take initiative and risk.

As one of the nations' founding fathers, Ki Hajar Dewantoro, said, Indonesia needs a type of leader who is hands-on, gets off the ground, gains trust and respect, and is able to motivate the nation's spirit to re-awake.

3.3 *Indonesia Today — Micro-Level Condition after* reformasi *Era*

Tuomi (1999) has stated that the learning organization is part of knowledge management (KM). Tjakraatmadja (2003) also concluded more explicitly that, implementation of KM should be preceded by the implementation of a learning organization, developed through interaction between people, the environment and the learning infrastructure.

Out of the above 3 elements of a learning organization, the worker is the living element and key success factor for the creation of a learning *organization*. Knowledge and experience sharing among members of an organization is the key activity in nurturing organizational knowledge. The effectiveness of this knowledge sharing process is very much determined by the quality of the learning environment, manifested in mutual trust and a conducive learning culture within the organization. Leadership to motivate members and to provide infrastructure should be directed at maximizing the effectiveness of the process of absorption, development, dissemination and application of organizational knowledge.

According to research by Tjakraatmadja (2003), there are some general challenges in the implementation of KM in micro-level organizations in Indonesia, namely:

- *Personal barriers*, especially related to the maturity of the workers. Immature members of the organization do not have the ability and courage to create supportive environment for learning, thus discouraging employees from improving their work effort.
- *Barriers due to the poor learning habitat*, especially the need for a better *learning* culture and higher degree of mutual trust

among members of the organization. This is showcased by (1) the lack of common values to bond members of the organization, (2) the lack of opportunity for dialogue as a mechanism to build common aspiration and synergy out of diversity of viewpoints. If the learning habitat is immature, the intellectual competence of individuals in an organization cannot yet be transformed into (in)tangible assets as organization capacity.

- *Barriers due to (1) poor organizational structure*, lead to the inability to capitalize on the potential for regional autonomy and decentralization. Thus, there is no capability to transform individual learning outcomes into organizational knowledge. In other words, individual learning does not have a positive impact on organizational performance. Many of the efforts to improve personal competencies were simply motivated by personal or group motives. *(2) The performance evaluation system* was unable to motivate learning and collaboration. Therefore, the result of individual learning has not enhanced the organization's human capital capacity.

3.4 *Institut Teknologi Bandung (ITB) as Indonesia's Leading Academic Institution — Micro-Level Context of Knowledge Management in Indonesia*

The problems faced by organizations at the micro level are also evident at ITB, the leading academic institution in Indonesia. Our research in 2007 entitled "Knowledge Sharing in Academic Insitutions: Case Study at School of Business and Management (SBM) ITB" (Tjakraatmadja and Martini, 2008) concludes that knowledge sharing between faculty members in SBM is facing considerable barriers similar to the factors mentioned above. We interviewed faculty members in SBM ITB to understand their perceptions of knowledge sharing, related catalysts, and barriers. Knowledge sharing is perceived by the respondents as knowledge transfer, problem solving and process to help each other. Learning spirit, learning habit, absorptive capacity and facility were found to be catalysts of the knowledge sharing process.

Barriers relating to the micro organization as discussed above also represent barriers for knowledge sharing in SBM ITB, which can be classified into three groups:

- *Personal barriers*: faculty members have not benefited from the knowledge sharing process, especially when the topic shared is out of his/her expertise or not one of his/her favourite topics. Thus, many faculty members do not feel the need to share their own knowledge with others because they believe that it is useless for others.
- *Environment*: faculty members admitted that the environment, consisting of other members, has not given them appropriate support for the knowledge sharing process. There is little appreciation for those who share knowledge and not enough support from leaders to engage in knowledge sharing. The organization's leaders have not functioned as enablers to form a conducive learning environment.
- *Systematic barriers*: organizational members have little time to engage in knowledge sharing due to a variety of time-consuming routine tasks which need to be performed. The performance appraisal system is not giving them sufficient credit/rewards for knowledge sharing activities. As a consequence, faculty members are not motivated to engage in such activities.

Despite the fact that SBM ITB currently has the most comprehensive facilities in ITB, for example, library, computer labs, study rooms, meeting rooms, syndicate, access to international journals, magazines, books, incl. a wide international network with some prominent universities abroad, knowledge sharing will not become a prominent knowledge process unless the three barriers outlined above are overcome.

4. ITB's Knowledge Innovation Management (KIM): Implementation and Challenges

4.1 *ITB as a Centre of Knowledge Innovation Management (KIM): Guiding Principles*

ITB has long discussed the concept of business and the need for business activities in ITB. In 1993, guiding norms and principles for ITB business activities were formulated, aimed at directing ITB's business activities so that it could achieve independence. ITB's academic senate formulated the concept for ITB's independence based on 3 principles of ITB's business:

- ITB as a public university in science, technology and arts should discharge its obligation in Indonesia as an agent for development.
- ITB through the principle of "Tridarma Perguruan Tinggi"[1] should accelerate and enhance national development with dedication.
- Businesses run by ITB utilizing public funds should aim at achieving the independence of ITB, i.e., for ITB to deliver its obligations and provide the best service to Indonesia. This will be achieved if there is improvement in the following indicators: the "Tridarma Perguruan Tinggi" program, employee's welfare, creation of works valuable for the society and science and technology, ability to maintain and renew academic facilities, and ability to utilize institutional capabilities.

4.2 *Bandung High-Tech Valley (BHTV)*[2]*: The Role of ITB*

In order to realize the above concept of independence, ITB has tried to pioneer the birth of Bandung High-Tech Valley (BHTV) from the late 1990s onwards and to develop the business incubator. However, this independence is still far from being fulfilled. The fundamental challenge toward realizing the independence of ITB is the low spirit for knowledge sharing, due to the similarity of the characteristics of ITB's staff with the micro-level situation as described in the previous section.

The concept of BHTV has been discussed since a decade ago, before the economic crisis hit Indonesia. At that time, several Indonesian academics wanted to implement the Silicon Valley concept in Indonesia, concentrating high tech business in one particular place. But due to the instability of the Indonesian economic and political conditions, the

[1] Tridarma Perguruan Tinggi is the triple fundamental principle guiding the activities of universities in Indonesia, consisting of teaching, research and community service.
[2] Source: www.bhtv.or.id

BHTV concept still does not physically exist. The BHTV community today consists of ITB academics and key persons in Bandung and high-tech industries. It tries to rebuild the concept to make it feasible so that it can be realized in a shorter time. The vision of BHTV is to create a supportive ecosystem where technology-based businesses can grow. The focus also has changed by not only concentrating on IT business, but also developing existing business in Bandung using the advantage of technology, for example the textile industry, as well as to translate business ideas gleaned from academia into real business.

Bandung is chosen as the location because of these reasons:

- It has a number of top universities in Indonesia, which attract talented students from all over Indonesia, supplying large number of knowledge workers.
- Research institutions are scattered in Bandung that can help to detect the latest trends and produce business-related knowledge.
- Government agencies dealing with technology are located in Bandung, making it easier to build connections with the government.
- There is an increasing number of small and medium businesses in high tech industry as potential tenants in BHTV, e.g., software houses, medical tools companies, interactive game developers, etc.
- There is support in form of facilities and infrastructure such as book stores, stable supply of electricity, water and other utilities, as well as highway access to Jakarta.

With the advantages above, BHTV has the opportunity to be a pioneer of knowledge clusters in Indonesia as a realization of linkages between academia, business and government. It can also contribute significant amounts of income to the country. To reach that ideal state, BHTV should consider these key questions:

- *Location*: Will it be localized in certain areas in Bandung? What are the criteria for selecting the areas?
- *Impact*: What will be the impact for the surrounding community?
- *Business selection*: What are the criteria for selecting tenants? Will there be any prioritization for certain businesses?

- *Legality*: What is the legal form of BHTV? What is the relationship between tenants, BHTV, government, academia and other related subjects?
- *Sustainability*: What will be the funding source? How is the pattern of funding?

In order to answer these questions, some steps have been taken, e.g., conducting a survey of industry clusters in Indonesia in Bandung. Regular internal meetings between stakeholders in BHTV community, as well as with external parties such as discussion fora with local government to develop regulations for IT businesses are also being held. Seminars and workshops with speakers from prominent IT firms are geared at knowledge sharing between specalists within the community. BHTV has also outlined plans to develop the Cimahi Cyber City concept and build the Cimahi Business Center in cooperation with the local municipalities in Cimahi, a satellite town of Bandung. In order to reap the benefits and to learn from successful global players, it has also attempted to open links with global companies in IT research and business, such as Synopsis in the USA and Infosys in India. Further ideas include the development of a cyber-colony.

Currently there are 6 core competences developed in BHTV, namely microprocessor design and implementation; design and manufacturing of internet phone (IP)-based telecommunication; GPS (Global Positioning System) and GIS (Global Information System) design; software design; game developing and computer animation; and security system design, biometric and internet security. Other than those core competences, BHTV is also looking to develop local potential such as the textile industry and medical equipment technology.

Big opportunities await BHTV and also huge challenges. The biggest challenge is to get full support from all parties that determine the existence of BHTV: government as regulator, experts and research institutes as knowledge providers, small industries as tenants, and financial institutions as funding resource. ITB has a big role to play in this, as a hub to link all parties and facilitate the development of BHTV to become the pioneer of knowledge cluster in Indonesia.

4.3 *An Example of an Innovation Zone from the Business Sector: Unilever's Greenest and Cleanest Village Program*[3]

An example of an innovation zone in an organization is a program by Unilever Indonesia called "the greenest and cleanest village program". Unilever's mission, vision and corporate purposes are:

- Its mission is to add vitality to life. Unilever meet everyday needs for nutrition, hygiene and personal care with brands that help people feel good, look good and get more out of life.
- Its vision is to be the first choice of consumers, customers and communities.
- Its corporate purpose is: "... to bring our knowledge and international expertise to the service of local consumers. Our long term success requires a total commitment to expectional standards of performance and productivity, to working together effectivelly and to a willingness to embrace new ideas and learn continuously. It's about knowledge and learning".

Unilever Indonesia has developed knowledge sharing culture programs with employees, consumers, customers, community and forum, and top leaders. How is the knowledge sharing culture with the community embedded in Unilever's business in Indonesia?

Unilever started to develop its environmental program in the city of Surabaya, where the company's factory is located in 2002. It has initiated an environmental campaign in collaboration with Jawa Pos, the leading newspaper in East Java since 2005. It also organized a

---

[3] Based on the result of MAKE (Most Admired Knowledge Enterprise) Award 2007, where Unilever Indonesia won one of the prizes (Source: http://www.unilever.co.id/id/ourcompany/beritaandmedia/siaranpers/_2007/UnileverMenangdalam2007IndonesianMostAdmiredKnowledgeMAKEAward.asp).

competition to build the greenest and cleanest neighborhood at the smallest unit of a community. The objectives of the program were:

- to create awareness about environmental issues, especially waste management and greening issues;
- to spread the environmental spirit among the community in Surabaya through intensive knowledge sharing in daily Jawa Pos and
- to call for action.

4.4 *Knowledge Innovation Zone (KIZ) in ITB: A Concept for ABG Smart-Collaboration*

The knowledge sharing process will be effective if it is supported by ABG collaboration which would include providing the domain, policies, and rules and regulations for members of the organization. To minimize conflict and uncertainty, there need to be clarity on the role, task, authority as well as agreement on the rules for learning and working in a wider sense, for the process of knowledge transformation in ABG organizations. Therefore, at the initial state it is important to review the definition, commissioning and support processes of KIM roles in the member organization.

Smart ABG collaboration is aimed at:

- Accelerating the commercialisation of technology
- Improving regional and national economy
- Achieving independence of ITB

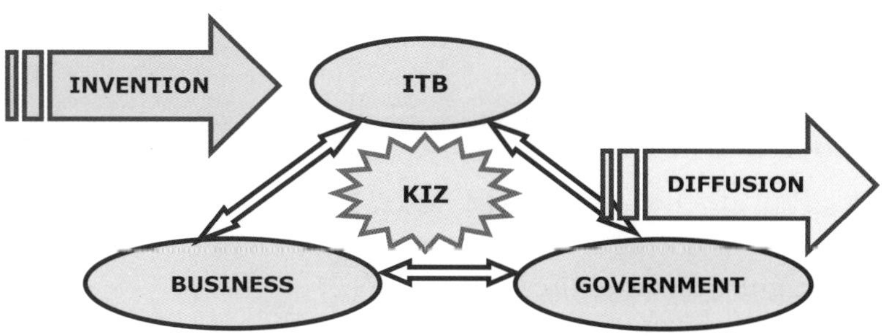

Figure 2: Concept of KIM between ABG organizations in ITB (based on Amidon, 1998)

Conceptually, the main role of each of the ABG elements can be defined as follows, based on Tjakraatmadja (1997):

i. **Roles of Academia (ITB):**

- Support R&D process by utilizing ITB facilities such as laboratorium, research institutes and centers, as well as ITB's academics.
- Provide for a Knowledge Innovation Zone (KIZ), as the center to coordinate activities of innovation and technology management of the members.
- Appoint a manager to oversee the daily operation of KIZ. Management of KIZ should act as *entrepreneurial manager*, function as coordination center, strengthen collaboration between ABG organization, and abide by the norms and targets of each of the organization based on the principle of fairness and mutual benefit.
- KIZ manager is tasked with the selection of tenants, prepares tenant to be future entrepreneurs and help tenants to achieve excellence in business.
- In the context of KIM, ITB is tasked with the development of KIM concept, KIM strategy and KIM tools to be implemented in KIZ.

ii. **Business Roles:**

- Provide funding and facilities for KIZ operation.
- Provide facilities, business information, internships, and professionally assist the training and business practice of the members of KIZ.

iii. **Government Roles:**

- Provide the land for KIZ area.
- Create attractive business environment through policy, regulation and bureaucracy.
- Monitor the process of socialization and security in the surrounding of KIZ.

iv. **KIZ Roles:**

- Accelerate commercialization of R&D result and product or process innovation (minimizing time to market).
- Minimize the risk of new business failure by training young entrepreneurs and integrating them with the sources for intellectual capital in campus, government, business and market.
- One stop service for business set-up, in the form of office space, administration, utilities, workshop and laboratorium.
- Space for ABG collaboration to conduct entrepreneurial management in developing regional and national economy.
- Space to implement Knowledge Innovation Management.

Critical issues to be faced by KIZ members during the tenant phase (requiring the support of academicians such as scientists, engineers and business/management experts) include:

- Accelerating and minimizing the risk of failure between the processes of invention (acquisition of new knowledge), innovation (application of new knowledge) and diffusion (acceptance and adoption of new knowledge).
- Feasibility study of the business (incl. the entrepreneur) — technically, financially, commercially (and mentally).

Critical issues to be faced by KIZ members as far as young entrepreneurs are concerned (to be assisted by business profesionals, experts, and government) are:

- Preparing legal set-up of the new company.
- Access to financing, technology and personel.
- Development of entrepreneurial and creative culture.
- Business plan of the company.

The role of ABG collaboration at each phase of the knowledge innovation process can be summarized by the following figure.

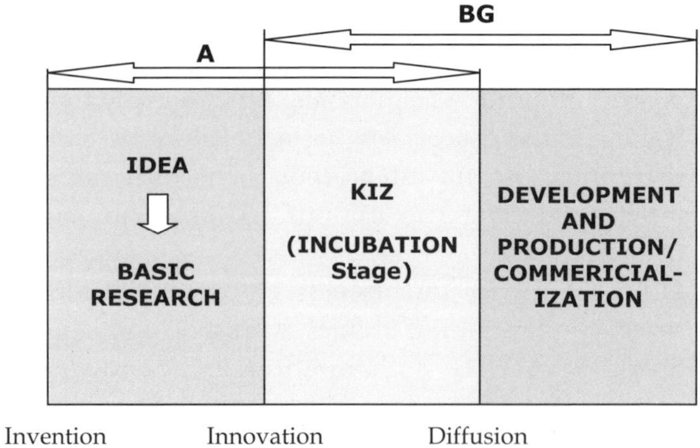

Figure 3: Knowledge innovation process

The presence of KIZ in ITB can hopefully help to achieve ITB's independence and prosperity for the surrounding community by:

- Job creation
- Transfer of knowledge/technology into businesses
- Substitute foreign technology
- Diversification of local business
- Revitalizing sunset industries
- Investment into new companies
- Application of knowledge into technology creation and commercialization to create new products or processes, or innovation to replace existing technology

Maximizing the benefit of ITB-KIZ should be based on the concept of environmentally-friendly Knowledge Innovation Zone, to be the 'greenest and cleanest industrial zone' as described in the Unilever example, as the prerequisites of building an industrial society described below.

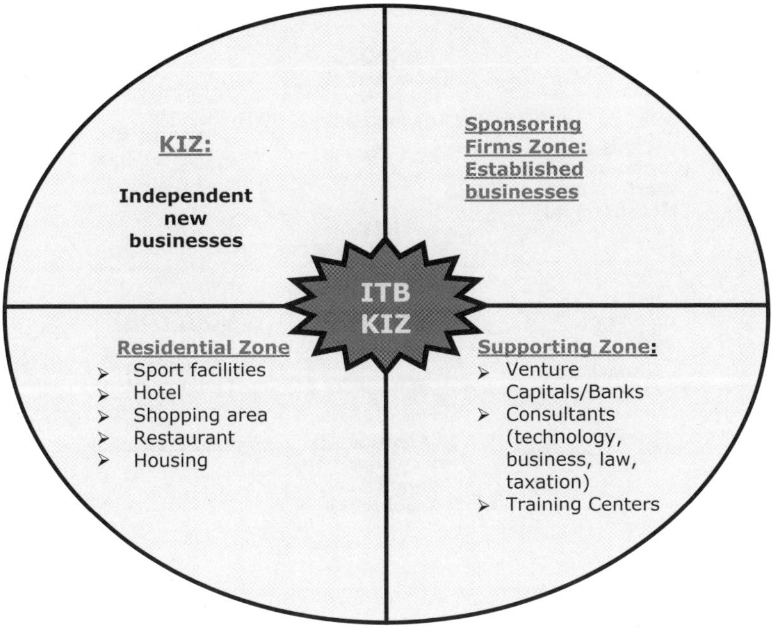

Figure 4: ITB Knowledge Innovation Zone (KIZ) concept: Greenest and cleanest industrial zone

5. KIM Framework: The Triple Helix Collaboration between Indonesian ABG (Academia-Business-Government)

5.1 *The Five Enablers of ABG Smart Collaboration in Indonesia*

Smart collaboration between ABG organizations could work if the society is mature enough. In the case of Indonesia, the immature state of Indonesian society as described earlier needs to a taken into account. Referring to the context of Indonesian macro and micro level state of development and the meaning of KIM for organization at micro and macro level, we can identify five enablers for KIM Framework between ABG organizations in Indonesia, as proposed by Lambe (2010) in Figure 5. Those five enablers should focus on its

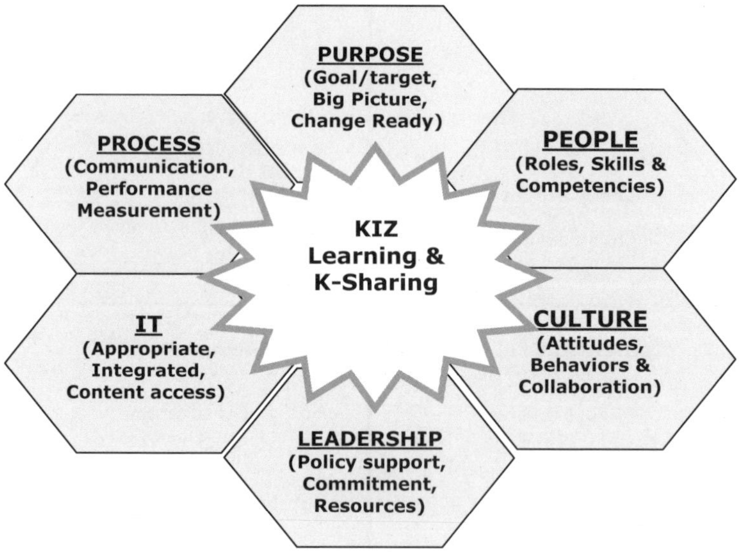

Figure 5: KIM framework for ABG collaboration (based on Lambe, 2005)

purpose, which is to achieve the goal and target. Purpose is defined as a consensus between all members of the ABG organizations, and describes the commitment of all organization for change.

The characteristics, roles and key activities that those enablers should have can be identified as follows:

i. People:

This is the main enabler that reflects the readiness of the members of the organization to play a role, willingness and ability to change, and specifically the quality of their skills and competence. The key aspects for development to enable knowledge sharing between members in ABG collaboration can be achieved by training with an emphasis on:

- Systematic problem solving skills;
- Experimenting with new methods;
- Learning from experiences, both from success stories and failures;
- Positive thinking and open-mindedness in listening to others

- Benchmarking from best practices
- Transforming and sharing knowledge in an effective and efficient manner

In order to serve those purposes, ABG collaboration should also formulate the mechanism — formally and informally — for effective sharing of knowledge, mental model and vision.

ii. Culture:

This enabler describes the attitude and behavior of the people in the ABG organization, who are ready for collaboration to realize successful KIM. Indonesian society can be shaped as a result of:

- *Trust*, between members of the ABG collaboration. Trust should serve as the foundation for sharing of vision, mental model and knowledge. Some of the traits to be achieved in gaining a high level of trust in an organization are: awareness of the role of each member; empathy; fighting spirit and confidence; open communication; honesty and friendship; ability to resolve conflict; adherence to work values and ethics; constructive criticism; positive thinking and patience under pressure.
- *Learning culture*, as the manifestation of values and trust into daily habits. This serves as the foundation for the attitude and behavior of members of ABG collaboration in perceiving the importance of knowledge sharing among them. Some of the traits to be achieved in gaining a high level of learning culture in an organization are: spirit for knowledge sharing; attentive and fondness to serve other members of the society; tolerance toward opinion differences and unintended mistakes or due to new innovation; spirit to learn and to change; putting common interests above oneself or group; positive thinking in order to focus on similarities than differences; reflect on personal as well as common vision and mental model.

iii. Leadership:

This enabler describes the roles of top leaders of ABG organization, who have the awareness and readiness to execute the policy for collaboration, as well as being fully involved in mobilizing resources. The maturity of the society can be achieved if the leaders are able to ignite the spirit and accelerate the birth of knowledge workers as human capital. Involvement of top leadership is an absolute prerequisite for successful implementation of KIM in ABG organization. Thus, the initiative to build learning organization in Indonesia must come from the top leadership. This is especially important given the condition of the majority of Indonesian society where the level of maturity is not high. Indonesian culture also holds a paternalistic behavior, where the role and personality of the top leadership is still very dominant, especially in setting organizational vision and strategy.

To lead an immature society, a leader must be willing to be hands-on and engage in a discussion with the members of the organization. This is especially important when it comes to discussing the future, defining the culture, and determining work plans as well as the indicators of success in the organization. In reference to one of the Indonesian founding fathers Ki Hajar Dewantoro, this is a type of leader who follows the motto "*ing madyo mangun karso*" (*building spirit from the middle*) and who integrates three traits of leadership:

- *Visionary leadership*, i.e., leaders who can provide clarity and spirit to drive the organization towards a common vision; motivate and build the sense of solidarity; build commitment and confidence for success; develop the willingness to learn and improve the competence and insight of the members.
- *Synergistic leadership*, i.e., leaders who can build openness, honesty and selflessness in collaboration; able to weigh and build synergy from differing opinions; build spirit and motivation to share knowledge; promote tolerance toward differing personalities; as well as able to understand *power and politics in the* organization.
- *Transformational leadership*, i.e., leaders who can be good listeners and trusted advisors; able to develop and empower member's competences and; able to approach with a good heart; promote

openness and honesty; become a role model for members to have ethics and moral commitment; and develop spirit to learn and improve competence.

The ultimate mission of the leadership enabler is to achieve the maturity state of all members of the organization. When this is achieved, the leadership type would evolve into one with focus on supporting from behind ("*tut wuri handayani*") as stated by Ki Hajar Dewantara). *Who should play the role of Leadership Enabler in the collaboration between organizations of ABG?*

Given the paternalistic character of the Indonesian people, the government would be in a better position to lead the smart collaboration among organizations of ABG. The government has a strategic position with its authority and access to power. However, many Indonesian bureaucrats are still trapped in politics, finding it difficult to rise above vested interests, unable to be selfless, visionary, synergistic and transformational as required. Hence Indonesian academia, especially ITB and other leading Indonesian universities, should take the role as pioneer in building smart ABG collaboration in Indonesia.

iv. Process:

This enabler is described as the process of collaboration which needs to be designed and maintained so that it is functioning well, especially with regard to connected information, supply of knowledge and work support. To ensure a smooth collaboration, an organizational system must be built as follows:

- *Develop communications and awareness programme*:
  Along the line of KIM role, communication and awareness programs are necessary for members to understand role and objective toward success of common KIM strategy.
- *Review rewards and recognition*:
  Collaboration of ABG organizations requires action plans that can be measured, including indicators, in order to stay within the common goals. Therefore, a performance measurement tool would be necessary, especially to reflect on the progress of every

plan/program. Gap analysis needs to be conducted and aggregate performance measures should be launched. This will provide an understanding of the current state of the collaboration to achieve its target based on a common vision.

- *Develop KIM evaluation framework*:
  Finally, ABG smart collaboration would be meaningful if the benefit can be continuously enjoyed and improved. Evaluation to monitor and provide an early warning system for program improvement to be sustainable is important, too.

v. Information Technology (IT):

This enabler is described as the technology that helps to facilitate the flow of knowledge, in order to be appropriate, integrated, and allow for easy access to content. IT functions as the tools for collaborative members to guide and facilitate the development of knowledge sharing culture, and finally, to build KBS (Knowledge Based Society).

The process of knowledge sharing is the key in smart ABG collaboration, whose effectiveness would be determined by openness and willingness of sharing amongst the members, as well as the quality of IT facility and system mechanism which can accelerate knowledge flows faster to be more accurate and transparent, so that individual knowledge can be institutionalized into common knowledge.

Moreover, ABG would work if there is a system that can guide and facilitate the development of a knowledge society and knowledge sharing culture. It should review the range of information platforms and seek to simplify and integrate the infrastructure through, e.g., information architecture and usability improvements, possible consolidation and migration of content to fewer platforms.

5.2 *The Role of KMSI (Knowledge Management Society Indonesia)*

Availability of expertise and experience in the implementation of KIM in Indonesia is still at the early stage in all ABG organizations. This shortcoming is more obvious when it comes to collaboration in implementing inter-organizational KIM with synergy. Therefore, the support of expertise from KMSI would be very helpful in the form of training, coaching and mentoring, as well as consultation for the design of KM organization, system, technology and tools according to the contextual needs.

Based on the KIM Framework above, and the role of ABG and KMSI, the following is Knowledge Innovation and Sharing Model for collaboration between organizations of ABG in Indonesia (Figure 6). This model of KIM ABG in Indonesia is developed from a combination of the KM models in Unilever, Sucofindo Indonesia and BP, with adjustment to the Indonesian context.

6. Conclusions

This conceptual paper attempted to answer the key research question of how a smart collaboration between ABG (Academia-Business-Government) in Indonesia can be developed and to find out what it

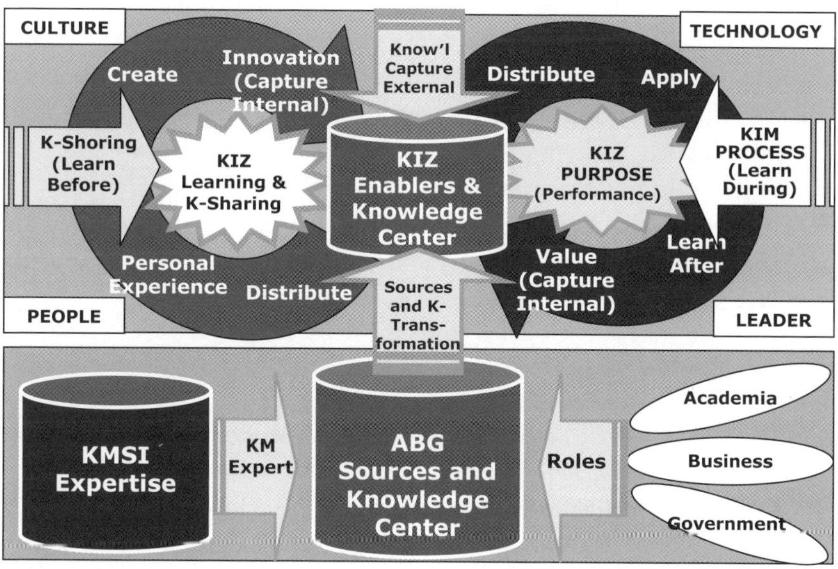

Figure 6: Knowledge innovation and sharing — ABG Indonesian model (Tjakraatmadja, 2007) based on the KM models of Unilever, Sucofindo and BP

takes in terms of framework and key success factors to enable such collaboration. Learning from its own experience of initiating and developing BHTV — as well as benchmarking of KM models in various businesses — and framing it in the macro and micro context of Indonesia, we argue that it is imperative for ITB as the nation's leading university being the pioneer in science, technology and arts to play a significant role in this collaboration. The proposed model by ITB is for a KIZ, where specific physical space will serve as the collaboration hub for the various ABG members.

It is evident that in the case of Indonesia where a paternalistic cultural pattern still holds, the difficulty lies on who should lead the collaboration. This kind of society puts high expectation on the government as the source of authority to take the lead. However, the current Indonesian political system — where democracy is flourishing but society has not reached full maturity — has not yet succeeded in creating the necessary governance. Bureaucrats are trapped in politics, which reduce its governance capacity, despite the strengthening elements of the private sector and civil society. Hence Indonesian academia, especially ITB and other leading Indonesian universities, should play a bigger role as pioneers in building smart ABG collaboration in Indonesia. Nevertheless, people would also remember well that it is the practitioners in the business community that can make this happen with their practical experience, resources, as well as technical and commercial competence.

Therefore, the key for success lies in the mutual respect and understanding among all entities in this smart ABG collaboration. Each has an important role to play to make this a success.

References

Amidon, D. M. (1998). The evolving community of knowledge practice: The ken awakening. *International Journal of Technology Management*, 16(1–3):45–63.

Amir, S. (2002). *Demokrasi dan Teknologi*. The Prospect. (http://www.theindonesianinstitute.org/gdemo02.htm).

Amir, S. (2004). The regime and the airplane: High technology and nationalism in Indonesia. *Bulletin of Science, Technology & Society*, 24(2):107–114.

Baker, K. A. and G. M. Badamshina (2003). Knowledge Management (http://www.science.doe.gov/sc-5/benchmark/Ch%205%20Knowledge%20Management%2006.10.02.pdf).

Bijker, W., T. Hughes, and T. Pinch (eds.) (1987). *The Social Construction of Technological Systems: New Directions in the Sociology and History of Technology.* Cambridge: MIT Press.

Campbell, N. D. (2000). *Using Women: Gender, Drug Policy, and Social Justice.* New York: Routledge.

de Geus, A. (1997). *The Living Company.* Harvard Business School Press.

Drucker (1969). *The Age of Discontinuity: Guidelines to Our Changing Society.* New York: Harper and Row.

Halal, W.E. (1998). Knowledge entrepreneur. *Executive Excellence*, 15(19):12–13.

Lambe, P. (2010). Collaboration and Competition: The knowledge research institute of Singapore as a model KM system. In *Governing and Managing Knowledge in Asia*, T. Menkhoff, H.-D. Evers, and Y. W. Chay (Eds.), pp. 267–280. Singapore: Word Scientific Publishing.

Leydesdorff, L. and H. Etzkowitz (1998). The triple helix as a model for innovation studies. *Science and Public Policy*, 25(3):195–203.

Leydesdorff, L. and M. Meyer (2003). The triple helix of university-industry-government relations. *Scientometrics*, 58(2):191–203.

Lim, M. (2003). *Civil Society, Public Sphere and the Internet in Indonesia*. Asia.com: Asia Encounters the Internet.

Nonaka, I. and H. Takeuchi (1995). *The Knowledge-Creating Company.* New York: Oxford University Press.

Sveiby, K. E. (1997). *The New Organizational Wealth: Managing and Measuring Knowledge-Based Assets.* San Francisco: Berrett-Koehler.

Tjakraatmadja, J. H. (1997). *Manajemen Teknologi.* Studio Manajemen TI-ITB.

Tjakraatmadja, J. H. (2001). *Manajemen Transformasi Pengetahuan dalam Organisasi Belajar* (Management of Knowledge Transformation in Learning Organization), PhD Dissertation, ITB.

Tjakraatmadja, J. H., F. M. Hartanto, and M. Aroef (2001). *Influences of Visionary Leadership and Transformation Learning Culture to Knowledge Transformation Process in Learning Organization Environment.* Global HRM Conference — Comparative HRM Learning from Diversity, Barcelona, Spain.

Tjakraatmadja, J. H. and D. C. Lantu (2006). *Knowledge Management in Learning Organization Context.* School of Business and Management — ITB.

Tjakraatmadja, J. H. (2007). *Knowledge Management dan Potensinya Untuk Membangun Ekonomi dan Masyarakat Berbasis Pengetahuan* (Knowledge Management and Its Potential to Build Knowledge Based Economy and Society), Pidato Ilmiah Majelis Guru Besar (Scientific Lecture before the Board of Professors), Institut Teknologi Bandung, 8 September.

Tjakraatmadja, J. H. and L. Martini (2008). *Knowledge Sharing in Academic Institution: Case Study at School of Business and Management — Institut Teknologi Bandung (SBM-ITB)*, Unpublished Master Thesis.

Tuomi, I. (1999). *Corporate Knowledge: Theory and Practice of Intelligent Organizations.* Helsinki: Meraxis.

Chapter

11

Achieving Knowledge Economy Status through Good Knowledge Governance

The Singapore KBE Story Revisited

Thomas Menkhoff, Hans-Dieter Evers, Chay Yue Wah and Solvay Gerke

1. Introduction

The effective governance of knowledge represents a key enabler of economic growth in an era where the creation of wealth from immaterial production (services, computer-assisted production etc.) exceeds that from material production (Drucker, 1994; Nonaka, 1994; Nonaka and Nishiguchi, 2001; Willke, 2007; Hornidge, 2007; Menkhoff *et al.*, 2010). The mega shift towards the increasing importance of intangible assets such as know how or R&D breakthroughs represents new challenges and opportunities for policy makers, development organizations and business leaders tasked with 'capacity development' and/or 'intellectual capital management' aimed at deriving benefits from talents, information and communication technologies (ICT), and global knowledge flows. This is evidenced by the increasing digital divide and impediments with regard to the implementation of sustainable knowledge-based strategies and systems at country, regional, and organizational levels.

One country which has successfully managed to leverage on knowledge is the Republic of Singapore, an island city-state located at the southern tip of the Malay Peninsula, south of the Malaysian state of Johor, and north of the Indonesian Riau Islands. With a population of about 5 million people it is one of the smallest countries in Southeast Asia. Colonised by the British during the 19th century, Singapore with its natural deep water port quickly became an important export outlet for Malayan primary products such as tin and rubber. After the Japanese occupation and the end of World War II, Singapore became a crown colony (1946), and was granted internal self-government in 1955. In 1963, it joined the Federation of Malaysia as an independent state from which it was expelled in 1965. When Singapore became fully independent in 1965, having no natural resources, it was faced with severe problems such as unemployment, labor unrest, poverty, poor housing conditions etc. As a consequence, the government, under the People's Action Party (PAP), embarked on a strategy of export-led growth driven largely by foreign multinational firms (MNCs) and government-linked companies (GLCs). During the last 40 years, fast-paced government-led industrialisation has created a robust and increasingly knowledge-based economy with a well-diversified manufacturing industry based on electronics, chemicals, mechanical engineering and biomedical sciences as well as a sophisticated service sector. Together with Hong Kong, South Korea and Taiwan, Singapore is one of the four original 'East Asian Tigers' (Menkhoff, 1998; Evers *et al.*, 2004).

2. The Importance of Good Knowledge Governance

The output of Singapore's emerging knowledge-based economy so far has been impressive as indicated by the increasing number of patents filed in part in cooperation with other countries, for instance, the USA (Figure 1). The number of patents filed increased from 1,096 in

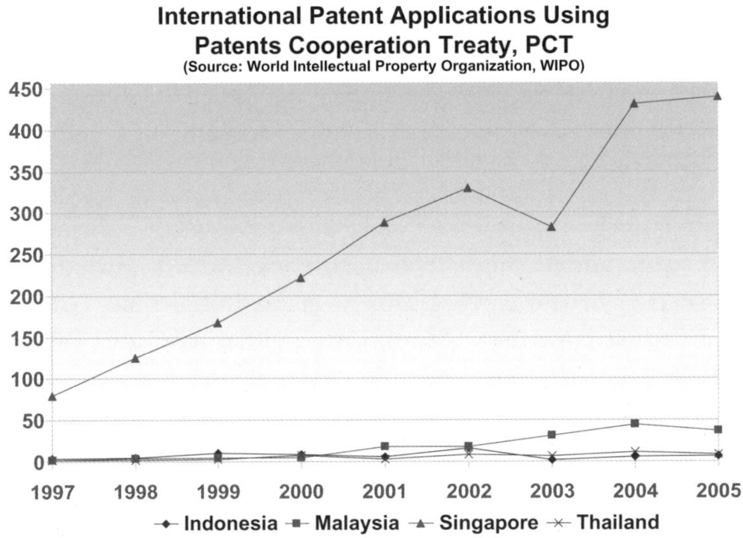

Figure 1: Number of PCT patent applications by Singapore, Malaysia, Thailand and Indonesia, 1997–2005

2001 to 1,739 in 2007. The number of patents awarded increased by 6 percent from 891 in 2006 to 941 in 2007. R&D expenditure per patent application increased from $2.7 million per patent application in 2006 to $3.6 million per patent application in 2007 (Menkhoff *et al.*, 2010:189–190). In 2006, R&D expenditure represented 2.31 percent of Singapore's GDP, in comparison with Germany's 2.51 percent, France (2.12 percent) and Korea (2.98 percent). Singapore is fast catching up with the OECD countries. In 2007, R&D as a percentage of GDP increased to 2.61 percent, a commendable ratio, compared to the US (2.62 percent in 2006). Singaporeans have more mobile phones per 1,000 inhabitants than countries such as South Korea or Malaysia (Evers *et al.*, 2004; Evers, 2010). It also has the highest number of PC owners per 1,000 inhabitants.

To shed some light on the question why Singapore has achieved such remarkable results, we propose two major hypotheses: (i) firstly, a premise which emphasizes the role of efficient knowledge governance in developing a knowledge-based economy and (ii) secondly, a supposition that emphasizes the importance of effectively managing diversity.

Proposition 1

> Singapore's capability to achieve the status of a knowledge-based economy is based on the ability to effectively apply and accumulate value-creating knowledge from various sources and to utilize such knowledge with great impact through good knowledge governance.

Singapore launched its state-led start into a knowledge society (based on high value added, high-tech industries and a knowledge-based service sector) in the early 1990s after an initial phase of export-oriented industrialization (Menkhoff, 1998). The idea of a rising k-society originated from academic circles in the USA and Japan in the 1950s and 1960s. This idea reached European nations around the beginning of the 1970s and the Asian tigers a decade later. In Germany, for example, the term 'information society' is mainly used as it has a long tradition in the political sphere which can be traced back to the translation of a Japanese report entitled "Japan's Technological Strategy" by the German Federal Ministry of Education and Science (1972); the term is also easier to convey to citizens. In Singapore, the terminological preference was 'knowledge-based economy', arguably justified by the island-nation's need to develop from 'third' to 'first world economy' based on export-led growth, access to foreign technology, capital, knowledge and a competent workforce.

In both Germany and Singapore the term knowledge society is defined procedurally, i.e., by programmes such as the 'Information Society Germany 2010 Action Plan' initiated by the German Federal Government or the 'Intelligent Nation 2015' Plan developed by Singapore's Infocomm Development Authority. While both countries place heavy emphasis on an ICT and knowledge infrastructure so as to foster economic prosperity, differences exist with regard to what kinds of knowledge topics are addressed and the defining elements of a k-society. Singapore today, in contrast to Germany, places more emphasis on fostering creativity and the development of creative industries while the traditional broad definition of knowledge in Germany (that allows for fine arts, craft and music education, critical discussion as well as time and space or personal development, hobbies and recreation) is slowly eroding due to the rising focus on knowledge and knowledge production (Hornidge, 2006).

Historically, Singapore's government-led drive to improve the knowledge and technology base of the economy can be traced back to the foundation of the National Science and Technology Board in 1991. In line with the national strategic clustering approach towards development, a new Strategic Economic Plan (1991) was launched to support the development of new value-added manufacturing and services thrusts. Two Science Parks were set up to facilitate the R&D activities of high-tech companies (Koh *et al.*, 2004; Menkhoff *et al.*, 2010). Various scholarship schemes were launched to train young scientists abroad. The National Information Infrastructure (NII) was further fine tuned and improved to leverage on broad-band technologies in line with Singapore's intelligent island vision.

The Asian financial crisis and increasing competition from China and India gave another strong impetus to concentrate on new, innovative technologies, science & technology as well as innovation to further diversify the knowledge base of Singapore's economy. During the last ten years, Singapore's planners in the powerful Economic Development Board (EDB) have continued to envision a broader defined knowledge-based economy as evidenced by the ten-year plan Industry 21 aimed at developing Singapore into a vibrant global hub of knowledge industries in manufacturing and services.

Several statutory boards responsible for the development of the country's knowledge and high-tech industrial base have been reengineered to adjust their organizational missions to the rapidly changing environment. An example is the National Science and Technology Board (NSTB) which was renamed the Agency for Science, Technology and Research (A*STAR) with its two research councils, The *Biomedical Research Council* and the *Science and Engineering Research Council*. These councils perform important functions to ensure that A*STAR's various organizational units support the various broader strategic objectives. The Biomedical Research Council, for example, oversees the development of core research capabilities within the various A*STAR research units such as the Genome Institute of Singapore (GIS). It promotes translational medicine and cross-disciplinary research in conjunction with the emphasis on human healthcare advancement and supports biomedical research in the wider scientific community (including public universities and hospitals). The Biomedical Research Council also takes care of human capital development in the biomedical sciences and promotes societal awareness of biomedical research.

Physically, Singapore's biomedical ambitions have led to the construction of the Biopolis complex which was completed in 2004. This "biomedical city" is situated at the Buona Vista Science Hub. Tenants include the Genome Institute of Singapore (GIS), the Singapore Institute of Molecular Biology, the BioTechnology Center, the Bioinformatics Institute, the Institute of Biomedical Engineering and other R&D organizations. Biotechnology represents one of the four pillars of a knowledge-based economy prioritized by the Singapore government. Besides industries in the areas of biotechnology/biological sciences, future growth is expected to be driven by microelectronics, robotics and artificial intelligence, information technology, laser technology and electro-optics, and communications technology. Next to the Biopolis, Singapore's R&D hub for Infocommunication Technology, Media and Physical Sciences and Engineering is emerging: Fusionopolis.

Decisive attempts have been made to further optimize the governance system of science and technology in line with the national vision to become a vibrant knowledge-based economy. In 2006, the new National Research Foundation (NRF) was set up as a department under the Prime Minister's Office to provide secretariat support to the Research, Innovation and Enterprise Council (RIEC), chaired by the Prime Minister; to coordinate the research of different agencies within the larger national framework in order to provide a coherent strategic overview and direction; to develop policies and plans to implement the five strategic thrusts for the national R&D agenda, and to implement national research, innovation and enterprise strategies approved by the RIEC, and to allocate funding to programmes that meet NRF's strategic objectives.

Singapore's national R&D Agenda is based on five strategic thrusts:

- To intensify national R&D spending to achieve 3 percent of GDP by 2010;
- To identify and invest in strategic areas of R&D;

- To fund a balance of basic and applied research within strategic areas;
- To provide resources and support to encourage private sector R&D; and
- To strengthen linkages between public and private sector R&D.

These thrusts reflect the need to increase the country's national expenditure on R&D so as to catch up with other innovative countries such as Finland and Sweden, to focus on key strategic clusters where Singapore has competitive advantages, e.g., electronics, chemicals, aerospace and logistics, marine engineering, and biomedical sciences, to achieve scientific excellence on the basis of a vibrant research ecosystem, to increase public sector research and synergies between R&D and industry (source: http://www.nrf.gov.sg).

Further strategic elements include the greater commercialization of know how and protection cum marketing of intellectual property and patents created, attracting and retaining foreign talent for strategic industries (e.g., digital media) and alignment of tertiary educational institutions with Singapore's long-term development needs. A key role is played by the International Academic Advisory Panel (IAAP) comprising eminent academic, research and corporate leaders. It is chaired by Dr. Tony Tan, Chairman of the National Research Foundation and provides broad guidance on the development of the university sector. Over the years Singapore's universities have become more autonomous. As corporate entities they now have their own board of trustees and management which enables them to decide their own strategic direction with greater autonomy over finances and human resources. These changes in how universities are managed are founded on the basis of the overall strategic direction and policies for the university sector as set by the Government.

Regular reviews of the higher education landscape ensure that the system is needs-based and relevant to the demands of industry. As a result of such reviews, a new institution was founded recently, the Singapore Institute of Applied Technology (SIAT). The measure is part of the so-called *Polytechnic-Foreign Specialised Institute* (poly-FSI) initiative aimed at providing more upgrading opportunities for polytechnic graduates to obtain industry-relevant and reputable degrees. SIAT was set up as a distinct entity within MOE and is responsible for planning, managing and implementing degree programmes offered by foreign universities in partnership with Singapore's five polytechnics. Participating foreign universities will award degrees in their own institution name. Strategic areas in which these new degree programmes will be offered include Applied Sciences and Engineering, Digital Media, and Health Sciences. Through SIAT, polytechnic graduates will have access to cost-effective, innovative full-time and part-time courses locally.

Another strategic project is the new Singapore University of Technology and Design (SUTD) which was established in 2010 in collaboration with MIT (USA) and Zhejiang University (PRCh). The aim of the SUTD is to advance knowledge and nurture "technically grounded leaders and innovators to serve societal needs" with particular emphasis on Design, through an integrated multi-disciplinary curriculum. SUTD's degree programmes will focus on Architecture and Sustainable Design, Engineering Product Development, Engineering Systems and Design, and Information Systems Technology and Design (http://www.su.edu.sg).

A key strategic initiative in building and sustaining the knowledge economy is the National Continuing Education and Training (CET) Masterplan, introduced in 2008 to develop a highly skilled, competent and competitive workforce across different and diverse industries. This National CET Masterplan was conceived for two important demographic shifts: (a) a highly qualified workforce of the future where about 60 percent of the resident workforce will have at least a diploma qualification by 2020 compared to 36 percent in 2007, and (b) nurturing emerging and growth industries by equipping Singaporeans with the skills for job opportunities in new growth industries, whether they are preparing for new jobs, changing careers, or acquiring new skills for their jobs.

2.1 *Critical Success Factors*

Key elements of Singapore's KBE strategy include heavy investment in information and communication technology (ICT) infrastructure, training programmes to increase computer literacy, Government support of R&D and R&D personnel, creation of research institutes in selected fields such as ICT, informatics, biotechnology etc. and gearing the education systems towards holistic education and the creation of centres of excellence. The Singapore story illustrates how important knowledge governance competencies are so that economic and developmental knowledge goals can be obtained (Menkhoff *et al.*, 2010; Willke, 2007). This is evidenced by the impact of the various development blueprints (e.g., IT 2000), organizational arrangements and regulatory bodies (e.g., Infocomm Development Authority of Singapore/IDA) which have transformed Singapore over the years from an entrepôt towards a modern knowledge-based economy (KBE).

Knowledge governance (Stehr, 2004; Willke, 2007; Menkhoff, Evers and Chay, 2010) is both a managerial-administrative process and a structure of authority relations; it involves the channelling of resources in building up knowledge management capabilities and improving the competitive advantage of a country in the world market by utilizing knowledge as a factor of production. Like the term 'good governance', good knowledge governance refers to the multiple tasks and capabilities of governmental units in embedding and creating appropriate institutions to leverage on global knowledge — from establishing high quality primary schools to the codification of national ICT plans aimed at creating an intelligent nation. Relevant governance factors in the Singapore case include:

- High steering power and efficiency of government, state and knowledge elites;
- Outstanding planning and organising competencies of units tasked with realizing national KBE goals. Examples include:
  - Strategic Economic Plan (1991) with cluster development goals for the manufacturing sector and services;
  - IT 2000 Plan (1992) aimed at developing national Internet-Broadband network;
  - Industry 21 (1999) with focusing on knowledge-intensive sectors and the various Science and Technology Plans (see Table above);
  - Intelligent Nation Masterplan iN2015 (see table above);

- Systematic creation of knowledge hubs, competence centres and centres of excellence;
- Development of knowledge clusters as learning regions;
- Transfer of knowledge to stakeholders in civil society, Government and business;
- Acquisition of knowledge through global producer networks;
- Use of glocal knowledge to build comparative advantages in niche areas.

2.2 *Networked Government*

Knowledge governance has come a long way in comparison to the emphasis on ICT and automation which prevailed during the early 1980s when the civil service computerization initiative was launched. The 'Civil Service Computerisation Programme' (1980–1999) is an important milestone in Singapore's journey towards an 'intelligent island' as indicated in Table 1. In 1980, the Civil Service Computerisation Group (CSCG) was established to improve the productivity of the civil service and to propel Singapore into the information age. In 1981, the National Computer Board (NCB) was set up to implement computerisation plans in the Civil Service, to

Table 1: Singapore's IT Masterplans (1980–2015)

Year	National Infocomm Plans	Infocomm Plans of Government
2010–2015	Next Gen NBN	
2006–2010	iN 2015	iGov2010
2003–2006	Connected Singapore	e-Government Action Plan II
2000–2003	Infocomm 21	e-Government Action Plan
1992–1999	IT2000	Civil Service Computerisation Programme (1980–1999)
1986–1991	National IT Plan	"
1980–1985	National Computerisation Plan	"

Source: The Straits Times, May 30, 2006, S8.

spearhead the development of information systems, to coordinate computer education and to promote the computer services industry. Reducing paperwork and automation were key goals during this phase. From 1985–1990, during the National Information Technology Plan (NITP), led by NCB, the focus shifted to the private sector and it was aimed at developing the IT industry and promoting the use of IT for productivity and competitiveness improvements in all economic sectors. In the 1990s, the launch of the 'IT 2000: The Intelligent Island' plan marked the beginning of the 3rd CSCP phase with its emphasis on IT, telecommunications infrastructure and sophisticated online and networking systems for Singapore's citizens (citizen-centric services).

In 1999, the Infocomm Development Authority of Singapore (IDA), a statutory board of the Singapore Government, was formed. This was a result of the merger of the National Computer Board (NCB) and the Telecommunication Authority of Singapore (TAS), driven by the growing convergence of information technology and telephony. In 2000, it launched the 'Infocomm 21' blueprint, a strategic plan to turn Singapore into a global infocomm capital for infocomm industries, research and development, venture capital, intellectual capital, education and thought leadership, as well as an international showcase and test-bed for infocomm applications and services.

Infocomm 21 has since evolved into 'Connected Singapore' so as to develop new sources of growth, including areas involving creative input, such as design and the arts: "Connected Singapore sees infocomm technology as a true connector, bringing together the power of computing, communications and content, to create new business opportunities, consumer value and cultural experiences" (Straits Times 30/5/2006, S8). 'Infocomm 21' and 'Connected Singapore' were instrumental in realizing the ambitious goals of the e-Government Action Plans I+II with its emphasis on providing effective public e-services. Currently efforts are underway to deploy the Next Generation Nationwide Broadband Network (in short: Next Gen NBN) by 2012 with broadband speeds from 100 Mbps to users in business, homes and schools as part of the iN2015 blueprint to drive Singapore's transition into a global knowledge city.

Key Strategies of Singapore's E-Government Action Plan

- Electronic service delivery
- Service-wide infrastructure
- Knowledge-based workplace
- Infocomm education
- Technology experimentation
- ICT governance and management

Singaporeans have access to more than 1,500 e-services with the ease of a mouse-click. One of the challenges in this area is in the delivery of quality services that the public values. As a consequence, government agencies collaborate across agency boundaries with the aim of providing citizens, businesses and other stakeholders with a seamless online experience when interacting with them.

3. The Importance of Robust Knowledge Cultures and Diversity

Proposition 2

Our second proposition is as follows:

> The development of robust organisational knowledge cultures and the capability to leverage on diversity are preconditions to create new knowledge.

As the literature about technology and innovation management suggests, the development of robust knowledge cultures (often comprising culturally diverse knowledge workers) is a precondition for achieving new value creation in an era of globalisation (Trompenaars and Hampden-Turner, 1997; Powell, 1998; Knorr-Cetina, 1999; Chow *et al.*, 1999; Koh *et al.*, 2004). As discussed by Evers *et al.* in chapter 1 of this volume, a spatial and/or regional concentration of firms, suppliers, research institutes and universities as well as other knowledge relevant institutions is a preconditon for creating new knowledge resources. Important in this context is the variety, i.e., the diversity of participating organisations and knowledge producers who need to supplement and also compete with each other on the basis of *coopetitive* relationships, i.e., a mix of competition and cooperation (Evers, Gerke and Menkhoff, 2006:8; Evers, 2008; Zouhaer M'Chirgui, 2005).

From the perspective of managers who have to deal with such processes, there are a couple of challenges because they do not always fully appreciate the unique features and characteristics of knowledge organisations and experts (for example, when colleagues belong to different ethnic groups). The production of new knowledge is not always based on rational action. Knowledge workers have certain behavioral traits, cultural values, customs, preferences and emotions. They have unique frames of reference such as a strong interest in their own career or the wish to be autonomous which can make managerial interventions with regard to leadership, innovation management or productivity improvement difficult. Sometimes there are insufficient cross-linkages between experts in an organisation due to territorial patterns and knowledge management ignorance which can prevent synergetic innovation qua interdisciplinary exchange.

Knowledge organisations have a specific culture of knowledge production which can be positive or negative. While an adequate technical equipment of offices, workshops, labs etc. is of great importance, efficient innovation and value creation processes depend on the existance of non-technical enablers such as (relative) discursive freedom, a certain risk orientation, the ability to come up with new ideas and an authentic, stimulating environment.

3.1 *What Makes Robust Knowledge Cultures Tick?*

As our studies in Singapore have shown (Chay *et al.*, 2009; Evers and Menkhoff, 2008; Menkhoff *et al.*, 2010), there are several factors which play a role in creating robust knowledge cultures:

- Ability to leverage on 'synergetic diversity effects' in order to innovate;
- Personality traits and competencies of knowledge workers (e.g., openness, intercultural empathy);
- Competition over resources and/or recognition but no open conflicts;

- Stringent rules of conduct but no undue regulation;
- Strong internal cooperation, diversity of knowledge producers and interdisciplinary knowledge exchange.

Singapore provides an interesting empirical basis for related studies because the cultural knowledge diversity in local organisations is relatively high due to the large number of foreign talents (ca. 80 percent of the knowledge workers in the government-linked R&D sector are foreigners). An interesting example is the Biopolis research team of Dr. Alan Coleman, a former colleague of Professor Ian Wilmut who was instrumental in cloning Dolly the sheep. He is now working in Singapore, in "ES Cell International (ESI)", a research and production company operating in Biopolis specializing on stem cells. His vision is to find a medicine for diabetes through the "engineering" of insulin producing stem cells. Another example is the team of Professor Axel Ullrich (Director, Molecular Biology, Max-Planck-Institute for Biochemistry) who is in charge of the Singaporean "Onco Genome Laboratory".

Usually, these team leaders are obliged to integrate local specialists into their teams in order to ensure knowledge transfer and learning. In this context, an interesting question which requires more empirical research emerges: How can a robust knowledge culture be created in multi-cultural teams? Besides strategic knowledge alliances, value creation in teams and knowledge-based organisations require knowledge sharing (Rulke and Zaheer, 2000). Our research has shown that the sharing of knowledge is of strategic importance for the creation of new knowledge and that success or failure of achieving economic goals depends on the respective epistemic culture (Menkhoff *et al.*, 2010). Institutions with low communication thresholds such as the German Fraunhofer institutes facilitate the transfer of experience-based knowledge from R&D to the private sector. Organisations without a sense of shared agreements on what the issues are, run the risk of foregoing synergetic value creation opportunities due to silos and/or latent prejudices of their members towards people who do not belong to the ingroup due to their ethnicity or other reasons such as mistrust (Tjosvold *et al.*, 2000;

Hutchings and Michailova, 2004). Especially in plural societies and/or ethnically heterogenous organisations, this can become a serious challenge. In Singapore such issues are tackled head on qua corporate training measures with a focus on multi-cultural awareness building.

3.2 *Social Capital as Catalyst of Knowledge Transfer*

What are the driving forces behind knowledge transfer processes? Our research suggests that social capital is a key driver (Lin, 2001; Yli-Renko *et al.*, 2001). Bourdieu defines social capital as "the aggregate of the actual or potential resources which are linked to possession of a durable network or more or less institutionalized relationships of mutual acquaintance or recognition" (1985:248). This definition focuses on the benefits of this resource. In our context this could mean that knowledge workers are motivated to cooperate with others on the basis of more intensive, trustworthy relationships so as to get access to other resources (Gupta and Govindarajan, 2000; Gefen, 2000; Hutchings and Michailova, 2004; Tan *et al.*, 2005; Joshi and Sarker, 2006). As we have further elaborated in Menkhoff *et al.* (2009), we interpret social capital and its various dimensions as a catalyst for knowledge exchange. The following factors can be distinguished:

1. Organisationel factors such as care and concern, incentives, recognition and so forth;
2. Individual motives in terms of pro-social values or impression management;
3. Relational dimension (competency and openess).

Organisational Factors. The organizational climate of an organisation plays an important role which can either break or make positive relationships between members. Among the most important facets of this dimension are *organizational care* (Krogh, 1998; Krogh *et al.*, 2001) experienced by employees. Contrary to conditions of low-care, high-care environments usually facilitate social exchange which open up new opportunities for interaction and knowledge creation. Furthermore, *recognition and rewards* (Bartol and Srivastava, 2002) for sharing in action can have a positive impact upon performance and knowledge outcomes.

Individual Motivation. The agency dimension of social capital examines the role of individual motives in engaging in social interactions that would enable them to acquire the resources available in such interactions (Archer, 2003; Cicourel, 1973). What motivates donors who are requested to make certain assets available without any immediate return? Among the key facets identified to explain motives in this dimension are *prosocial motives* (Rioux and Penner, 2001), *impression management*, *altruism* (Jensen, 1998; Conte and Paolucci, 2002), and *shared values* (Cicourel, 1973).

Relational dimension. In this essay we look at the relational dimension of social capital though the concept of relational embeddedness, which has been described by Granovetter (1992) as the kind of personal relationships people have developed with one another through a history of interactions. This concept focuses on the building of trust into the relations individuals have that influence their behavior (Putnam, 1993; Fukuyama, 1996, 1999; Cohen and Prusak, 2001). Among the key facets of this dimension are *competence* (Blau, 1964; Schurr and Ozanne, 1985), *integrity* (Hosmer, 1995; Luhmann, 1979) and *open-mindedness* (Tjosvold, Hui and Sun, 2000).

Which of these factors are decisive and what are the implications? Our research findings suggest that contemporary organizations, which engage in knowledge-intensive and knowledge-generating activities, need to institute an environment conducive to the development of all three dimensions of social capital in order for effective knowledge sharing to take place. Particular emphasis needs to be placed on organizational climate variables such as recognition and rewards, which turned out to be very critical predictors of knowledge sharing.

As we showed in our research (Chay *et al.*, 2009), the structural dimension of social capital matters and so does the relational dimension. The criticality of open-mindedness as another predictor of knowledge sharing implies that organizations need to implement proper recruitment and screening processes so as to attract a particular type of person who has the required demographic and personality traits (Truch *et al.*, 2004) which may make sharing easier. Voluntary interactions between human actors aimed at exchanging information and experiences often occur when people are comfortable with each other, e.g., social similarities. The plausible assumption that personal compatibility predicts knowledge sharing will have to be examined in a future study.

Organizational members consider the possible costs of knowledge sharing and hoarding very carefully before they act. Prosocial motives or altruism do not matter much in the context of our sample which might be a function of the fact that many of the respondents were highly qualified knowledge workers who are known to have a unique orientation (e.g., they are loyal to their own profession but not necessarily to their employer). Individuals who are highly competent in their work abilities turned out to be less likely to share what they know (in contrast to individuals who are low on competency) when they perceive that there are few rewards or when sharing is not recognized by the organization.

Overall, our research provides evidence for the importance of an effective performance management system as well as authentic places for social interactions if an organization wants to successfully manage the transition from a 'knowledge is power culture' to a high-performing organization where knowledge sharing is seen as a key enabler of improved business performance, innovation and value creation (Nonaka *et al.*, 2001).

4. Knowledge Sharing Challenges in Multicultural Societies

In multi-cultural societies additional problems can arise because behavioral patterns and cooperation motives of knowledge workers as well as possible barriers can not always be fully anticipated and understood. As indicated above, there is a shortage of studies which examine the impact of inter-cultural peculiarities and differences with regard to national cultural dimensions of science and technology (S&T) on the one hand and knowledge sharing processes on the other. The following table refers to very broad differences between "Asian" and "Non-Asian" cultures with regard to S&T as discussed by experts of the Word Bank (http://info.worldbank.org/etools/docs/library/238718/1.1%20JeanEric%20Aubert.pdf).

The 'Asian' way

- Technology followers (technology transfer through and licenses)
- Limited technology and innovation capacity (largely product-oriented innovation)
- Relationship-based systems/use of diasporas
- Communitarian societies
- Increasing investments in ICT and local innovation capabilities (gradual)
- Relatively weak domestic SMEs

The 'Western' way

- Technology leaders
- Science-based innovations through systematic research & development
- Relatively high level of system trust (rule of law)
- Strong indigenous S&T landscapes
- Individualist societies
- Technologically strong SMEs

Research in Southeast Asia has shown that the effective management of inter-cultural differences is the key for the development of sustainable knowledge structures (for example the establishment of a multi-cultural "community of interest" in the area of vocational training aimed at adjusting curricula and syllabi to new goal dimensions). Especially during the synergetic cooperation of experts from different national cultural traditions such as foreign experts from individualistic countries and their counterparts from more collectivistic societies, knowledge traps can occur which can theoretically hamper the achievement of project goals and mutual learning. As Berrel *et al.* (2001:88–89) have stressed in their analysis of a Malaysian-Australian joint venture in higher education, there are several problems at the intersection of both cultures which they conceptualise as "different ways of knowing":

> "Different ways of knowing manifested themselves in a variety of organisational contexts. For example, while the unequal distribution of power and paternalistic management styles sat comfortably with Malay sensibilities, Australian managers less readily accepted power differentials and domineering styles of management within the organization... Other manifestations of the ways of knowing were observed in the positive attitudes of Australian managers towards recognizing achievements as well as the rights of the individual, views that are supported by universalistic social philosophies. In contrast, Malay managers extolled collective action and the rights of the group, views motivated more by particularistic social philosophies... This distinctly homogenous group felt more comfortable among their compatriots. In contrast, the low-context and heterogeneous group of Australian managers appeared more unconstrained in their behavior, discriminating less between insiders and outsiders."

The quotation implies that there are significant country-specific differences with regard to work and organizationally-relevant values (defined as semi-conscious feelings developed during childhood about what is right and what is wrong) as discussed by Hofstede *et al.* While members of individualistic cultures tolerate that people pursue their own personal interests, members of collectivistic cultures do not always appreciate such behavior due to emphasis placed on the importance of the group, group goals and group well-being (Apfelthaler, 1999:54).

As these statements show, it is important in the context of future empirical studies to examine the antecedents and critical success factors of knowledge sharing processes in multi-cultural contexts. Such an approach would allow the evaluation of the synergetic potential of cultural diversity in the context of efficient value creation and innovation processes. A possible hypothesis could be that members of groups with a collectivistic orientation, due to shared values and reciprocity pressure, tend to share knowledge resources with other members more easily while individualistic actors might perceive that "knowledge hoarding" is the way to go in developing their careers. That would point to a deep gap which would require good inter-cultural management on the basis of inter-cultural cooperation competence.

The overall positive experiences of international firms in Singapore imply that the tiny city-state has successfully managed knowledge governance challenges. This positive experience is a result of the know how transfer and learning as well as the achievements of the Singapore government in leveraging on global knowledge to achieve local development goals in education, business and the public sector (Toh *et al.*, 2002; Hornidge, 2007).

5. Conclusion

In this essay, we have attributed Singapore's knowledge economy (KBE) success to the power of good knowledge governance and the ability to effectively draw upon the diversity of knowledge producers and knowledge consumers in a multi-cultural, globalised environment. By now Singapore has a very well developed knowledge infrastructure in terms of ICT, research institutes and diverse knowledge workforce. Considerable research is being conducted by scientists and researchers in institutions of higher learning and research centers in line with the blueprints of the Singapore government aimed at staying economically competitive in the knowledge-based economy era. Given the rise of both China and India, Singapore has to derive its growth increasingly from the production, dissemination and application of knowledge.

Besides its academic value, the Singapore story sends out a powerful message to policy-makers as it draws attention to the significant role of visionary knowledge elites as drivers of sustainable development (which are not always in adequate supply if one compares the different development trajectories of Asian countries). The continuous legitimacy of Singapore's strategic groups around the People's Action Party (PAP), government planning agencies and technology conglomerates is based on many factors such as their track record, competencies, command over resources as well as knowledge and evidence-based (good) management. The official discourse about the urgent need to build up and to attract new growth drivers such as the biomedical sciences or digital media in view of the rapidly changing environment and the emergence of new competitors makes a lot of sense. It seems that there is no other alternative development path, i.e., respective policy measures are perceived as rational, legitimate and value-added. Singapore's survival and national interests serve as ideological justifications of KBE-related policy measures and provide meaning for those in charge and Singapore's citizens. This raises several questions one of which is, who is supposed to assist Asian latecomers in knowledge-based development in an era of globalization, regional integration and latent knowledge conflicts. In these turbulent times of increasing ignorance, where an 'explosion of despair' beyond the shores of Singapore can not be ruled out, one might argue that knowledge for development (K4D in short) assistance is of utmost importance to ensure further growth and stability in Asia.

References

Apfelthaler, G. (1999). *Interkulturelles Management.* Köln-München-Wien: Manz Verlag/Fortis.

Archer, M. S. (2003). *Structure, Agency, and the Internal Conversation.* Cambridge and New York: Cambridge University Press.

Bartol, K. M. and A. Srivastava (2002). Encouraging knowledge sharing: The role of organizational reward systems. *Journal of Leadership and Organizational Studies,* 9(1):64–76.

Berrel, M., M. Glöt, and P. Wright (2002). Organisational learning in international joint ventures — Implications for management development. *The Journal of Management Development,* 21(2):83–100.

Bourdieu, P. (1985). The Forms of Capital. In *Handbook of Theory and Research for the Sociology of Education,* J. G. Richardson (ed.), pp. 241–258. New York: Greenwood.

Blau, P. (1964). *Exchange and Power in Social Life.* New York: Wiley.

Chay, Y. W., T. Menkhoff, B. Loh, and H.-D. Evers (2009). Social capital and knowledge sharing in knowledge-based organizations: An empirical study. In *Knowledge Management, Organizational Memory and Transfer Behavior: Global Approaches and Advancements,* Murray E. Jennex (ed.). Hershey, PA, USA: IGI Global.

Chow, C. W., G. L. Harrison, J. L. McKinnon, and A. Wu (1999). Cultural influences on informal information sharing in Chinese and Anglo-American organizations: An exploratory study. *Accounting, Organizations and Society*, 24:561–582.

Cicourel, A. V. (1973). *Cognitive Sociology: Language and Meaning in Social Interaction.* Harmondsworth: Penguin.

Cohen, D. and L. Prusak (2001). *In Good Company: How Social Capital Makes Organizations Work.* Boston: Harvard Business School Press.

Conte, R. and M. Paolucci (2002). *Reputation in Artificial Societies: Social Beliefs for Social Order.* Boston: Kluwer Academic Press.

Davenport, T. H. and L. Prusak (1998). *Working Knowledge: How Organizations Manage What They Know.* Cambridge, MA: Harvard Business School Press.

Drucker, P. F. (1994). *Postcapitalist Society.* New York: Harper Business.

Evers, H.-D. (2010). Transition towards a knowledge society: Malaysia and Indonesia in global perspective. In *Governing and Managing Knowledge in Asia*, T. Menkhoff, H.-D. Evers, and Y. W. Chay (Eds.), pp. 93–115. Singapore: World Scientific Publishing.

Evers, H.-D. (2008). Knowledge hubs and knowledge clusters: Designing a knowledge architecture for development. ZEF working paper series 27. Bonn: Center for development research (ZEF), University of Bonn.

Evers, H.-D. and T. Menkhoff (2008). Strategic groups in a knowledge society: Knowledge elites as drivers of biotechnological development in Singapore. In *The Straits of Malacca: Knowledge and Diversity. ZEF Development Studies* Vol. 8, S. Gerke, H.-D. Evers, and A.-K. Hornidge (Eds.), pp. 85–98. Berlin, London, Penang: LIT Verlag.

Evers, H.-D., S. Gerke, and T. Menkhoff (2006). Knowledge and development — Strategies for building a knowledge society. ZEF Policy Brief No. 6, Centre for development research, Dept of political and cultural change, University of Bonn (April).

Evers, H.-D., S. Gerke, and R. Schweisshelm (2004). Malaysia, Singapur, Indonesien: Wege zur Wissensgesellschaft. Southeast Asian studies working paper No. 20, University of Bonn.

Evers, H.-D., M. Kaiser *et al.* (2003). Entwicklung durch wissen — Eine neue globale wissensarchitektur. In *Entwicklungszusammenarbeit in der Weltgesellschaft*, M. Kaiser (ed.), pp. 43–78. Bielefeld: Transcript Verlag.

Fukuyama, F. (1996). *Trust: The Social Virtues and the Creation of Prosperity.* London: Penguin.

Fukuyama, F. (1999). *The Great Disruption: Human Nature and the Reconstitution of Social Order.* London: Profile.

Gefen, D. (2000). Lessons learnt from the successful adoption of an ERP: The central role of trust. In *Decision Making: Recent Developments and Worldwide Applications*, S. H. Zanakis, G. Doukidis, and C. Zopounidis (Eds.), pp. 17–30. Dordrecht & Boston, MA.: Kluwer Academic Publishers.

Gerke, S., H.-D. Evers, and R. Schweisshelm (2005). Wissen als produktionsfaktor: Südostasiens aufbruch zur wissensgesellschaft. *Soziale Welt*, 55:39–52.

Gerke, S., H.-D. Evers, and A.-K. Hornidge (eds.) (2008). *The Straits of Malacca: Knowledge and Diversity*. Berlin, London, Penang: LIT Verlag/Straits G.T.

Goffman, E. (1969). *The Presentation of Self in Everyday Life*. London: Allen Lane.

Granovetter, M. S. (1985). Economic action and social structure: The problem of embeddedness. *American Journal of Sociology*, 91(3):481–510.

Gupta, A. K. and V. Govindarajan (2000). Knowledge management's social dimension: Lessons from nucor steel. *Sloan Management Review*, 42(1): 71–80.

Hofstede, G. (1980). *Culture's Consequences: International Differences in Work-Related Values*. Beverly Hills, CA: Sage.

Hofstede, G. (2001). *Culture's Consequences: Comparing Values, Behaviors, Institutions and Organizations Across Nations* (2nd Edition). Thousand Oaks, CA: Sage.

Hornidge, A.-K. (2007). Deutschland und Singapur: Zielgerade wissensgesellschaft. von überraschenden parallelen und unterschieden. *Südostasien*, 1:53–56.

Hornidge, A. K. (2006). Knowledge society. Vision and social contruction of reality in Germany and Singapore. ZEF development studies, Vol. 3. Münster: LIT-Verlag.

Hosmer, L. T. (1995). Trust: The connecting link between organizational theory and philosophical ethics. *Academy of Management Review*, 20(2):379–403.

Hutchings, K. and S. Michailova (2004). Facilitating knowledge sharing in Russian and Chinese subsidiaries: The role of personal networks and group membership. *Journal of Knowledge Management*, 8(2):1367–3270.

Jensen, M. C. (1998). *Foundations of Organizational Strategy*. Harvard, MA: Harvard University Press.

Joshi, K. D. and S. Sarker (2006). Examining the role of knowledge, source, recipiwnt, relational, and situational context on knowledge transfer among face-to-face ISD teams. Paper presented at the 39th Annual Hawaii international conference on system sciences (HICSS-39), 4–7 January 2006 (published in the HICSS-39 2006 Conference Proceedings ed by Ralph H. Sprague).

Knorr-Cetina, K. (1999). *Epistemic Cultures: How the Sciences Make Knowledge*. Cambridge, Harvard University Press.

Koh, F., W. Koh, and T. Tschang (2004). An analytical framework for science parks and technology districts with an application to Singapore. *Journal of Business Venturing*, 20(2):217–239.

Lin, N. (2001). Building a network theory of social capital. In *Social Capital: Theory and Research*, N. Lin, K. Cook and R. S. Burt (Eds.), pp. 3–29. New York: Aldine De Gruyter.

Luhmann, N. (1979). *Trust and Power*. London: John Wiley & Sons.

Menkhoff, T. (1998). Staat, markt and modernisierung — The Singapore experience. In *Länderbericht China — Politik, Wirtschaft und Gesellschaft im chinesischen Kulturraum. Schriftenreihe der Bundeszentrale fuer politische Bildung*, No. 351, H.-P. Carsten and M. Lackner (Eds.), pp. 240–257. Bonn, Germany.

Menkhoff, T., H.-D. Evers, and Y. W. Chay (eds.) (2010). *Governing and Managing Knowledge in Asia*. New Jersey and London: World Scientific Publishing (2nd edition).

Menkhoff, T. and S. Gerke (2002). *Chinese Entrepreneurship and Asian Business Networks*. London and New York: RoutledgeCurzon.

Menkhoff, T., W. L. Tan, and Y. W. Chay (2007). The effects of entrepreneurial growth orientation on organizational change and firm growth. *Small Enterprise Research*, 15(2):88–99.

Menkhoff, T., Y. W. Chay, H.-D. Evers, and B. Loh (2008). Leadership in knowledge sharing: Creating value through collaboration. *Journal of Asian Business*, 23:265–282.

Menkhoff, T., E. F. Pang, and H.-D. Evers (eds.) (2008). The power of knowing: Studies of Chinese business in Asia. *Special Issue of the Journal of Asian Business*, 22 & 23.

Menkhoff, T., H.-D. Evers, M. Meyer, and L. Lim Meng Huat (2010). Building vibrant science and technology parks with knowledge management — Trends in Singapore. In *Governing and Managing Knowledge in Asia, Series on Innovation and Knowledge Management*, Vol. 9, T. Menkhoff, H.-D. Evers, and Y. W. Chay (Eds.), pp. 187–219. New Jersey: World Scientific Publishing.

Nahapiet, J. and S. Ghoshal (1998). Social capital, intellectual capital, and the organizational advantage. *Academy of Management Review*, 23(2):242–266.

Nonaka, I. (1994). A dynamic theory of organizational knowledge creation. *Organization Science*, 5:14–37.

Nonaka, I., N. Konno, and R. Toyama (2001). Emergence of 'Ba': A conceptual framework for the continuous and self-transcending process of knowledge creation. In *Knowledge Emergence: Social, Technical, and Evolutionary Dimensions of Knowledge Creation*, I. Nonaka and T. Nishiguchi (Eds.), pp. 13–29. Oxford, New York: Oxford University Press.

Persaud, A. (2001). The knowledge gap. *Foreign Affairs*, 80:107–117.

Porter, M. (1990). *The Competitive Advantage of Nations*. New York: Free Press.

Powell, W. W. (1998). Learning from collaboration: Knowledge and networks in the biotechnology and pharmaceutical industries. *California Management Review*, 40(3):223–240.

Putnam, R. D. (1993). *Making Democracy Work*. Princeton, NJ: Princeton University Press.

Putnam, R.D. (1995). Bowling alone: America's declining social capital. *Journal of Democracy*, 6(1):65–78.

Rioux, S. and L. A. Penner (2001). The causes of organizational citizenship behavior: A motivational analysis. *Journal of Applied Psychology*, 86(6): 1303–1314.

Rulke, D. L. and S. Zaheer (2000). Shared and unshared transactive knowledge in complex organizations: An exploratory study. In *Organizational Cognition: Computation and Interpretation*, Z. Shapira and T. Lant (Eds.), pp. 83–100. Mahwah, NJ: Lawrence Erlbaum.

Schurr, P. H. and J. L. Ozanne (1985). Influences on exchange processes: Buyers' preconceptions of a seller's trustworthiness and bargaining toughness. *Journal of Consumer Research*, 11(4):939–953.

Scott, W. R. (1995). *Institutions and Organisations.* Thousand Oaks: Sage.

Stehr, N. (1994). *Knowledge Societies.* London: Sage.

Stehr, N. (2001). Wissen und Wirtschaften. Die gesellschaftlichen Grundlagen der modernen Ökonomie. Frankfurt am Main: Suhrkamp Verlag.

Stehr, N. (ed.) (2004). *The Governance of Knowledge.* London and New Brunswick: Transaction Publishers.

Subramaniam, S. (2001). The dual narrative of "Good Governance": Lessons for understanding political and cultural change in Malaysia and Singapore. *Contemporary Southeast Asia*, 23.

Szulanski, G. and R. Cappetta (2003). Stickiness: Conceptualizing, measuring, and predicting difficulties in the transfer of knowledge within organizations. In *The Blackwell Handbook of Organizational Learning and Knowledge Management*, M. Easterby-Smith and M. A. Lyles (Eds.), pp. 513–534. Malden, MA and Oxford: Blackwell.

Tjosvold, D., C. Hui, and H. Sun (2000). Social face and open-mindedness: constructive conflict in Asia. In *Asian Management Matters: Regional Relevance and Global Impact*, C. M. Lau *et al.* (Eds.). London: Imperial College Press.

Triandis, H. C. and M. J. Gelfand (1998). Converging measurement of horizontal and vertical individualism and collectivism. *Journal of Personality and Social Psychology*, 74(1):118–128.

Triandis, H. C. (1988). Collectivism vs. individualism: A reconceptualization of a basic concept in cross-cultural psychology. In *Cross-Cultural Studies of Personality, Attitudes and Cognition*, G. Verna and C. Bageley (Eds.), pp. 60–95. New York: St Martins Press.

Trompenaars, F. and C. Hampden-Turner (1997). *Riding the Waves of Culture — Understanding Diversity in Global Business.* Nicholas Brealey Publishing.

Truch, A., D. Batram, and M. Higgs (2004). Personality and knowledge sharing. In *Leveraging Corporate Knowledge*, E. Truch (ed.), pp. 131–143. Aldershot: Gower.

Von Krogh, G. V. (1998). Care in knowledge creation. *California Management Review*, 40(3):133–153.

Von Krogh, G. V., K. Ichijo, and I. Nonaka (2001). Bringing care into knowledge development of business organizations. In *Knowledge Emergence: Social, Technical, and Evolutionary Dimensions of Knowledge Creation*, I. Nonaka and T. Nishiguchi (Eds.), pp. 30–52. Oxford and New York: Oxford University Press.

World Bank (1999). *Knowledge for Development: World Development Report 1998/99.* New York: Oxford University Press.

Willke, H. (2007). *Smart Governance. Governing the Global Knowledge Society.* Frankfurt/New York: Campus.

Yli-Renko, H., E. Autio, and H. J. Sapienza (2001). Social capital, knowledge acquisition, and knowledge exploitation in young technology-based firms. *Strategic Management Journal*, 22:587–613.

Zeleny, M. (2010). Knowledge of enterprise: Knowledge management or knowledge technology? In *Governing and Managing Knowledge in Asia. Series on Innovation and Knowledge Management*, Vol. 9, T. Menkhoff, H.-D. Evers, and Y. W. Chay (Eds.), pp. 23–57. New Jersey and London: World Scientific Publishing.

Zouhaïer M'Chirgui (2005). The economics of the smart card industry: Towards coopetitive strategies. *Economics of Innovation and New Technology*, 14(6): 455–477.

Chapter

12

Knowledge for Poverty Alleviation

A Framework for Developing Intangible Assets of Low-Income Communities

Serafin D. Talisayon and Jasmin P. Suministrado[1]

1. Introduction

Two compelling and recent paradigms, and two corresponding discourses, have emerged and developed quite separately during the last two decades: ***knowledge-based management*** (KBM) and ***sustainable development*** (SD). The term "knowledge-based management" is used here instead of the more commonly-used term "knowledge management" (KM) because there are valid criticisms that it is not

[1] The authors wish to acknowledge Daan Boom, former Head of the ADB Knowledge Center, and Maria Socorro Afable, Managing Director of CCLFI. Philippines, for their valuable comments and suggestions during the preparation of this paper.

knowledge *per se* that is managed.[2] We define KBM as management or governance premised on the observation that intangible assets such as knowledge have become more important than tangible assets in the creation of wealth. While the term KM has been used largely in the corporate sector where it started in the late 1980s,[3] KBM can be generically applied to knowledge-based approaches in any sector.

SD on the other hand was developed by national development planners and field level practitioners, with initial impetus from environmental advocates starting in the 1970s. SD is premised on the observation that economic growth, if pursued at the expense of social capital or natural capital, is not sustainable.

This chapter proposes to contribute to bringing together the hitherto distinct mental models behind these two schools of thought and action, and to apply it to a pressing development problem: alleviating or eradicating poverty at the community level. Specifically, this chapter will propose a framework, the Knowledge for Poverty Alleviation (KPA) framework, show how it provides additional explanation why successful projects are successful, and

---

[2] Nonaka, I. (2008). Strategy as distributed phronesis: Knowledge creation for the common good. In *Knowledge Management: From Brain to Business*, Proceedings of the International Productivity Conference 2007, Bangkok, Thailand, 18–19 January 2007. Tokyo: Asian Productivity Organization. In the still evolving field of knowledge management, the nomenclature still varies. Besides "knowledge-based management", the term "intellectual capital management" is also employed. Intellectual capital consists of three categories: human capital, structural capital (sometimes referred to as process capital or internal capital), and stakeholder capital (sometimes referred to as relationship capital, external capital or customer capital). These are also referred to as "knowledge assets". The term "intellectual capital management" and its categories are less prone to misunderstanding compared to the term "knowledge management" because of the many different meanings that the word "knowledge" denotes.

[3] Chase, R. (2007). Innovation and intellectual capital management set the agenda. In *Knowledge Management: From Brain to Business*, Proceedings of the 2007 International Productivity Conference, Bangkok, Thailand. S.D. Talisayon (ed.). Tokyo: Asian Productivity Organization.

point out next directions for its operationalization by development organizations.

The KPA framework consists of recognizing and leveraging various forms of intangible assets of a community to sustainably develop both its tangible and intangible assets.[4] From the KPA perspective, sustainability is pursued along the three value domains of sustainable development, where growth in economic, natural and social capital[5] is pursued in such a manner that growth of one is not at the expense of any other.[6]

The development of the KPA framework, processes and toolkits is work in progress. We are reporting in this paper the results from an on-going study jointly conducted by two non-profit foundations in the Philippines: CCLFI. Philippines which is a leading knowledge-based management and organizational learning advocacy and service

[4] We adopt here the standard meaning of the term "intangible asset" from the accounting profession, namely, that intangible assets are non-physical assets that generate value for an organization. However, intangible assets are not always entered in the books of account of an organization.

[5] The meanings of "social capital" include the goodwill, fellowship, cohesiveness, mutual trust and the quality of relationships among members of a social group that enable them to work together productively. Some authors include membership in civic organizations, political participation, and trust in government as part of social capital (Putnam, R. 1995. Bowling alone: America's declining social capital. *Journal of Democracy*, 6(1)). The OECD defines it as "networks together with shared norms, values and understandings that facilitate cooperation within or among groups" (Cote, S. and T. Healy, 2001. The Well-being of Nations: The Role of Human and Social Capital. Organisation for Economic Co-operation and Development, Paris). See also Office of National Statistics (2001). Social Capital: A Review of Literature. Social Analysis and Reporting Division (October), U.K.

[6] In this paper, the term "capital" is meant in the economic sense of factor of production, which is implicitly consistent with the common usages of the terms "intellectual capital" among KBM practitioners, "social capital" among sociologists and some economists, and "natural capital" in ecology and environmental management. A broader concept — "metacapital" — is proposed in Section 3.3 below. By the term "assets" we also mean particular entries in an accounting system but we adopt here the broader concept of "asset" proposed by Kiyosaki as anything that generates regular income (Kiyosaki, R. and S. Lechter, 2000. *Rich Dad Poor Dad*. Warner Books.)

provider in the Philippines, and Peace and Equity Foundation (PEF) which is a leading loan and grant-giving institution in the Philippines.

In April 2008, PEF and CCLFI.Philippines introduced the KPA framework to the Philippine Development Innovation Marketplace or *Panibagong Paraan* 2008, a World-Bank led and initiated program to seek out innovative ideas to address development challenges.[7] The two NGO partners constructed a "KPA website" to share results of on-going studies with the development community.

2. Review of Literature

This review of literature draws from two separate strands in development literature: the KBM discourse and the SD discourse.

Development literature does point to some beginnings in cross-applications of these two discourses. For example, there are community-level frameworks that put the spotlight on intangible assets of communities. At the level of national development policy, the two discourses started to meet with the Asian Development Bank's concept of knowledge-based development (KBD).

2.1 *Asset-Based Frameworks at the Community Level*

Community development had long been approached from a needs-based perspective, where development institutions focus on identifying the needs of the local community with the goal of presenting solutions that will address such needs. Recognizing the limitations that this approach presents in terms of community capacity building, as well as the undesirable impact of focusing on weaknesses rather than strengths, there came a shift in international development agency practice to the asset-based and strength-based approaches.

[7] See: Panibagong Paraan 2008 at http://go.worldbank.org/FU3ZT73ZU0.

Among the frameworks that emerged from this approach are (i) the Asset-Based Community Development (ABCD) proposed by the Institute for Policy Research at the Northwestern University, (ii) the asset-building framework now employed by the Ford Foundation, (iii) the Sustainable Livelihoods Approach (SLA) developed by the Department for International Development (DFID), U.K., (iv) the Community Capitals Framework developed by the North Center Regional Center for Rural Development, (v) Hernando de Soto's[8] proposal that various forms of sanctioned access to public goods is a form of community asset, and (vi) Jeffrey Sachs'[9] identification of various forms of tangible and intangible assets that local communities either own or can use for poverty alleviation. These frameworks share many similarities, including the recognition of both tangible and intangible community assets, the approach of building from strengths, and the emphasis on social capital to build social and economic enterprises.

The ABCD, proposed by Kretzmann and McKnight,[10] highlights the existing assets of the community and emphasizes the recognition and mobilization of these strengths to create economic opportunities. The approach utilizes appreciative inquiry methods for building a community history and for community dreaming, and mapping of often unrecognized community assets, particularly social capital. It also draws on participatory approaches to development, collaborative economic development, and strengthening civil society to develop them as citizens charting their own direction as opposed to clients of development projects.[11] The building of assets can be done by linking

---

[8] De Soto, H. (2000). *The Mystery of Capital: Why Capitalism Triumphs in the West and Fails Everywhere Else*. Basic Books.

[9] Sachs, J. (2005). *The End of Poverty: Economic Possibilities for Our Time*. Penguin Press.

[10] Kretzmann, J. and J. McKnight (1993). *Building Communities from the Inside Out*. Chicago, IL: ACTA Publications.

[11] Mathie, A. and G. Cunningham (2002). From Clients to Citizens: Asset Based Community Development as a Strategy for Community Driven Development. Occasional Paper Series, no. 4. Antigonish, Nova Scotia: St Francis Xavier University. http://www.stfx.ca/

existing initiatives in the community, creating community-wide organizations and engaging partners.[12]

The asset-building framework of the Ford Foundation guides its Asset Building and Community Development Program.[13] The basic premise is that building individual and community assets increases resilience and provides long-term stability and security that are requisites for permanent poverty reduction. The assets focused on are financial assets, natural resources, marketable skills, public assets, social assets, and effective community services and institutions.

Compared with the ABCD, a framework that puts explicit emphasis on sustainability is the Sustainable Livelihood Approach or SLA. Recognizing that promoting income-generating activities is not the same as enhancing livelihoods, the SLA takes into account other factors such as the vulnerability context of the poor, strategies used by the poor to cope with economic shocks, the asset pentagon of households and the community consisting of financial, human, social, physical and natural assets, and the structures and processes affecting livelihoods in the community.[14] The core concepts of the SLA are: people-centered, holistic, dynamic, builds on community strengths, explores macro-micro link, and considers sustainability.

The Community Capitals Framework,[15] on the other hand, was based on a research that looked into the characteristics of entrepreneurial and sustainable communities revealing that successful communities paid attention to seven types of capital: natural, cultural, human, social, political, financial and built. The application of this framework range from measuring what the community currently has,

---

[12] Roehlkepartain, E.C. (2001). An Asset Approach to Positive Community Change. Minneapolis, MN: Search Institute. (ED 462 461)

[13] See: http://www.fordfound.org/programs/assets

[14] Hocking, G. (2003). Oxfam Great Britain and Sustainable Livelihoods in the UK. *Community Development Journal* 38, no. 3, July: 235–242.

[15] For more information, see http://www.ncrcrd.iastate.edu/projects/commcap/7capitals.htm#resources

to identifying what potentials exist, and determining what capitals are needed in order to bring about a desired state.

One of the strengths and unique propositions of the Community Capital Framework is its emphasis on the interaction and interplay of the community capitals to build the other forms of capital. This interplay is based on the theory of cumulative causation formulated by Gunnar Myrdal[16] which states that, "the place that loses assets, for whatever reason, will continue to lose them through systems effect". Conversely, once assets are acquired, more assets will be acquired because everything is inter-connected as a system. Evidence of this interplay of assets (see Diagram 1) that can be characterized as "success leading to success" is found in applications of the Community Capital Framework in Chimapala, Mexico[17] and the Home Town Competitiveness (HTC) Program.[18]

In addition to the frameworks and approaches discussed above, certain frameworks that focus on unique and specific contexts have emerged. One such framework is the Peace and Conflict Impact Assessment (PCIA)[19] most useful in conflict-prone areas. This assessment framework anticipates and evaluates the potential and actual impact and unintended consequences of a proposed or completed development project on the state of peace in the area. It is based on the premise that projects cannot bring about positive impact and change in a community if relationships (social and stakeholder capital) are poor or deteriorate.

---

[16] Myrdal, G. (1957). *Economic Theory and Underdeveloped Regions.* London: Duckworth.

[17] Gutierrez-Montes, I. (2005). Healthy Communities Equals Healthy Ecosystems? Evaluation (and Breakdown) of a Participatory Ecological Research Project towards a Community Natural Resource Management Process, San Miguel Chimalapa (Mexico). Ph. D. Dissertation. Iowa State University, Ames, Iowa.

[18] Emery, M. and C. Flora (2006). Spiralling up: Mapping community transformation with community capitals framework. *Community Development*, 37(1), Social Science Module.

[19] Bush, K. (1998). A Measure of Peace: Peace and Conflict Impact Assessment (PCIA) of Development Projects in Conflict Zones. Working Paper No. 1, The Peacebuilding and Reconstruction Program Initiative & The Evaluation Unit.

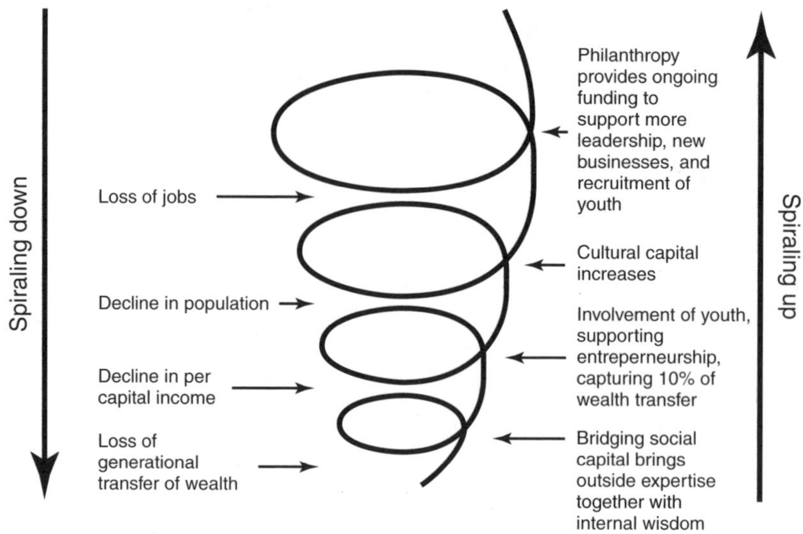

Diagram 1: The spiralling of capital assets

Source: Emery and Flora, 2006.[20]

2.2 *Knowledge-Based Development at the National Policy Level*

The World Bank proposed a knowledge-based framework at the national level, namely, the Knowledge-Based Economy (KBE) and a corresponding methodology for measurement, the Knowledge Assessment Methodology (KAM).[21] KBE consists of four pillars, education, science & technology and innovation, ICT infrastructure and economic incentive regime. The first three are factors that have been observed to significantly influence national factor productivities.[22]

[20] Emery, M. and C. Flora (2006). Spiralling up: Mapping community transformation with community capitals framework. *Community Development*, 37(1), Social Science Module.

[21] See http://www.worldbank.org/kam

[22] Chen, D. and C. Dahlman (2004). Knowledge and Development, a Cross-Section Approach. World Bank Policy Research Working Paper 3366. Washington, D.C.

KAM consists of corresponding national data elements that are readily available for longitudinal and cross-sectional comparisons.

In a technical note from the Asian Development Bank, the SD dimension was added to KBE, resulting in the Knowledge-Based Development (KBD)[23] framework (see Table 1, particularly the columns on "Social" and "Natural"). KPA contributes to the operationalization of KBD by applying intellectual capital concepts to a very pressing global concern, namely, eradication of poverty[24] at the community level. KPA is designed for application of KBD in the design, implementation and evaluation of community-based anti-poverty projects.

Table 1: Knowledge-based development model

	Economic (KBE)	Social	Natural
Education (development of **human capital**)	Education for a skilled workforce	Education for total human development	Education for sustainable development
Innovation (development of **structural capital**)	Systems, processes and technological innovations	New institutions and protocols for peace, equity and human rights	Environmental technologies, e.g., renewable energy technologies
Building Networks (development of **stakeholder capital**)	Financial and physical networks, e.g., IT infrastructure	Social networks, social trust, cultural integrity	Agreements protect and sustain planetary life support systems

[23] Asian Development Bank (2007). Moving Toward Knowledge-Based Economies: Asian Experiences (a Technical Report). Asian Development Bank, September.

[24] See for instance discussions on the Millennium Development Goals in the UN Millennium Project 2005. Investing in Development: A Practical Plan to Achieve the Millennium Development Goals. Overview. United Nations Development Programme.

2.3 *Nurturing Personal and Community Engagement in Development*

A number of studies across many disciplines corroborate each other towards the conclusion that intangible human and social qualities do produce tangible development outcomes. This causal link can be the basis for nurturing lasting personal and community engagement in the development process.

- An analysis of seventeen successful and innovative development projects in the Philippines revealed that their underlying success factors were not technical or tangible factors but largely "internal changes in the participants" such as a "we" feeling among the community (cohesiveness, or high social capital), a strong feeling that "this project is ours" (sense of ownership), and strong commitment (the motivational aspect of human capital).[25]
- In the corporate sector, the tangible impacts of intangible human qualities have also been recently demonstrated well by some authors. Covey and Merrill argued that when trust is low, performance is low and business costs go up.[26] Marcum and Smith marshalled research data to show that executives' ego can work both ways: increase competitiveness as well as pose a hindrance to further greatness or excellence.[27] Ego has substantial impacts on the corporate bottom line. There is accumulating evidence that emotional intelligence contributes more than technical knowledge towards excellent work performance and effectiveness.[28]

---

[25] Talisayon, S. (1991). Lessons. In *Innovative Development Processes in the Philippines: Case Studies*, S.D. Talisayon (ed.). University of the Philippines: Asian Center.

[26] Covey, S. and R. Merril (2008). *The Speed of Trust: The One Thing that Changes Everything*. Free Press.

[27] Marcum, D. and S. Smith (2007). *Egonomics: What Makes Ego Our Greatest Asset* (or *Most Expensive Liability*). Simon & Schuster.

[28] Newman, M. and G. Ainsworth (2004). The Business Case for Building Emotional Capital. Byzedium Pty Ltd. See: http://www.byzedium.com.au/cpa/htm/htm_mod_link.asp?id=63

- At the national level, the intangible quality of "social trust" had been correlated by Fukuyama[29] with lower transaction costs and higher level of economic development. In a U.K. government review, various studies noted that high social capital is associated with better health, improved longevity, better educational achievement, lower rates of child abuse, and less corruption in government.[30]

At the other extreme, the "development disasters" that largely energized the SD movement were of two types: environmental and human. The second type of disaster led to discourses on "participatory" development modes, "empowerment" of disadvantaged groups, "inclusive" development frameworks, "appreciative" inquiry instead of interventionist or "top-down" models, and similar approaches.

The SD discourse that culminated in the 1992 Rio Summit where 118 countries adopted Agenda 21 is rooted in widespread development experiences that sustainability depends on the preservation and enhancement of all three forms of capital: social capital (community cohesiveness, trust, relationships, local values, local support systems, etc.), natural capital (biophysical support systems in the natural environment), and economic or financial capital (economic benefits, participation in management and control, ownership). The conventional SD wisdom is that the third, if pursued at the expense of the first and/or the second, is a kind of development that eventually will not be sustainable. As a result, a wide variety of group and social processes has emerged which we can label as "empowerment"[31] and has become integral in the SD discourse as an end, as well as a means to the end of authentic personal and community engagement in development.

---

[29] Fukuyama, F. (1995). Social capital and the global economy. *Foreign Affairs*, 74(5):89–103.

[30] Social Analysis and Reporting Division (2001). Social Capital: A Review of Literature, Office of National Statistics, U.K., October.

[31] Unfortunately, the term "empowerment" from the transitive verb "to empower" connotes a solely top-down or externally-driven process. However, this may be the best well-understood label we can use for the moment.

Empowerment can generally be taken to mean providing individuals and groups the right, authority, and ability to make decisions and choices within a particular domain. Three aspects of empowerment are put forward by Papineau and Kiely[32] — perception of self-efficacy and control; acquisition of knowledge, skills, and resources; and participation in collective action.

At the heart of empowerment is citizen or community participation. In the words of Sherry Arnstein, "participation" is "the redistribution of power that enables the have-not citizens, presently excluded from the political and economic processes, to be deliberately included in the future".[33] There are many types of "participation", from the genuine to the more superficial ones. Arnstein's "ladder of participation" is useful for distinguishing among different types of participation (see Diagram 2).

Talisayon[34] observed that the productivity and stability of a project revolves around three interrelated questions which pertain to the extent to which all or most of the affected and concerned parties perceive an interest in, identify with, or are engaged in, the project. A project that is for, by and of the people would be a truly participatory project.

1. Ownership and control: Of whom is the project?
2. Management, from conceptualization to design and implementation of the project: By whom is the project?
3. Benefit: For whom is the project?

---

[32] Papineau, D. and M.C. Kiely (1996). Participatory evaluation in a community organization: Fostering stakeholder empowerment and utilization. *Evaluation and Program Planning*, 19(1):79–93.

[33] Arnstein, S. (1971). A Ladder of Citizen Participation. Journal of the Royal Town Planning Institute. Also: Arnstein, S. (1969). A ladder of citizen participation. *JAIP*, 35(4):216–224. Available: http://lithgow-schmidt.dk/sherry-arnstein/ladder-of-citizen-participation.html.

[34] Talisayon, S. (1989). Designing for Consensus: The ASEAN Grid. Institute of Southeast Asian Studies, Singapore, p. 105. See also: Talisayon, S. (1991). Lessons. In *Innovative Development Processes in the Philippines: Case Studies*, S.D. Talisayon (ed.). University of the Philippines: Asian Center.

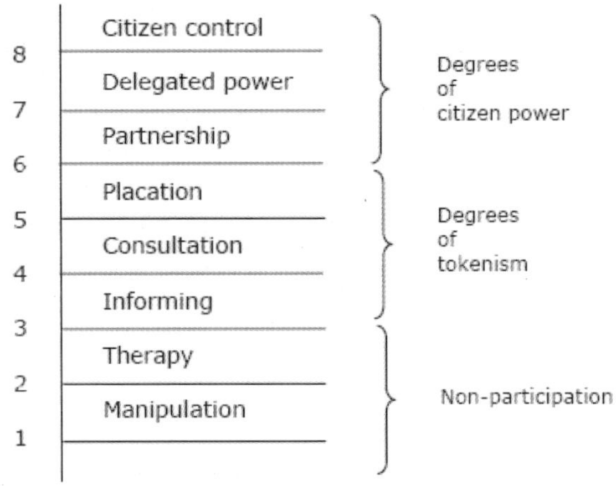

Diagram 2: Eight steps on a ladder of citizen participation

Source: Arnstein (1969).[35]

At the corporate level, practitioners of KBM frequently encounter the issue of more fully engaging knowledge workers. The issue is frequently labelled as a "change management" problem that often accompanies implementation of an organization-wide knowledge management (KM) program. A cross-case analysis of 22 case studies of good KM practices in nine Asian countries reveals a common theme: the use of a variety of approaches to motivate knowledge workers:[36]

- Rewards and recognition schemes: Knowledge Dollar (K$) and the Joint President's and CEO's Knowledge Management Award in Airtel (India); the Learning Award for knowledge transfer and an Enterprise Award for intrapreneurship in Unilever Indonesia; various corporate awards by Wika (Indonesia).

[35] Arnstein, S. (1969). A ladder of citizen participation. *JAIP*, 35(4):216–224.

[36] Talisayon, S. (2008). Concluding observations. In *Knowledge Management in Asia: Experiences and Lessons*, S.D. Talisayon (ed.), pp. 349–355. Tokyo: Asian Productivity Organization.

- Measures to show the benefits and rationale for KM: Infosys (India) and Goldsun (Vietnam).
- Optimization of the convergence between personal and organizational goals: Department of Health (Philippines).
- Mix of informal and formal communication modes to strengthen buy-in from employees and customers, e.g., "floor walks", tea sessions and informal gatherings: Qian Hu (Singapore); balance of face-to-face meetings and virtual interaction: SCG Paper (Thailand).
- Redesign of physical spaces for interactions and to foster openness and trust among employees: SCG Paper (Thailand); redesigned library environment to be more reader friendly, using ergonomics furniture and encouraging a more cheerful mood using paintings and appropriate color scheme for walls and furniture: Bank Negara Malaysia.
- Top management commitment or executive sponsorship: result of surveys in Thailand and Malaysia; selection of Chief Knowledge Officer on the basis of commitment, leadership ability and recognition from other staff: Siriraj Hospital (Thailand); creation of a motivational organizational culture characterized by caring leadership which supports active questioning and allows for mistakes: JTC Corporation (Singapore).
- Learning as a win-win activity for employees and the company: Multimedia Cyber College at CAPCO (Taiwan); learning through face-to-face interaction in a team or community of practice: Unilever Indonesia, SCG Paper (Thailand), Siriraj Hospital (Thailand), Samsung Advanced Institute of Technology (Korea); cross-functional teams, benchmarking projects and study visits or attachments: Bank Negara Malaysia.
- The honor and prestige of being a mentor or coach: Airtel (India), Wika (Indonesia); the role of "begawan" or sage: Bank Indonesia.
- "Praise Ground," a practice of peer-to-peer public compliments for exemplary KM behavior at Samsung Advanced Institute of Technology (Korea).

3. The KPA Framework

A phenomenon observed about the global economy is the growing contribution of intangible assets over tangible assets in market value of corporations.[37] In other words, knowledge assets, which constitute most of intangible assets, have become the major creator and repository of corporate value. A similar phenomenon can be observed about wealth creation at the community level.

When CCLFI.Philippines and PEF reviewed the most successful projects among more than 950 anti-poverty projects funded by PEF, it is apparent that project success can be explained in part to (a) use of available intangible assets of the community and (b) sustainable build-up of the community's tangible and intangible assets. These were not conscious decisions made at the time these projects were conceived and designed; rather, these were ex post or hindsight observations.

Again, the term "sustainable" is used here in the same sense of sustainable development, which can be reframed as simultaneous growths of economic, natural and social capital which are not at the expense of one another.

To illustrate (a), Diagram 3 summarizes and explains why a water and sanitation project for an upland cultural community in Mindanao, Philippines was among the more successful projects. Here, "success" was judged from the granting agency or PEF's criteria of having achieved preset project objectives.

The plus (+) factors on the left side are mostly intangible assets, with the exception of the entries under the "Technology and Infrastructure" category. Although adjudged as successful, this particular project does have negative (–) factors (see the right side). These factors can alert the development planner/evaluator that the project success may not be permanent or may be at risk if certain vulnerabilities are not addressed.

---

[37] For example, see Blair, M. and S. Wallman (2001). *Unseen Wealth, Report of the Brookings Task Force on Intangibles.* Brookings Press.

ANATOMY OF SUCCESS AND VULNERABILITY

Assessment of various forms of capital used in the water/sanitation project for Tbolis

SUCCESS FACTORS

Stakeholder/Social Capital

+ Support from NGO
+ Skills and track record of NGO in development, including grant sourcing
+ Support from NGO founder who has become Vice Mayor
+ Trust of the people on the NGO and its founder
+ Support from two corporations
+ Goodwill from adjacent barangay who donated labor

... Political factions linked to ethnicity, Ilonggo vs Tboli

Technology and Infrastructure

+ Technology of water system design
+ Technology of toilet bowl and system design
+ Water and sanitation infrastructure

Human Capital

+ Organizational skills of NGO
+ Labor donated by community
+ Leadership of some project participants
+ Prior capabilitiy building efforts by NGO for PO

+ ...Management skills of PO which took over water system (mixed)

Access rights

+ Access rights to spring water, donated by owner of spring

RISK FACTORS

Cultural Capital

+ *Dagang* practice or cooperative self-help

... Prevailing worldview that water is a free public good, including practice of open access

... Some weakness in cooperative and sharing culture, e.g. failure of some families to share cement

... 40% unwilling or late in payment of water fees

Structural Capital

+ Control given to PO instead of barangay council
+ Prior needs assessment

+ ...Clarity of project benefits to beneficiaries (uncertain)

+ ...Unclear about actual "felt needs" by community (NGO-centric needs assessment?)

... Water fees introduced only after water system was started

... No connection between fee paid and amount of actual consumption

Natural Capital

+ Relatively clean water in identified spring from hill

... Denudation of watershed above spring

... Some consumers at higher elevation

Diagram 3: Anatomy of success and vulnerability of water/sanitation project for Tbolis

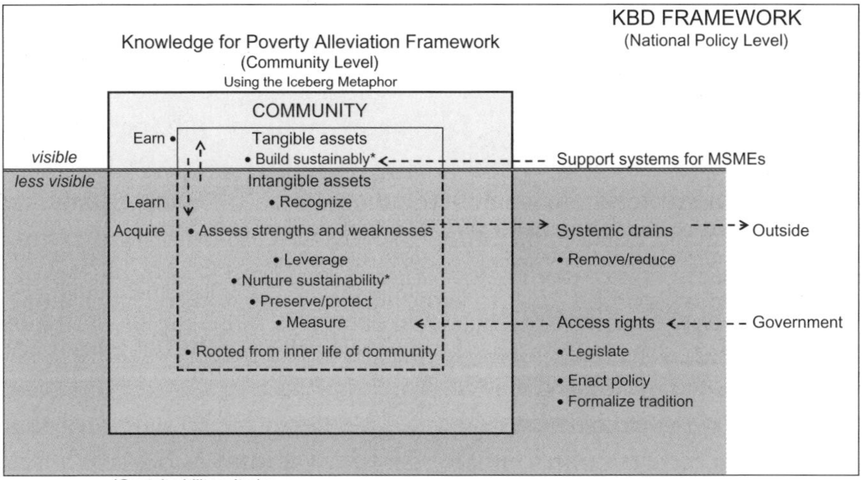

*Sustainability criteria:
- Economic: viable and competitive in the global knowledge economy
- Environmental: continuing stability, reproducibility and productivity of life support systems
- Social: project is of, for and by the community
- Growths of economic, natural and social capital are not at the expense of one another

Diagram 4: Building assets within the KBD framework

3.1 *Recognition and Use of Intangible Assets of the Community*

The crucial starting point in the KPA framework is *recognition* of various forms of intangible assets a community may possess or has access to, something that is not readily visible to many development planners and workers. Once recognized and appreciated, a set of management actions, both old and new, comes into view (see bullet points in Diagram 4 and Section 3.6). The framework can be visualized using the iceberg metaphor, to stress the perceptual reality that most of the important community assets are less visible to most people (see Diagram 4).

3.2 *Reframing "Poverty"*

From the KPA perspective, what was referred to by many as "poverty" is very often lack only of tangible assets: financial capital and local infrastructure. Once the development planner realizes that a "poor" community may have other forms of wealth that can be leveraged, accessed or built upon, then the process of poverty alleviation becomes a broader process of (a) using external and internal assets to sustainably generate both tangible and intangible assets of the community (horizontal arrows in Diagram 4) and (b) using intangible assets to generate more tangible assets, and vice-versa (vertical arrows in Diagram 4).

3.3 *Broader Concept of Capital*

Increasingly, researchers from many disciplines are recognizing that many other factors contribute to wealth creation besides the traditional factors of production — land, labor and capital. Development practitioners have recognized the contribution to national development of other factors such as technology, education and public infrastructure. The World Bank's KBE model acknowledges the importance of four "pillars": education, ICT infrastructure, innovation systems including science and technology programs, and a favorable policy and regulatory environment.[21] This concept was expanded into the KBD model by the Asian Development Bank, by incorporating the three elements of intellectual capital employed by practitioners of knowledge-based management (or knowledge management) and three elements of sustainable development.[23]

Proponents of asset-based frameworks applied to community development (see Section 2.1) argue for the usefulness of a broader concept of "asset" and "capital" to include social-political, cultural and natural dimensions.

There is a growing realization among policy makers and decision makers that knowledge assets are the prime creators of value or wealth in the knowledge economy. The intellectual capital school draws from corporate experience to show that a firm's intangible assets cover three factors: human capital, structural capital (also called process capital or internal capital) and relationship capital (also called stakeholder capital or more narrowly as customer capital).[38]

Other concepts of capital are gaining currencies in other disciplines. In sustainable development, ecology and environmental management, the concepts of "natural capital", "biological capital", "biodiversity capital" and "energy capital" are appearing in academic literature and in practitioners' discourses. The concept of "social capital" is now used more widely outside the social sciences: economic development, community development and business management. Less common but notable are new concepts in the behavioral sciences such as "integrity capital", "positive psychological capital" and "emotional capital". We saw examples (see Section 2.4) demonstrating how intangible human and social qualities produce tangible outcomes.

To summarize, there are increasingly many factors becoming recognized as contributing to wealth or value creation. One of the skills that policy makers and decision makers need to learn to make better decisions in the context of the global knowledge economy is the skill of recognizing and managing intangibles. This skill requires a shift in mental model. To facilitate this paradigm shift, we propose the use of a new term — "metacapital" — to refer to those factors, whether tangible or intangible, and whether public or private, that contribute to production or wealth creation.

3.4 *KPA: Building Assets within the Larger KBD Framework*

A basic thesis in KPA is that intangible assets do generate tangible returns (see upward arrow in Diagram 4 labelled "Earn"). This is

---

[38] This school of thought is best represented by the following well-read authors: (a) Sullivan, P.H. (2000). *Value Driven Intellectual Capital: How to Convert Intangible Corporate Assets Into Market Value.* Wiley. (b) Stewart, T.A. (1998). *Intellectual Capital: The New Wealth of Organizations.* Doubleday Business. (c) Sveiby, K.E. (1997). *The New Organizational Wealth: Managing and Measuring Knowledge-Based Assets.* Berrett-Koehler Publishers.

consistent with the broader meaning of "asset" proposed by Kiyosaki as anything that can generate regular income.[39]

The reverse is also happening. Tangible assets such as grants from development funding institutions, loans for a microenterprise and members' contributions to a cooperative can be converted to intangible assets through training and capacity building (downward arrow labelled "Learn, Acquire").

The KPA framework clarifies how addition or erosion of community metacapital is taking place in relation to external agents and factors (horizontal arrows in Diagram 4) that are influenced more by KBD approaches. It brings out how systemic drains can contribute to worsening of poverty (in metacapitals, both tangible and intangible), and how local and national governments can act to prevent that erosion or to contribute to the build-up of community assets.

For example, systemic socio-economic and legal arrangements do continuously drain local communities of their metacapitals:

- Drain in natural capital. Based on the Regalian Doctrine, resource extraction companies deplete local natural resources and transfer them outside, but little of the economic proceeds returns to local governments and to the local people.
- Drain in human capital. Drawn by better opportunities, valedictorians (top performers in secondary schools) and other local talent migrate to big cities and to developed countries.
- Drain in fiscal resources. Only a small part of taxes collected locally and remitted to the central government comes back in the form of either cash or new public infrastructures or services.
- Drain in private financial resources. Local branches of commercial banks based in Manila are more deposit-takers than lenders,

[39] Kiyosaki, R. and S. Lechter (2000). *Rich Dad Poor Dad*. Warner Books. For example, a family car is normally considered as an "asset". However, Kiyosaki argues that a family car generates regular expenses and therefore should not be classified under "asset". Consistent with Kiyosaki's redefinition of asset, indeed intangible assets can and do generate regular income. This redefinition is more aligned with the meaning of the term "capital" as a factor of production.

resulting in net flow of private savings from local communities to Manila.

- Other drains can come from: incursion of formalized land titles/patents and informal occupants into ancestral domains of indigenous peoples, policy changes accompanying periodic changes in political leadership, premature exposure to foreign competitors due to reduction of tariff barriers and other trade liberalization measures, etc.

On the other hand, the government can contribute to the build-up of community metacapitals:

- Legislation. Example 1: Republic Act 8371 or the Philippine Indigenous Peoples Rights Act (IPRA) set the ground rules for recognizing and operationalizing the rights of indigenous cultural communities to their traditional or ancestral domains, such as the requirement of "Free and Prior Informed Consent" of the community before a mining permit is issued over part of their ancestral domains. Example 2: Republic Act 7586 or the Philippine law on National Integrated Protected Areas Systems (NIPAS) defines municipal (fishing) waters as 15 kilometers from the shoreline; this is the exclusive fishing ground for municipal fishermen; commercial fishers are not allowed to fish within this zone.
- Policy. Example 1: the Philippine Department of Energy adopted a policy in 1994 (Energy Regulations I-94) requiring power companies to allocate PhP 0.0025 per kilowatt-hour of total electricity sales to a fund a local "development and livelihood fund" and a similar amount for "reforestation, watershed management, health and/or environment enhancement." Example 2: the Philippine Office of the President enacted several Executive Orders in the 1990s which deregulated and de-cartelized many industries such as the airlines, domestic shipping, telecommunications, insurance, etc. The result of this set of policies is lower costs and higher quality of services, with widespread benefits to low-income households.

- Programs. Example 1: Village drugstores (called *"Botika sa Barangay"* in the Philippines) which were designed, franchised, resupplied and supported by the national government to distribute cheaper generic drugs, are relatively risk-free business ventures available to local communities. Example 2: Local government funded feeding programs aligned with the national government's nutrition promotion policy.
- Formalization of traditional rights. Example: harvesting of some designated forest products such as rattan is illegal in the Philippines, but the Department of Environment and Natural Resources granted exception to indigenous cultural communities such as Agtas who had been traditionally depending on the product for their livelihood.

The KBD concept introduced by the Asian Development Bank is yet to be operationalized. Here, an important issue is how a national government, as the national rule-setting institution, can create a roadmap of legislations, policies and programs so that the systemic drains that exacerbate poverty at the community level can be removed or reduced, and affirmative action can be taken to assist local communities build up their metacapitals, both tangible and intangible. Addressing these KBD concerns will provide a favorable context and environment for successful and sustainable KPA applications at the local level.

3.5 *Nurturing Personal and Community Engagement*

As we have seen, there is probably a causal link between intangible human and social qualities on the one hand, and tangible outcomes, on the other hand. Hitherto, focus of development attention on tangible assets tends to ignore or fail to appreciate the intrinsic values of intangible assets, values which are rooted in the inner life and values of the community. This perspective may be the basis for designing and applying tools — old and new — for effectively sensing and nurturing of personal and community engagement in the development process. Nurturing community engagement and

Table 2: Inner foundations of community intangible assets

Intangible Assets	Foundations on Inner Life and Values of the Community (*Negative in Italics*)
Social capital	Trust, loyalty, reputation, sharing, *factionalism*, *ethnic or religious conflict, rich-poor gap*
Human capital	Commitment, leadership, previous trainings, skills and expertise, entrepreneurship, indigenous knowledge and wisdom, *dysfunctional egoism of leader, corrupt managers*
Cultural capital	Community and family values, beliefs and practices that enhance development, *animosity to outsiders, apathy, fear of innovations*
Structural capital	Community organization and procedures, tools and formal/informal systems, *inequitable distribution of managerial roles, project created new local elites*
Stakeholder capital	External support systems, linkages to NGOs and civil society organizations (CSOs), goodwill, *animosity to outsiders, bad community reputation, presence of criminal syndicates*
Access to natural capital in the public domain	Proper utilization of the natural resources, *overharvesting and destructive harvesting, mismanagement of wastes and waste disposal*

building community assets then become mutually reinforcing processes.

One must understand that the intangible assets are connected to each other in a dynamic and evolving nature, such that intervention on one asset will have certain effects, intentional or unintentional, on the others. This interplay of capital is important, as the synergy it creates may contribute significantly to the development of the different dimensions of well-being of a community.

3.6 *Participatory and Appreciative Approaches*

To help energize, nurture and sustain community engagement, KPA processes must be rooted within the worldview and local social context of the community.

There are many variants of participatory and appreciative community-based processes that can be adapted to help community members and development workers alike recognize the intangible assets of the community, and to clarify the community members' collective development vision or goals. The energy typically generated in appreciative group inquiry and clarification of group vision can be marshalled for subsequent planning, problem-solving and addressing any weaknesses in their intangible assets.[40]

The output of a facilitated community self-discovery process, namely an emic identification and assessment of community intangible assets, is a good input for a pre-project community assessment and ensure that it is grounded on the social reality of the community.

3.7 *Sustainability Criteria*

SD criteria encompass three value domains: natural, economic and social. To assess the extent that a community project builds tangible and intangible assets in a sustainable manner, checklists or scales will have to be constructed and validated: environmental, economic and empowerment scales, respectively.

There is substantial literature on economic viability and sustainability of micro, small and medium-scale enterprises (MSME).[41] What needs to be pointed out here is another important link between KPA and KBD, or between the community operational level and the national policy level. The success of anti-poverty projects is very dependent on a variety of governmental and other external support systems (see Diagram 4) such as laws that facilitate access to financing; linkages of MSMEs to larger firms, and external markets and value chains, business incubation and business advisory facilities, relatively

[40] For example, a workable cyclical process combining the energy-generating appreciative inquiry of David Cooperrider and the problem solving modes in organizational development is described by Richley, B. and T. Lingham (2007). A time to build: Strategically linking positive and conventional change methodologies to develop leadership capacity. *OD Practitioner*, 39(1).

[41] For examples, see USAID's microLINKS (Microenterprise Learning, Information and Knowledge Sharing) at http://www.microlinks.org/ and the Socio Economic Research Portal for the Philippines at http://serp-p.pids.gov.ph/details.php3?tid=746.

risk-free including government-supported MSME-ready franchises, government programs such as one-village-one-product (OVOP), and other forms of stakeholder capital from the community perspective.

A desirable goal in MSME development is to develop means and models for effectively linking local enterprises to global e-commerce and value chains, i.e., to fully participate in the global knowledge economy.

The operationalization of the KBD framework will also require the conversion of current knowledge about sustainability in the social and natural domains into validated checklists useful for development planners and managers for designing and monitoring/evaluation of anti-poverty projects.

3.8 *Management Actions*

By intent, the KPA framework is oriented to action. The actions cover: (a) recognizing, assessing, leveraging and sustainably building community tangible assets by using its intangible assets, and vice-versa, (b) accessing or seeking access rights to external including governmental resources, and (c) working for government legislations, policies, regulations and programs that remove or reduce systemic drains in community assets. Many actions are already being undertaken; what the KPA framework can do is to situate those actions within a coherent KBM framework and to identify action areas that require further R&D (Table 3).

Diagram 5 summarizes the KPA framework in a single diagram. Factors that are beyond the control of the community and are influenced more by KBD approaches are presented in the flesh box at the bottom.

4. Application of KPA in Analyzing Project Success[42]

An initial application of the KPA framework to six of the most successful projects funded by the PEF provides some very interesting

[42] This section is informed by the results of two case study workshops — the Case Study Presentation: SRRE and KPA Framework Validation Workshop attended by officers of PEF and CCLFI. Philippines held between February 8–9, 2008 in Quezon City; and the Case Study Presentation to Partners attended by partner institutions, and officers of PEF and CCLFI. Philippines held between February 19–20, 2008 in Mandaluyong City.

Table 3: KPA actions for a country project

	KPA ACTIONS FOR A COMMUNITY PROJECT		
	Project Identification and Design	Project Implementation and M&E	Ensuring Post-Project Sustainability
Local Actions	• Identify, assess and leverage community intangible assets	• Develop intangible assets in relation to inner life and values of the community	• Evaluate and plan for continuous growth of intangible and tangible assets
	• Identify risks and vulnerabilities in community intangible assets	• Manage risks and vulnerabilities in community intangible assets	• Evaluate outcomes of risk management of intangibles
	• Plan build-up of sustainable community intangible and tangible assets: use intangibles to build tangibles and vice-versa	• Implement sustainable and synergistic build-up of new community intangible and tangible assets	• Design and implement complementary and follow-through projects
National Actions	• Work for access rights to needed resources from the government or from other property owners	• Access support systems from the government, civil societies and international development agencies	• Work for government actions to institutionalize the project, and government reforms to remove/ reduce drains in community capital

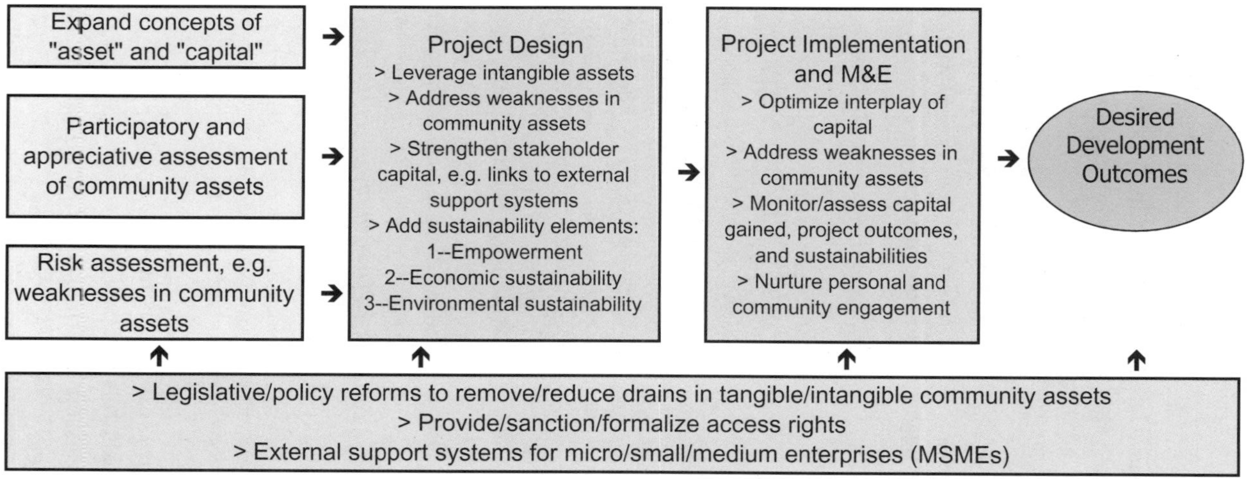

Diagram 5: KPA Framework

Table 4: PEF projects analyzed using KPA

Project Title	Location	Implementing Organization (Partner Institution)
1. Tubig Para sa Barangay at Mga Impormal na Grupo (TUBIG) (*Water for the Barangay and Informal Groups*)	Mandaluyong City and Pasig City (Luzon)	Foundation for Entrepreneurial Development, Inc. (FEDI)
2. Uplifting and Normalizing Malnourished Children of Lamut, Ifugao through Macro and Micro Nutrient Supplementation and Advocacy	Lamut, Ifugao Province (Luzon Island)	Lamut Women's Organization (LAWO)
3. Health Plus Social Franchising	Antique Province (Visayas Islands)	Antique Federation of NGOs (AFON)
4. Bridge Financing for Education in Northern Samar	Mondragon, Northern Samar Province (Visayas Islands)	Sacred Heart Institute for Transformative Education (SHIFT) Foundation
5. Micro-Livelihood Support Program (MLSP) and Expanded Micro-Livelihood Support Program	Clarin, Misamis Occidental Province (Mindanao Island)	Gata Daku Multi-Purpose Cooperative (GDMPC)
6. Integrated Health Initiatives for Two Indigenous People (IP) Communities	Barangay Payapag, Bacauag and Barangay Gacepan, Sison both in Surigao del Norte Province (Mindanao Island)	Surigao del Norte NGO Coalition for Development (SUNGCOD)

insights on the usefulness of this new framework in explaining project success (see Table 4).

These cases assessed various community assets along different phases of the project:

1. Pre-project in terms of capital and vulnerabilities existent in the community even before the project;

2. During project implementation in terms of actual capital used, leveraged, and liabilities addressed; and
3. Post-project in terms of capital or benefits gained.

The examination of the pre-project picture provided indications of how the project was able to succeed, as it made explicit the different assets that were used, and the different liabilities or vulnerabilities that needed to be addressed for project success. This first step provided insights on the cases. As the officer-in-charge of Foundation for Entrepreneurial Development (FEDI) stated during the presentation of the cases to partner institutions, "*I would like to affirm the KPA as a good tool and framework very akin to resource mapping done before the project. (It is) clearly stating many resources in various forms that are often overlooked, particularly cultural capital*".

Although not formally or intentionally included in the project design, the existing community assets surfaced in the pre-project assessment were employed as key components of project strategies. Among the most important assets utilized in the different projects are certain types of human capital. From the side of the implementing and/or grassroots organizations, this human capital takes the form of leadership capabilities, skills and qualities that served as the driving force behind the organization. From the side of the community, this takes the form of common aspirations for better lives with respect to having land tenure in the case of the indigenous communities in Surigao del Norte, access to water in the case of the informal settlers in Mandaluyong and Pasig, and better nutrition and nourishment in the case of Ifugao. The combination of these two types of human capital, from the side of the community and from the side of the implementing organization, seemed to have propelled project success, as it resulted in cooperation, participation, and commitment from both ends.

Stakeholder capital, particularly good relationships with local government agencies and units, is another form of capital that was leveraged across all the projects. This was strongest in the case of the malnutrition project in Ifugao Province, where the leveraging of stakeholder capital was used as the primary strategy by the implementing organization LAWO.[43] They took an active role in collaborating with the Convergence Team of the LGU (local government unit), composed of the Municipal Health Office, Municipal Social Welfare Office, the Municipal Agricultural Office, and the Office of the Mayor, starting with project conceptualization and development. Several prospective partners from various sectors were also involved by LAWO in order to generate more resources in cash or in kind. The AFON-implemented project on social franchising also heavily involved the Antique local government and the local Department of Health as partners in the project, with the Rural Health Offices housing the franchised Health Plus Outlets where the generic drugs are sold, and 43 community-based organizations managing such outlets. The project providing land tenure to IP (indigenous peoples) communities by SUNGCOD also cited the involvement of the government units as the critical success factor of the project.[44] Land was acquired for the IPs through the barangay LGUs, while the Department of Education was tapped to provide adult literacy program.

In terms of vulnerabilities that had to be addressed, two forms of capital emerged very strongly. These were the need for greater structural capital in the form of organizational capability building and strengthening of implementing organizations, and certain elements in the communities' cultural capital such as prevailing beliefs, norms and mindsets that had to be addressed so that the community can take full advantage of the benefits offered by the project.

The building of structural capital, in the form of improved organizational policies, processes and capabilities, served as a key element in ensuring that the social franchising concept was successfully implemented across Antique. Although AFON, a confederation of NGOs in Antique, already had a wide network of organizations, it is the enhanced capability and strength of these member community

[43] Fuellos, F. *Feeding the Future: The Case of Lamut*, Ifugao. (unpublished)
[44] Morallas-Basarte, A. *Integrated Health Initiatives for two IP Communities in Surigao del Norte.* (unpublished)

organizations which established the 43 outlets in Antique that contributed to the success of the Health Plus model.[45] The Gatu Daku[46] also had to set up its policies and systems in relation to microfinance in order for it to successfully implement the MLSP (micro-livelihood support program). They also had to establish self-help groups patterned according to the famous Grameen micro-lending system.

Projects also had to be sensitive to the existing cultural assets and liabilities of the community. For example, the project on Health Plus had to contend with the predominant thinking that "branded is better" (compared to cheaper generic drugs). The malnutrition program in Ifugao had to consider the Filipino taste in the preparation of the micronutrient supplement called Vita-Meal. Innovations in the preparation and nutritious recipes had to be introduced in order to sustain the feeding program to make the food appealing to children. The microfinance project of Gata Daku had to face the prevailing practice of borrowing from usurers. In the case of the water project of FEDI, community organizations had to act on the threat of syndicates puncturing water pipes by guarding the portion of the pipes running through their homes 24 hours a day.[47] Syndicates who sell water at extremely high prices are common among informal communities in the urban areas.

These cross-case findings showing assets that were leveraged and liabilities that were addressed demonstrate that the KPA provides a more comprehensive picture of a development project and the stories behind its performance. This was affirmed and appreciated by the fund manager of one of the projects when she stated: "*It's only now... and thank you because what you saw, we did not see... It gives me now a broader and clearer perspective of how the project succeeded*".

---

[45] Garcia, F. *AFON Social Franchising Revisited.* (unpublished)

[46] Morallas-Basarte, A. *Microfinance: Building Financial Capital through Structural Capital, the Case of Gata Daku Multi-Purpose Cooperative of Misamis Occidental.* (unpublished)

[47] This was shared and validated by FEDI during the case presentation and validation workshop among PEF and partner institutions.

When it comes to project results, the KPA framework recognizes that project gains and benefits are more than what is initially indicated in the project objectives. Gains are likewise realized with other capital and assets of the community, with improvements in one asset contributing to the enhancement of other assets. The cases provided evidence of how this build-up of assets happens. As Sister Digna Dacanay of SHIFT noted, "*It is true; this interaction of capital generates more capital, more synergy*". SHIFT saw how the financial capital provided by the project for bridge financing of a community high school translated to improved human capital of the teachers who found more time to perform their work which in turn enhanced the structural capital of the community as the parent-teacher association was strengthened.[48] The case of SUNGCOD showed how good stakeholder capital in the form of strong relationships with the LGU allowed for the acquisition of a physical asset (land for the IP communities) which in turn led to improved human capital in the form of increased self-esteem and which also led to enhanced structural capital as the IPs were formally organized to manage their dwelling places.[49]

The cases also provided evidence of how the macro context, which is the domain of KBD, affects development projects at the community level, which is the realm of KPA. Three cases are worth noting here. The first is the Health Plus social franchising project which provides a clear example of how enabling policies and laws at the national level facilitate development projects at the community level. In this case, the Law on Generic Drugs and an Administrative Order allowing legitimate community organizations to manage drug outlets called *Botika ng Barangay* (Village Pharmacy) provided the legal foundation for the social franchising concept.

The second is the LAWO feeding project which was anchored on the Municipal Nutrition Action Program of the LGU which was in turn contributing to the realization of the Philippine Nutrition

---

[48] Garcia, F. *Bridge Financing for Education in Northern Samar* (unpublished manunscript).

[49] Morallas-Basarte, A. *Integrated Health Initiatives For Two IP Communities in Surigao del Norte* (unpublished manuscript).

Action Plan. The presence of a national program guaranteed the involvement of the LGU in this community-initiated project.

The third is the SHIFT project which demonstrates how laws of the national government can both serve to facilitate and hinder community development. The project was operating in the context of Republic Act 6728 that mandated the Educational Service Contracting[50] scheme of the Department of Education. However, while this provided the venue for the government to support the education of its youth, the bureaucracy of the government places undue pressure on private schools because of constantly delayed subsidies. It was in fact this perennial problem that the bridge financing project attempted to address.

5. KPA and Project Sustainability

One of the unique benefits of using the KPA as a lens in examining the development projects is its ability to provide insights on the likelihood of sustainability of project gains. In the six case studies, the results are quite striking. While certain sustainability elements common across the projects were determined to be present, many other elements were identified to be missing. The potential role of these missing elements was affirmed during two case validation exercises that involved key officers of PEF and CCLFI. Philippines and the partner institutions.[51]

In terms of social empowerment, the elements that were present were the following: the increased sense of confidence and self worth of the community, together with the organization of community members into formal groups; and the institutionalization of

[50] Under this scheme, private secondary schools were contracted to provide free education through the provision by the government of student subsidies paid directly to the private school.

[51] These are the Case Study Presentation: SRRE and KPA Framework Validation Workshop attended by officers of PEF and CCLFI. Philippines held between February 8–9, 2008 in Quezon City; and the Case Study Presentation to Partners attended by partner institutions, and officers of PEF CCLFI. Philippines held between February 19–20, 2008 in Mandaluyong City.

programs, through agreements with the LGUs, for instance, to ensure that the programs can be continued even after the end of the project.

The elements related to social empowerment that were lacking and that should have been addressed by the projects as early as project conceptualization include:

1. Developing leaders and second-liners;
2. Continuous organizational strengthening, capacity building, and developing organizational self-reliance;
3. Help the community attract and work with various stakeholders who can assist them in their development needs; and
4. Helping the community determine their development needs, participate in planning and charting directions.

In terms of economic sustainability, the absence of an income-generating or sustainable livelihood component was evident in some projects. It was noted that many of the projects were on the expense side (providing access to services and resources that had to be bought) and were not matched on the income side. There were no clearly defined activities on how the community will be able to afford such services on a sustained basis. This was particularly noticeable in the case of FEDI where access to water was accompanied by monthly water bills and SUNGCOD's project with the indigenous people where settling them in a permanent location would expectedly create the need to purchase basic services and utilities.

6. Concluding Observations

The wedding of two powerful development paradigms — knowledge-based management and sustainable development — is a continuous and challenging task. This paper on the Knowledge for Poverty Alleviation or KPA framework attempted to contribute to this task in several ways.

Firstly, building on previous proposals and studies, the KPA framework outlines how intangible assets of communities can be recognized, assessed, leveraged, developed and managed in ways that are sustainable. Action items are identified to help bring theory to application.

Secondly, in extending the intellectual capital discourse to community-level anti-poverty projects, it was found necessary to reframe prevailing concepts of "asset", "capital" and even "poverty." To facilitate this reframing process, this paper proposes a new and broader concept: metacapital. KPA development is a work in progress that invites cross-disciplinal communication and collaboration among economists and sociologist, among ecologists and policy researchers, and among corporate KBM practitioners and rural SD practitioners. There is a need to test and validate operational tools for application at the field level.[52]

Thirdly, because the problem of poverty at the local level is embedded within the national policy context, the developments of the KPA and KBD frameworks should proceed together. This paper points to important linkages between KPA and KBD, and corresponding action items to address national policy issues and other systemic factors that may at times work to exacerbate poverty at the local levels.

Finally, the attention to intangible assets underscores a new insight: that authentic personal and social engagement in development is rooted in the inner life and values of the community — an

---

[52] One of the tools developed so far is "Anatomy of Success and Vulnerability", a summary table (Diagram 3 is an example) which quickly organizes and shows the strengths and vulnerabilities in a community's intangible and tangible assets, and in external assets it has access to. Development workers who are familiar with a project, when shown its "Anatomy of Success and Vulnerability", readily appreciate the value of the tool and the new perspective behind it, for presenting new or additional insights why the project is succeeding and cues where it may fail in the future. A tool for pre-project participatory community assessment process is also being developed, which would provide useful inputs to the project design stage. PEF and CCLFI. Philippines also is developing two other tools: a table on Interplay of Capitals which shows assets leveraged, vulnerabilities addressed, and interventions made during project implementation, and a table on Changes in the Assets/Capital which compares pre- and post-project situations.

insight that is corroborated by similar or parallel findings in the corporate sector. Applying KPA starts with a participatory and appreciative community self-assessment to surface the community's hitherto unrecognized intangible assets. This process produces inputs to the pre-project assessment prior to project design.

Chapter

13

Knowledge Loss

Managing Local Knowledge in Rural Uzbekistan

Hans-Dieter Evers and Caleb Wall

1. Introduction: Towards a Theory of Knowledge Production and Local Knowledge Loss

Currently knowledge is produced in great quantities and spread around the globe at what appears to be an ever increasing rate. The law of diminishing returns does not seem to apply to the production and dissemination of knowledge. The more is known in a society or community, the more valuable new knowledge becomes. There is not only an abundance of knowledge but also an increasing literature on the creation of new knowledge, on the sharing of knowledge and the productivity of new knowledge. So far an important aspect has been largely absent from this debate, namely the loss of knowledge.

In the context of this chapter, we shall limit our scope and refer to knowledge as a human resource that is used to guide social action. In this limited perspective, all factual knowledge is derivable from experience but knowledge is needed to produce, share and acquire new knowledge. We shall concentrate our attention on indigenous and on local knowledge. Both terms are often used interchangeably but we distinguish between the two. Indigenenous knowledge is passed on from generation to generation and is firmly grounded in the tradition of a group, community or society. Local knowledge is acquired through learning and adoption to local conditions. Knowledge that has been brought back from studying abroad, from reading, viewing or listening to mass media or adopted through other channels has become local knowledge, as soon as it has been adopted to local social and cultural conditions and is ready to be applied locally. Local knowledge constitutes a 'milieu' within a network of social interaction (Evers, 2010:63). Indigenous as well as local knowledge may be lost. The dialectics of knowledge production leads both to knowing and not-knowing. The more knowledge is produced the more we know what we don't know (Evers and Menkhoff, 2010:145). Solving a research problem usually opens up new research questions and thus increases our ignorance. Another form of creating ignorance is knowledge loss. This process does not indicate that old knowledge is up-dated or replaced by new knowledge but the attrition of a stock of available and usable knowledge. In the following section we will show how knowledge loss has happened and how knowledge seepage, in some cases knowledge destruction, has occurred in rural Uzbekistan. Based on empirical field research we shall be able to systematise forms of knowledge loss and discuss its consequences for rural development.

2. Knowledge Loss in Practice

What is not known is as important as that which is known. Whilst Evers and Menkhoff (2010:145) discuss 'the growth of ignorance' in terms of a relative growth of ignorance as a corollary to a growth of knowledge, we wish to discuss here the growth of ignorance in terms of knowledge loss in Uzbekistan during and after the Soviet period. This is based on one year (in 2005) of field research in the Khorezm region, in the western periphery of

Uzbekistan.[1] That is to say that indigenous knowledge has in some ways been 'lost' during the period of Soviet colonialism of Khorezm. Likewise there has been some attrition of indigenous knowledge in the post-Soviet era. This is pertinent in terms of knowledge of livestock production and post-harvest processing. In addition there is the simple, static, ignorance of new technologies and farming methods that are available in other parts of the world, yet which are not known in Uzbekistan. This is largely caused by the knowledge control approach adopted by the government of Uzbekistan. Also relevant is the role of ineffective linkages between farmers and local (i.e., Uzbek, non international) research institutes and universities, for failing to combat knowledge attrition in rural Khorezm. Whilst indigenous knowledge is dynamic, in that it is constantly evolving and changing, it is not always in the ascendancy. Instead what we show in this paper is that whilst the Soviet period introduced a considerable amount of new agricultural knowledge, which was adapted to local conditions and thus made 'local', there was also a considerable growth of real ignorance (distinct from relative ignorance, see Evers, 2000) which is manifest in Khorezm today. This is not to totally discount the level of indigenous knowledge growth that has occurred in the 15 years since Independence. What is worthwhile mentioning is that a significant amount of knowledge was simply 'lost' during the Soviet period, mainly due to the collectivisation of certain agro-economic activities. With de-collectivisation and the break down of existing collectives, post-1991 this attrition of knowledge continued. This is particularly relevant for the rural economy of Uzbekistan, as this 'lost' knowledge could potentially play a large developmental role in promoting new livelihood strategies, such as through post-harvest processing.

---

[1] Field work was carried out in 2005–06 by Caleb Wall within the framework of the interdisciplinary project "Economic and Ecological Restructuring of Land and Water Use in the Region Khorezm (Uzbekistan)" of the Center for Development Research (ZEF).

2.1 *Livestock Production*

With collectivisation in the 1920s, livestock production was transferred from an almost purely domestic affair into a collectivised and specialised industry of the state.[2] Whilst post-WWII reforms within agricultural production allowed for limited domestic production of livestock within the household economy, large amounts of indigenous knowledge had already been lost. For instance chickens, which were allowed for much of the Soviet period yet were reared from eggs centrally and distributed to the households. The healthcare of these chickens was centrally managed, with a *kolkhoz* veterinarian being responsible for ensuring regular inoculation (Interview with key informant, 17 August, 2005). With the demise of the Soviet Union, these structures collapsed, leaving rural households without their pre-collectivisation knowledge. This is manifest in many ways, for instance the ignorance of how to treat sick chickens (Field notes, 5–6 April, 2005). This same lack of knowledge is also the case with cattle production. We conducted a survey on farmer's knowledge of cattle health and milk production in October and November of 2005, including 50 in-depth interviews with farmers. The findings of this survey confirmed the issue of knowledge loss, which was even identified by a number of respondents themselves, noting the decline in cattle rearing post-de-collectivisation.[3] The education level of those involved in livestock tasks, most importantly feeding and milking, was

---

[2] We are conscious here that pre-1920 knowledge on livestock was far from static. Rather the Russian Imperial history has bought with it considerable amounts of new knowledge and different animal breeds. Likewise the gradual shift away from nomadic and pastoral livestock production towards centralised rearing should not be seen in an historical context. Rather we are discussing a phenomenon of post-Soviet knowledge loss which is very different from these knowledge 'transitions' between different modes of production, because it was the shock event of decollectivisation that destroyed one system of knowledge whilst not fully developing the new system. This is thus a study of knowledge in dynamic transition.

[3] For instance, over 80 percent of respondents had cattle, sheep or both, yet only 17 percent of these respondents claimed to have had any formal or informal training in livestock health of foddering practices.

limited. Very few respondents expressed any knowledge of sanitary and hygiene rules associated with milking, the one woman who did employ a strategy for sanitation possessed this 'specialised' knowledge because of Soviet era training. Whereas two of the respondents (both men) were owners of a large number of cattle and possessed a superior level of knowledge about the anatomy and feeding requirements of cows, yet professed that they applied little of this. We were unable to fully understand why this was the case, but our suspicion is that the cause is tied up with the ceiling on entrepreneurship discussed later in this paper. Another interesting aspect of livestock knowledge is the gendered issue of knowledge transmission within the family structure. To quote directly from a research assistant:

> "People learn farming mostly from family members, from their childhood taking care of cattle is one part of their life. There is no special age or time to teach farming for children, they learn farming as a one part of their daily lives. According to the survey, male children learn farming from their fathers, 18 men from the 27 male respondents learnt farming from their fathers. The other learnt farming from the everyday life practice, this could have been the absence of their father or they lived with their mother in the childhood. The female farmers learn farming both from female members of the family and the male members of family. For example, in the survey, 10 female farmers learnt farming from their mothers and 11 women learnt from their father and both from mother and father."

This male dominated knowledge transmission process, combined with the issue of knowledge loss reinforces the importance of understanding how cultural norms find their expression in knowledge sharing processes, even in cases where this knowledge is diminishing. What it also illustrates is the state of "relative ignorance" and indeed the post-1991 growth of ignorance, which has occurred in rural Uzbekistan. This is not to say that no knowledge exists, on the contrary rural Khorezm illustrates the ways in which farmers respond to challenges in creative ways. For instance how social networks are used to promote the breeding of cattle and sheep with those from other *kishlaks* (villages). Here farmers are demonstrating that they understand the risks of inbreeding (and indeed vocalised this understanding in interviews) and are acting upon this knowledge in a culturally grounded manner. That is using existing social and community linkages with other *kishlaks* to mutual benefit. This is another instance where knowledge takes a form and function which mirrors the cultural context in which it operates. Equally we should remember the political power function that knowledge plays in agriculture and realise that Soviet centralisation of production and the specialisation that this entailed were not politically neutral, rather the process served to further the centralisation of control and to reinforce central power. With the collapse in agriculture, there have been masters and large scale farmers who have been able to profit from superior knowledge (or political connections) to build businesses based on the knowledge deficit of others. Thus the knowledge loss is dynamic, it is changing and local solutions are being developed to confront the post-1991 collapse or livestock knowledge.

2.2 *Post Harvest Processing*

During the Soviet period almost all industrial processing of raw agricultural materials occurred outside of Uzbekistan and Central Asia (Spoor, 1999:5). Cotton, wool, leather and other agricultural commodities were 'exported' to other Soviet Republics for processing, in accordance with the Soviet doctrine of division of labour.[4] This created a system of agricultural production and dependency, reminiscent of European colonialism, with uneven development and reciprocal differentiation (Wall, 2004; Kandiyoti, 2002a) and one that had a deleterious effect upon indigenous knowledge in Khorezm. Without entering into the discourse on whether Soviet rule of Central Asia constituted a colonial relationship, it is worthwhile noting that the impact on the indigenous knowledge regarding commodity processing bears much in common with colonial experiences from India and elsewhere. So just as colonial India was created as a dependent, vassal,

[4] Interestingly, this was taken as much from descriptive work of Marx on the capitalist labour process as it was from the normative writings of the American management scientist Taylor.

state by way of moving all processing of cotton towards England (Baran, 1957) a similar case arguably occurred in Uzbekistan. This is because the knowledge associated with how to process agricultural commodities, such as cotton and wool, was simply 'lost' or destroyed between 1917 and 1991.

Cotton prior to the Soviet period was also grown to order for Tsarist authorities and processing was centralised towards Moscow (Peachy, 2004:3). The attempts at creating post-harvest facilities in the period after 1991 have had mixed success, whilst the industrialisation of the cotton industry remains a state priority, numerous formal and informal barriers are erected. The economic and political problems behind these barriers are the topic for another study. Rather, we focus here on a common commodity, wool. Whilst cotton is economically, ecologically and socially the most important crop in Khorezm and Uzbekistan; post-harvest processing remains slight (Kandiyoti, 2002a, 2002b). Likewise wool plays a minor role in the economy of Khorezm, yet holds considerable potential, as explored below.

Case study: Wool processing

We know from historical writings that there was a well developed wool industry in Khiva during the 'Kushan' period (Tolstov, 1948) and it is reasonable to assume that domestic processing occurred until the imposition of soviet rule (circa 1917–1924). However, there is little indigenous processing of wool occurring in Khorezm today. Whilst small domestic production of woollen socks and gloves does occur, this is a rather specialised activity; we were able to find only two households in the research village who were engaged in this trade.[5] Likewise, there are two carpet factories in Khorezm which process large amounts of wool, yet these operate outside of the indigenous knowledge system. Economically, wool products make up only a fraction of internal trade and are negligible in terms of exports

[5] It is of course possible that the kishlak was exceptional or that we simply missed a form of processing, however the experiences were of a considerable level of ignorance about wool processing.

(Ruzmetov *et al.*, 2004:8–10). Yet sheep rearing is quite common in Khorezm, with 16 percent of respondents to our rural survey (N = 457) reporting that they kept sheep.[6] Thus we attempted to explore why it is that wool is not being processed and found that knowledge loss is a significant contributing factor. For instance, one informant, the owner of over one hundred sheep and forty goats, which he grazes in the desert, is an eloquent example of how far the wool processing industry has declined. Despite having such a large flock, it is simply not economically worthwhile to sell his fleeces "I get 50 cym[7] per kilo of wool — it costs that much just to shear the wool — I am not interested" (Polvon, Interview, 13 May, 2005). Thus each year the wool is composted in the desert and goes to waste. This is an understandable reaction to problematic economic conditions and is perhaps little related to knowledge loss. However, the astounding aspect of the Polvon case study is that he is interested in making a profit from his wool and sees it as potentially valuable. Yet he admitted to be unaware how he could make a profit from this latent resource. One option involved turning the wool in ropes, with which to tether his sheep at night. Yet even for this he was going to have to consult an *usta*, or 'master'[8] (in this case an agricultural engineer) to access this knowledge.

This stands in some contrast to the Khiva carpet factory, which was established in 1972, producing wool-synthetic blended carpets for around the Soviet Union (Interviews with factory manager and head of work brigade, 7 October, 2005). Whilst there were obvious disruptions in the immediate period after 1991 and the disintegration of the Soviet Union, the factory has been able to continue production, indeed in 2001 a large investment was made in buying new German technologies (Field notes, 7 October, 2005). In discussions with the various 'brigade leaders', the heads of each manufacturing

---

[6] An average of 5.94 sheep per household which reported having sheep, with a maximum of 25 and a minimum of one.

[7] Approximately US$ 0.05 / EUR 0.04 in 2005.

[8] An emic term from Khorezm which denotes 'mastership' in a particular realm of knowledge. For more discussion see Wall (2006).

process in a form of labour organisation imported from the Soviet era, each demonstrated how they were able to apply their knowledge and adapt to new realities of operating post-1991 (Field notes, 7 October, 2005). Indeed, the investment in the German technology meant that new knowledge was acquired, and adapted to suit local conditions, reversing this trend of knowledge loss. It is worth noting that now they use much less wool in their carpets, explained both in terms of cost and in the difficulties of sourcing quality wool domestically. Given the low price for wool explored above, this disconnect may suggest that there is a crucial need for development in the wool post-harvest sector, and one that certainly involves knowledge as a central point.

As explored elsewhere in greater detail (Wall, 2006; Wall and Lamers, 2004; Kandiyoti, 2003), the state's control over the labour process (direct power) and indirect control impeding innovation, both lead to this knowledge loss and the failure of the local knowledge system to innovate. By restricting (through power relationships) the development of new technologies, and continuing to control labour in a manner which favours large scale 'mechanised' industries the local knowledge system has been unable to develop or rediscover the knowledge needed to process wool. Rather a set of economic barriers, discussed later, combine with a simple lack of alternative sources for knowledge. The state retains a monopoly on agricultural knowledge, controlling through direct and indirect means what knowledge can be developed. In the case of wool processing, it is not a state priority and thus the local system remains in ignorance of how to process this potentially valuable product.

2.3 *Post-Soviet Knowledge Loss*

Knowledge loss in Khorezm was not restricted to the Soviet period. In the years since 1991, when the Soviet Union collapsed, farming in Khorezm has gone through significant reorganisation, which has at times led to even more knowledge loss. Whilst the Soviet era system of knowledge governance was imperfect, it did provide a well resourced agricultural research infrastructure (Morgunov and

Zuidema, 2001). The examples of knowledge loss discussed above refer to processes which were either related to Soviet-era agricultural organisation or which were on-going at the time of independence. We discuss here two instances of where the demise of the USSR led directly to knowledge loss in Khorezm.

Case study: Kolkhoz communism cattle farm

An eloquent example of knowledge loss is that of a large private cattle farm just outside of *kolkhoz Communism*[9] in Gurlan. It consists of 200 cows, 12 pigs, 70 hectares of cropped land and 15 employees. Privatised from the collective in 2001, it was sold to the local animal expert who had worked at it previously and who held high esteem within the village. He had studied animal sciences in a Moscow Institute and was a 'master' in animal health and milk processing. It seems that the specialised knowledge of this 'master' was a key rationale for why it was privatised to him and not to another individual, the purchasing process remained opaque. This was described by the current farm manager in that the 'master' had understood how to make excellent cheese, how to care for the animals when they were sick and was an 'expert' on all issues of farm and livestock management (Interview, 19 May, 2005). When we were shown around the farm, the current manager spoke of all the challenges that they now face as a business because the 'master' died the previous year at the age of 43 — leaving only young sons (the oldest being in the 8th grade, circa 15 years old) and married daughters. Thus it was left to his wife to continue as manager — even though she had some training whilst living in Moscow, she was not a 'master'.[10] Thus all the accumulated knowledge of the 'master' was lost, with very little evidence that those left behind were able to continue. "We carry on doing our own jobs as before, but do not know how to do many things that the master knew" (Interview, 19 May, 2005). The

[9] Now officially going by another name, but locally referred to in the old manner.

[10] How much of this was because of her gender and how much was because of a lack of knowledge, we are unsure of.

decline in the farm was palpable. Only 14 of the 200 cows gave milk anymore, there was a lack of knowledge about breeding and encouraging milk production. Likewise, the business no longer produced cheese of any sort, instead selling (less profitable and less transportable) cream on the local market. The most post-harvest processing that occurred on site was the boiling of cream to make baby food, utilising only a Chinese separator (to separate the cream) and a wood fired *kazan* (large pot) to boil the cream in. In 2005 the business was, for the first time, growing cotton — largely because they have been unable to continue making money from the cows.

Other efforts at diversification, which had been started by the 'master', were flagging for instance pig production. It was initially the idea of the master, but after he died the herd is being slowly culled. One example of knowledge loss within the farm became evident at an occasion when a first time mother pig crushed all but one of her piglets. The manager did not know if this was normal or what to do about it. They had only two sources for knowledge to replace that which was lost. Either from the Farmers Union in Gurlan which provided booklets, or from the wife of the deceased master, with few other options apparently open to them. In either instance this external knowledge could not replace the knowledge that was suddenly lost with the early death of the 'master', and with only young sons and no other knowledge reproduction strategy; this knowledge was lost to the cattle farm. This incidence clearly shows the importance of a succession of generations for knowledge loss in case there are no institutionalised avenues to pass on local knowledge.

What this illustrates in the case of post-Soviet Uzbekistan is that the agricultural knowledge system has been unable to adapt to the economic and social disruptions of post-1991 independence. There are insufficient levels of knowledge within the local system and, constrained by the state, it is not possible for the local system to innovate and create new knowledge internally, or to access external knowledge sources outside of the state system. As a corollary of this, we see how state control and interests are actually enhanced by the situation, with the farm turning (voluntarily, if for want of other choices) towards cotton production which is a central state interest. Thus the states monopoly on agricultural knowledge is reinforced, with this 'privatised' farm reverting to the centralised knowledge of the state for cotton, precisely because of knowledge loss from the local knowledge system.

Case study: Seed selection

Seed selection has, for some species, deteriorated rapidly in the post-Socialist period. The loss of improved varieties, especially for maize, has been caused by a break down in the collective systems of seed breeding, selection and distribution. Whilst the GoU has been largely effective in ensuring the supply of improved varieties of cotton and wheat (the strategic crops) there has been a decline in the availability and quality of improved seeds for maize and some other cultures, including potatoes (Nasriddin, Interview, 26 April, 2005). So whilst state attention is focused on the two strategic crops, which command their own breeding centres, other crops lack centralised seed selection centres. In many ways seeds are a physical expression, an artefact, of knowledge. The ability to select and reproduce improved varieties involves a complex set of knowledge, for example procedural knowledge in how to select seeds and dynamic knowledge in constantly improving strains, and the end result of improved seeds are an expression of this knowledge chain.

Seeds are symbols, invested with knowledge, which illustrate how indigenous knowledge is created, shared and used. Yet what we observed in Khorezm was that this knowledge chain had been broken. Seed improvement techniques that we know existed during the Soviet period (various archival sources; Nasriddin, Interview, 26 April, 2005) have subsequently broken down. For example, the *sovkhozes* and *kolkhozes* previously conducted a lot of their own seed selection and storage for non-strategic crops whereas cotton and wheat were generally the concern of higher institutes or specialised academies (Truth in Khorezm, 22 August, 1959). This was conducted by trained specialists, within the ambit of their work at the collective farm, and these seeds were then also passed horizontally and vertically upwards through the network of collective farms

(Unknown, 1988:308–310). What improved seed did exist for varieties such as Maize in the form of imported hybrids, often labelled locally as '*Ulughbek*', was distributed through the *kolkhoz* farm system. In post-Socialist Khorezm non-strategic crops are increasingly grown from heritage seeds that the farmers collect themselves. This reversion to heritage seeds has been significant, necessitated by the break down in the former *kolkhoz* farms and systems of seed selection and improvement (Van Dusen, 2006). If we accept that seeds are the physical manifestation of a knowledge chain, then it is fair to discuss the quality of these seeds (in terms of harvest quality and yield) as an expression of the knowledge inherent in these seeds. This is where knowledge has been lost in post-Soviet Uzbekistan.

The varieties of seeds available for crucial fodder crops such as maize and sorghum are inferior to those previously available in collective farms (Nasriddin, Interview, 13 April, 2005). The same is even true, to a lesser extent, with wheat. Whilst specialised breeding centres do exist within Uzbekistan, it would appear that high quality wheat seeds are not distributed through the former *kolkhoz* system. Whether for lack of infrastructure, finance or political will, state plan farmers do not always receive improved wheat seeds (ibid.). Indeed, our interviews identified that it was necessary for a farmer to travel to the Jizzax or Samarkand rayons in order to buy improved wheat seeds. The same is not true for all crops. Indeed imported European seeds for various kitchen vegetables and cash crops such as watermelon; cucumber and tomatoes are available in the bazaars of Khorezm.[11] Yet this importation is exactly the point, there has been a loss of knowledge of improved seed varieties and how to develop these within Uzbekistan. The increased reliance on seed sources from outside of Uzbekistan is emblematic of the knowledge lost in the post-Soviet period. Although it should be noted that this situation is complex, as vegetable production has recovered to almost pre-1991 levels (Ali *et al.*, 2003:21). Yet this has been because of indigenous knowledge creation and local knowledge sharing, rather than because

[11] Vegetable seed distribution is discussed in more depth in a case study in the next section.

of any explicit state assistance (see next case study, also Van Dusen, 2006). So whilst the state has continued to invest in cotton and wheat production, this somewhat myopic policy has lead to a marked reduction in the quality of genetic material for other crops, especially those which provide nutritive fodder for livestock. This is because the state does not profit from areas of agriculture outside of cotton and wheat, yet exercises control over the entire agricultural production process. This restricts the development of local knowledge because innovation is not encouraged or even really allowed.

With the break down of Soviet era capacity in seed selection, there has not been investment from the state in non-strategic crops. In some cases the knowledge to do so has 'leaked' back to Russia, in other cases the knowledge potentially exists to select better seeds but the physical infrastructure to allow this knowledge to be used, is not present. Because seeds are a carrier of knowledge, the knowledge on seed selection needs to be used to be effective, in the absence of use this knowledge is being lost, and it is only being recovered because of the growth of indigenous knowledge in Khorezm (see next section) rather than because of any state assistance. This demonstrates how knowledge loss is more fluent than might otherwise be assumed, as we see clear evidence in the case study on seed selection, of how local people (especially women) are actively creating and sharing knowledge through their selective breeding (and sharing) of vegetable seeds, using both indigenous and introduced varieties to deliver improved nutritional and economic outcomes.

2.4 *Limits on Knowledge*

> "If I get more I will have to give all to the kolkhoz, there are a lot of taxes... everyone wants a tax. The environmental protection department, the customs, everyone... so it is not worth having more sheep" (Polvon, Interview, 13 May, 2005).

There are also limits on individual farmers and on agriculture in general that contribute to this phenomenon of post-Soviet Knowledge loss. It has been said that "there are no medium sized businesses in Uzbekistan, only large and small ones" (Rasanayagam, 2002:55). This reflects the dominance of state sponsored companies in all spheres of the economy. Whether these are official monopolies such as the cotton sector, or businesses which are officially private yet are controlled by the same political class that controls the rest of society and the economy, so called 'minister millionaires'. At the farm level these restrictions are played out in a 'ceiling' that is placed upon individual ambitions and entrepreneurialism. Wealth building and value adding to commodities is possible only to the extent that ones' political capital allows, which in the case of most '*kolkhozniks*' (rural farmers) is very low. Without going into the details of the economic system, the effect that this has on knowledge creation is stifling. Farmers remain unwilling to expand their production (see above quote) because of a real concern that they will end up worse off. Processors of raw products express a similar concern (Interview, 11 May, 2005). Likewise, because the labour process remains state rather than enterprise controlled, insufficient surpluses are being generated to allow experimentation and greater knowledge flows. There is also a more direct restriction on innovation, with state norms and mandated methods preventing the development of local knowledge. Thus the preconditions for knowledge creation, to replace knowledge that has been displaced with the fall of Communism, do not exist in Khorezm, nor is the current government allowing such a condition to develop because of the link between power and knowledge, which is central to state control over agriculture. These limitations and restrictions on economic life, which have direct consequences for knowledge creation and loss, should be considered as part of the system of knowledge within Khorezm, as a key constraint to indigenous development. This is because knowledge would be able to develop indigenously were it possible for producers to profit from further developing their production and labour processes. Yet this is not possible under the current system of state economic and knowledge control.

3. Theoretical Implications

What is clear from the examples in rural Khorezm is that there is a phenomenon of 'Knowledge Loss'. This is when knowledge within the local knowledge system disappears or ceases to be available. The displacement of old knowledge with new knowledge is a natural, indeed essential, aspect of a dynamic knowledge system (cf. 'normal science', Kuhn, 1972). This is not the type of loss we wish to discuss here; indeed we distinguish between this process of 'normal science' where new knowledge displaces the old, and genuine knowledge loss. Rather when we conceptualise knowledge loss, it is the destruction or leakage of knowledge from a discrete knowledge system. This form of knowledge loss is not replaced and thus is distinguished from displaced knowledge. So when old knowledge is improved upon and new lessons learnt this form of loss is not of concern, it is only when knowledge is lost and not replaced, that we consider this the concern of knowledge management. We theorise here that knowledge loss is an area of crucial interest for knowledge management and knowledge governance literature, because it demonstrates a failure in a knowledge management (or governance) system. Parallel to 'market failure' we can therefore speak of 'knowledge system failure' or in short 'knowledge failure'. Whilst some level of knowledge loss is probably unavoidable (through leakage into other systems) it demonstrates a failure to adequately manage and utilise the knowledge of a community, organisation or nation state. Thus it merits greater attention in the literature. Moreover, the phenomenon of knowledge loss would not appear to be restricted to any one system of knowledge, examples abound in universities, companies and indeed nation states.[12] The local knowledge systems studied in this paper certainly suffered some extreme instances of systematic knowledge loss, related to the downfall of the Soviet Union and the impediments to innovation that the governance structure enforces.

We theorise that there are various causes for the loss of this knowledge, ranging from the death of a knowledge broker, through to an attrition of knowledge due to misuse. These drivers of the knowledge loss phenomenon are explored in greater depth in the subsection below. However, we would caution that the term

[12] For reference to the dominant definitions of Knowledge in Singapore and Germany see Hornidge (2006).

'knowledge loss' does not necessarily mean that the knowledge is destroyed or is irretrievable, as there are cases in which knowledge is not so much lost as 'leaked' to another knowledge system. Thus the knowledge still exists, just not in the same knowledge system. This 'leakage' is discussed as one of the drivers of knowledge loss, along with the other drivers of displacement and misuse. Finally, we discuss the theoretical aspects of knowledge loss, attempting to establish some criteria for how to classify knowledge loss as a phenomenon worthy of study in knowledge management (KM) and knowledge governance. Knowledge loss is certainly an important area of study for KM theorists, as we attempt to better manage existing knowledge resources.

3.1 *Drivers of Knowledge Loss*

Knowledge is lost from a community in a number of different ways. Detailed here are the most common forms of knowledge loss as experienced in the field research presented above. Whilst some of these modes or drivers of knowledge loss may be unique to the field setting of rural Uzbekistan, many are not. Likewise, different communities of knowledge (e.g., the corporate world) may experience different manifestations of knowledge loss, although we suggest that the drivers may be similar, even if the form they take is different.

Death or displacement

Because knowledge is communicated between and carried by humans, when an individual dies or leaves a group, the knowledge they once held can be lost to the group or community. We saw in the case of the kolkhoz communism cattle farm, how the death of one key individual, especially a 'master' who held unique knowledge, had a deleterious impact on knowledge in the community. Certainly in local knowledge systems, where most knowledge is held personally and in an unwritten form, the risk of a 'master' dying has implications on the knowledge of the community. Yet it is not only death that can cause knowledge loss, but also simply an individual leaving the group, say, to take up alternative employment. It is thus predictable that knowledge is 'lost' to a system of knowledge because most knowledge is held by individuals, not collectively. Yet we find it surprising that the issue of key staff moving (or dying, somewhat more dramatic) is considered more of a human resource management than a knowledge management issue. Thus we would argue for a greater recognition in the literature of the risks of death and displacement for knowledge loss.

Misuse and misplacement

Knowledge can also be lost if it is misused or misplaced. That is, if knowledge is not applied and utilised, then it can be lost altogether. Certainly practical know-how, *aptitude*, is lost if it is not used and transmitted (shared) in its use. We see in the local knowledge system how collectivisation and the centralisation of post-harvest processing meant that local knowledge was lost precisely because it was not able to be utilised. This was not necessarily through any attempts at indoctrination or through wilful misplacing of the knowledge. Rather the indigenous knowledge on these topics was not used, thus it was not passed on from generation to generation, and it has been 'lost' to the knowledge community. The issue of unique indigenous knowledge being 'destroyed' is discussed in the literature, especially in cases of the medicinal uses of plants and of indigenous methods of conservation (cf. Stevens, 1997; Benz *et al.*, 2000). However, in the case of Khorezm this indigenous knowledge has been lost or destroyed for some time and it is only now (with the paucity of knowledge in the post-Soviet agricultural system) that this loss is acutely felt. As lamentable as the loss of indigenous knowledge is, the more important issue now is preventing the continued knowledge loss which is not being matched by developments in the rural economy or local knowledge system. The converse aspect of this is how indigenous innovation can be used to recreate knowledge which has been misused and misplaced, whilst preventing further knowledge loss.

Leakage

Knowledge loss does not always mean that the knowledge has been destroyed or lost entirely from the universe of knowledge. Rather from a systems perspective, all it denotes is that knowledge has been lost from the community or knowledge system. Thus a farmer moving from Khorezm and migrating to Russia (a common occurrence) carries with them a great deal of knowledge which is 'lost' to the community in Khorezm, yet which contributes to agricultural knowledge in the recipient locality (thus it is displaced). We label here this as 'leakage', a form of knowledge loss which is perhaps less dramatic than knowledge 'destruction'. The impact on the community (or potentially corporation) is perhaps similar in that their access to the knowledge is lost. Yet leaked knowledge is potentially retrievable (though it may not be) and does continue to exist in another knowledge system, with which knowledge sharing should still be possible. Thus leakage is the least dramatic form of knowledge loss, yet it is still an important driver.

Another form of knowledge leaking refers to the change of knowledge from one domain into another. Medical knowledge in cattle farming may no longer be used in that particular domain but may find acceptance in small animal farming.

4. Conclusion: Towards a Theory of Knowledge Loss

Knowledge loss is not a remote phenomenon, unique to one knowledge system. Rather we argue that the loss of knowledge is an issue for other knowledge systems as well. Knowledge loss is certainly a concern for anthropologists working on indigenous knowledge, fearful of 'losing' indigenous knowledge entirely as a result of modernisation (Cox, 2000). Equally, staff movements within the corporate world probably lead to a large amount of knowledge displacement, yet staff (and thus knowledge) retention is more often seen as a human resource than a knowledge management issue. Similarly in academia, which thrives on the wide interchange of knowledge and ideas and openly promotes the exchange of knowledge, much of this knowledge can be 'leaked' (i.e., it leaves academia for another knowledge community, say, a corporation) or it can be 'lost' altogether. Thus we attempt here to explain in theoretical terms how knowledge loss operates, what are the drivers of knowledge loss and how these can be ameliorated. We suggest that knowledge loss is a failure of knowledge management insofar as it demonstrates a lack of knowledge sharing, dissemination and use. The central argument being that knowledge must be reproduced (or stored in a repository) for it to be used and to continue to exist. Because local knowledge resides in individuals, who are apt to move to different knowledge systems (leakage) their doing so carries with them a considerable amount of knowledge. Key to reducing this is effective knowledge sharing during the time they are within the community or organisation. This provides the inherent benefit of greater knowledge utilisation through greater knowledge sharing, as well as reducing the risks of knowledge loss. Yet, individuals do not always share knowledge, when they do this sharing can be partial. In many cases this is because of the high transaction cost (and risk) associated with sharing their knowledge. We argue that knowledge management and knowledge governance theory needs to inform institutions (informal and formal policies) which can introduce better protections for individuals to share knowledge, in order to reduce the transaction costs of knowledge sharing. These transaction costs can be lowered by guaranteeing continued ownership of intellectual property, by establishing a proper policy framework for academic honesty and by enforcing these rules in a transparent manner. In the case of local knowledge the transaction costs are somewhat reduced by knowledge sharing within the family, shown in generational transfer of mastership. In the same way should projects, corporations and ultimately nation states develop structures which allow for enhanced knowledge sharing, by reducing the transaction cost of sharing this knowledge. Part of these systems must allow for knowledge which is no longer relevant, which is not useful or which is simply wrong, to be replaced by more appropriate knowledge. In this regard simple databases are somewhat counterproductive as they do not encourage the dynamic displacement and replacement of knowledge, which whilst it involves some knowledge 'loss' is actually a knowledge creation and sharing process. Thus we theorise knowledge loss as a phenomenon to be evidence of poor knowledge management. In its own right it is a failure of management and governance to allow knowledge resources, expensively produced within the community, to be lost. On a wider level it evidences a lack of knowledge reproduction and retention, which can be seen as a result of excessive transaction costs and risks to knowledge sharing.

References

Ali, M., T. A. Lumpkin, and U. Farooq (2003). Vegetable research and development in central Asia: A guideline for setting Priorities under data scarcity. *AVRDC — The World Vegetable Center*, 1–21.

Baran, P. A. (1957). *The Political Economy of Growth*. New York & London: Monthly Review Press.

Benz, B., E. J. Cevallos, M. F. Santana, J. Rosales, and S. M. Graf (2000). Losing knowledge about plant use in the sierra de manantlan biosphere reserve, Mexico. *Economic Botany*, 54(2):183–191.

Cox, P. A. (2000). Will tribal knowledge survive the millennium? *Science*, 7:44–45.

Evers, H.-D. (2000). Globalization, local knowledge, and the growth of ignorance: The epistemic construction of reality. *Southeast Asian Journal of Social Science* 28(1):13–22.

Evers, H.-D. (2010). 'Knowledge' and the Sociology of Science. In *Governing and Managing Knowledge in Asia*, T. Menkhoff, H.-D. Evers and Y. W. Chay (eds.), pp. 59–70. Singapore: World Scientific.

Evers, H.-D. and T. Menkhoff (2010). Expert knowledge and the role of consultants in an emerging knowledge society. In *Governing and Managing Knowledge in Asia*, T. Menkhoff, H.-D. Evers and Y. W. Chay (eds.), pp. 141–161. Singapore: World Scientific.

Hornidge, A.-K. (2006). The construction of K-Societies: Germany and Singapore. Ph.D. Thesis TU Berlin, Institute of Sociology.

Kandiyoti, D. (2002a). Post-colonialism compared: Potentials and limitations in the middle east and central Asia. *International Journal of Middle Eastern Studies*, 34:279–297.

Kandiyoti, D. (2002b). Agrarian Reform, Gender and Land Rights in Uzbekistan. United Nations Research Institute for Social Development.

Kandiyoti, D. (2003). Pathways of farm restructuring in Uzbekistan: Pressures and outcomes. In *Transition, Institutions, and the Rural Sector*, M. Spoor (ed.), pp. 143–162. Maryland: Lexington Books.

Kuhn, T. (1996). *The Structure of Scientific Revolutions* (3rd Edition). Chicago: University of Chicago Press.

Morgunov, A. and L. Zuidema (2001). The legacy of the soviet agricultural Research system for the republics of central Asia and the Caucasus, *ISNAR Research Report.* www.isnar.cgiar.org/publications/pdf/rr-20.pdf

Peachey, E. J. (2004). The aral sea basin crisis and sustainable water resource managementin central Asia. *Journal of Public and International Affairs*, 15, Spring.

Rasanayagam, J. (2002). Spheres of communal participation: Placing the state within local modes of interaction in rural Uzbekistan. *Central Asian Survey,* 21(1):55–70.

Ruzmetov, B., Z. Rahimov, and I. Rudenko, (2004). Analysis of farmer enterprises and agricultural markets in 2, N. (ed.), ZEF Work Papers for Sustainable Development in Central Asia.

Spoor, M. (1999). Agrarian transition in former Soviet central Asia: A comparative study of Kazakhstan, Kyrgyzsrtan and Uzbekistan, *Working Paper 298*, Institute of Social Studies.

Stevens, S. (1997). *Conservation through Cultural Survival: Indigenous Peoples and Protected Areas.* Washington D.C.: Island Press.

Tolstov, S. P. (1948). Following the tracks of Khorezmian civilization. Academy of Sciences of the USSR (VASKhNIL).

Unknown, A. (1988). *The Agricultural System and the Adoption of New Technologies.* Tashkent: Fan Publishing House.

Van Dusen, E. (2006). Evolving land tenure in rural Uzbekistan — A case study of smallholders and diversity in home gardens. Paper presented at the Workshop: "Assessing the linkages between reform, Land Tenure and Land Use". Humboldt University, Berlin, Germany, May 25, 2006.

Wall, C. (2004). Multiple barriers to technology change in rural Uzbekistan: A development perspective. *Masters Thesis — Massey University, New Zealand.* www.zef.de/download/articles/ wallMultiple_Barriers_technologychange.pdf

Wall, C. and J. Lamers (2004). Farmer priority setting: Issues and research needs for Khorezm, Uzbekistan. *Central Asian Journal,* 4(1): 5–25.

Wall, C. (2008). *Argorods of Western Uzbekistan: Knowledge Control and Agriculture in Khorezm. ZEF Development Series.* Berlin: LIT Verlag.

Chapter

14

Local Knowledge as a Strategic Resource

Fisheries in the Seasonal Floodplains of the Mekong Delta, Vietnam[1]

Solvay Gerke and Judith Ehlert

1. Introduction

The Mekong Delta is vibrant. People have lived alongside its watercourses for centuries, adapting to the natural environment and structuring their everyday life around it (Tran and Few, 2006:141). The flood season is traditionally the time of year during which the local population base an even greater amount of their nutrition on

[1] This article is based on field research undertaken in the Mekong Delta in 2008/09 as part of the WISDOM project (Water-Related Information System for the Sustainable Development of the Mekong Delta in Vietnam) at the Center for Development Research (ZEF), financed by the German Ministry for Education and Research (BMBF).

natural fish,[2] which has been an abundant source of subsistence in the past. However, where there is an abundance of water, there is not inevitably one of fish. Hence, ecology in the Mekong Delta is changing; the relationship between society and water is altering as humanity appropriates its surrounding natural resources. In the process of developing and modernising a rural area under pressure from a growing population, natural water products are decreasing. Industrialisation and agricultural intensification are triggering episodes of increasing environmental pollution, which affects the local water resources on which rural and urban livelihoods depend upon (Kelly *et al.*, 2001; Adger, Kelly and Huu Ninh, 2001:3).[3]

This article will discuss the relevance of local knowledge as a strategic resource in the fierce arena of competition for natural fish products. The importance of local knowledge, especially for the development of local communities, has become increasingly recognised by research worldwide (Antweiler, 1998; Evers, 2000; Neubert and Macamo, 2002). The broader term "local knowledge" includes any "locally and culturally situated knowledge that was and still is produced in local communities (Antweiler, 1998:490). Nonetheless, local knowledge does not necessarily result in sustainable or socially just activities, nor is it always shared by all members of the community. In this article, it is argued that local livelihood-based knowledge on water is neither collective nor integrative per se; rather, it is a strategically applied resource in the process of sharing and closing-off livelihood opportunities. Under the constraining conditions of livelihood insecurity, landless people, who have traditionally based their household economies on natural fish resources and fishing during the flood season as one component of wider livelihood strategies (DFID 2002), cautiously calculate the risk on whether and with whom to share their knowledge. This article will illustrate the measures that people apply and the actions they pursue deliberately or

[2] For regional aquatic resource consumption patterns, see DFID (2000:19).

[3] For a discussion on the impact of the dyke system in Can Tho City on natural fish resources, see Pham, Ehlers and Subramanian (2009:19).

unconsciously in order to guarantee that their knowledge on fish resources remains an exclusive and competitive asset, now and in the future.

We will place emphasis on the perspective of landless people and paddy farmers, since it is these actors' 'life-worlds' (Schütz and Luckmann, 1973) that serve as a reference for local rationalities of action. Our focus will concentrate on the changing human-water environment in order to better 'understand' — in the Weberian sense — the application of fishery knowledge as an adaptive process in the given environmental context. The underlying theme of all the interviewees' answers indicates a sharp decrease in natural fish stocks, the growing scarcity of which has been perceived as a qualitative change in the local natural environment, which has had an impact on people's actions pursued in order to adapt to this change.

In the following, we want to discuss two factors that impede knowledge from being a collective asset in a certain locality: First, we examine the tendency to control actively the transfer of knowledge in the struggle for natural resources growing in scarcity; and second, we look at the tacit character of local knowledge functioning as an immanent barrier to knowledge sharing. These two factors undermine the free movement of knowledge, instead sensibly determining to where it actually travels.

We will first give an overview of seasonal migration during the flood season to depict the context in which the encounter between land-owning paddy farmers settled in the area and landless people migrating to the area for fishing activities took place and were observed in the field.[4] The following elaboration draws on observations and interviews in a farming community in Can Tho City Province during the flood season. The flood season in the research site is characterised by an interesting dualism of rest and peace for the paddy farmers and dynamic action for the landless people engaging in

[4] The argument of this article is based on seventeen semi-structured interviews with mostly paddy farmers and landless people in a north western district of Can Tho City Province.

fish harvesting. How this encounter is structured on the basis of knowledge is an exciting aspect of local knowledge research.

2. 'Like an Ocean': Seasonal Migration and Knowledge Flow

The green fields of the Mekong Delta are harvested in August — before the seasonal rise in water levels. In October, the rice fields are covered with water and the scenery reminds one of an ocean, which is often how it is described by the local population.[5] All paddy farmers in the research site — a commune in the North-West of Can Tho City Province along the Cai San channel — cultivate two crops per year and let the floodwater into the fields after the summer-autumn crop. The flood season allows the soil time to rest and to take on the essential nutrients required for a successful winter-spring crop. By contrast, the area bordering this commune, belonging to another province — An Giang — follows a different flood management regime, as illustrated in Picture 1.

Farmers in this borderland keep the floodwaters out of their fields in order to help with triple cropping. However, this difference in flood management regimes stimulates seasonal migratory processes. Landless people in An Giang Province cannot harvest fish in the rice fields because the seasonal waters and, as a consequence, the fish are irrigated directly downstream without entering agricultural land. In the pursuit of accessible fishing grounds, landless people from the bordering Kien Giang Province, as well as those from the surrounding communes belonging to Can Tho City, temporarily migrate to the research site. The commune chosen as the research area is the focal point for the observation of such seasonal migration. The

[5] Flooding in the Mekong Delta is an annual event characterised by slow water onset, long inundation and slow withdrawal. The flood season in the delta inundates large areas in the period from July to November, with water levels constantly rising up from one to three metres, and affects about 8.5 million people, of which 17 percent live in urban and 83 percent in rural settings (Huu Ninh, 2007:6).

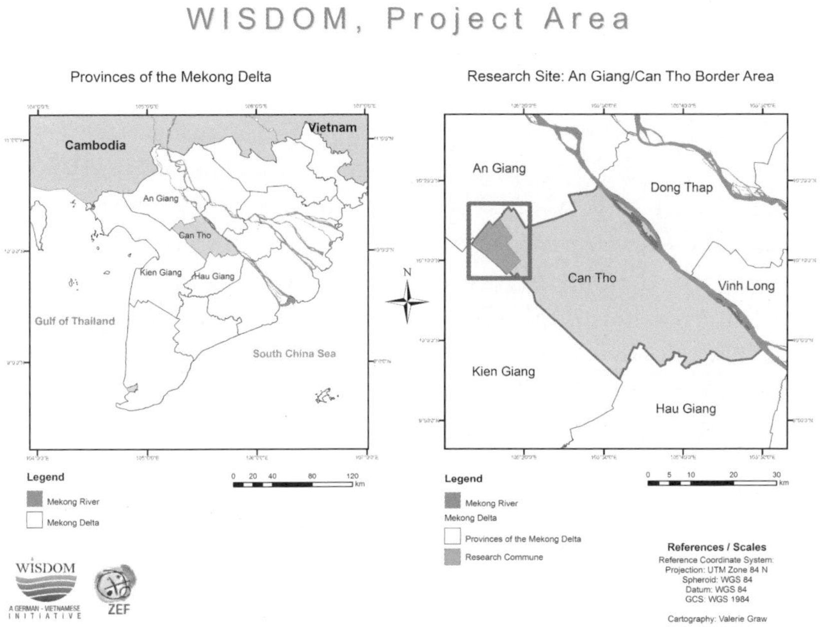

Figure 1: WISDOM, Project area

Photo 1: Flood management border lands: Green fields (3rd crop) in neighboring province An Giang and seasonal floodplains in Can Tho City province (H.-D. Evers)

natural and socio-cultural conditions in this commune make it a prominent destination for landless people coming from these neighbouring areas, because the flooded fields are wide and comparably rich in fish resources. Furthermore, land-owning farmers — Northern Catholic Vietnamese who came to the delta in the 1950s — base their livelihood exclusively on rice cultivation and thus leave the fishing niche open for exploitation by the landless. The Northerners' community in the North-West of Can Tho City forms a specific case of an 'internal diaspora' of Northern Vietnamese in the Mekong Delta. The Northerners possess larger landholdings than the 'local people', or 'Southerners', in the neighbouring communes of Can Tho City.[6] 'Local people' or 'Southerners' are terms used in the Northerners' community to refer to the people who lived in the area before they settled there; the term 'Northerners' is applied by the local population to denote the cultural and geographical origin of the new settlers. These emic categories serve to demarcate cultural 'otherness', often fuelled by mutual prejudice in terms of habits and character traits between the 'Northerners' and the 'local people'. Since it is only the people without any, or little, agricultural land among the 'local people' who migrate for fishing activities during the flood season, the Northerners use the term 'landless' interchangeably with 'Southerners' or 'local people' to refer to the people who annually visit to their lands. Although landowners in the Northerners' community use the flood season as a time to rest, the flood season is

[6] The termination of the first Indochina war in 1954, along with the foundation of a communist state in the north (Democratic Republic of Vietnam/North Vietnam) and a separate state in the south (Republic of Vietnam/ South Vietnam), induced internal migration. Between 1954 and '56, more than one million people moved south. Among them about 800,000 Catholics escaped the communist takeover of the north (UNHCR 2000:80). Catholic refugees arriving in the southern zone supported the succession of the anticommunist regime in Saigon, President Ngo Dinh Diem himself being a Catholic. Large-scale government support was given to the Catholics and the northern refugees strategically settled in rural development centres such as at Cai San channel in the western Mekong Delta between Long Xuyen and Rach Gia (Rambo, 2005:83).

an extremely important time for landless people from around the area, since it offers them natural fish for subsistence and small-scale business. The flooded fields in the area are openly accessible to anybody for fishing activities.[7] This is not only the case in the Northerners' community, but also common, informal practice. However, flooded fields in the Northerners' commune are wide and less crowded in the flood season because the commune itself mainly refrains from fishing on a larger scale. As soon as agricultural land is covered by a layer of floodwater, individual land property rights are redefined because the local populations as well as the seasonal migrants regard the flooded areas as common property. This is the period when seasonal migration for fish harvesting sets in. Landless people either come to the flooded fields every day by boat — if they live nearby — or temporarily settle down for the whole flood season on the residential land of rice farmers if they come from further away.

If people come from neighbouring provinces, they have to register with the commune's police department if they want to stay for the duration of the flood season.

Landless people migrate and carry with them knowledge on fishing tools, skills and experiences of how and where to make the biggest catch of fish. On their way, when temporarily settling down in farming communities for their fishing activities, they encounter rice farmers whose expertise lies in farming and who do not possess knowledge on fish harvesting. Further, they meet other landless people with comparable knowledge, who compete for the same fish resources on the paddy farmers' lands. This context poses questions

---

[7] Besides triple cropping, another exception for open access to flooded land, however, is culture fish raising. The modernisation of agri- and aquaculture will improve the opportunities for raising culture fish. This certainly is a reasonable measure to compensate for decreasing natural water products. However, the model of culture fish only benefits the landowners while flooded fields are literally fenced off and thus closed to landless people, who can neither participate in such models nor access the land for traditional floodplain fishing. During the flood season in 2008, however, the floodplain fish culture was not implemented and, in general, was not perceived as a successful model by the farmers in the research commune.

Photo 2: Seasonal migration: Landless people living on their boats for wild capture fisheries in the floodplains (H.-D. Evers)

about knowledge access: How is the transfer of knowledge controlled in the case of direct competition between landless people; and how is knowledge exchange structured in the context of non-interfering livelihood strategies between paddy farmers and fishermen?

2.1 *Local Social Spaces and Knowledge Encounters*

The concept of local knowledge suggests dynamism and flexibility. The stock of local knowledge is capable of integrating new knowledge and practical experience generated in the process of people actively engaging with and changing their natural environment (Antweiler, 1995:42). The integration of knowledge in this way, nevertheless, is depicted as a one-way endeavor of the recipient only, neglecting the position of the knowledgeable actor, who might hold back certain knowledge for strategic reasons. Furthermore, it underestimates the tacit component of local knowledge as something to be 'integrated' by the recipient; rather, one should talk of 'internalisation', which is achieved only through a long-term simultaneous process of practice and cognitive reflection. Consequently, this openness to 'external' influences might falsely imply the assumption that when people with different (or even the same) livelihood expertise and niches meet in a local space, 'new' knowledge is mutually 'interrogated', and then integrated through this kind of static exchange. Instead, such knowledge encounters constitute social spaces in which "communities of practice" (Gerke and Evers, 2006:5) meet beforehand to negotiate the rules and practices of knowledge transfer. Hence, not only might their knowledge as such enable people to cope with current situations or future challenges (Schröder, 1995:4) such as the growing competition for natural fish, but also the way they regulate its access might give them advantages in coping with growing livelihood insecurity. Since local knowledge is not something 'outside' or abstract from actors but instead their "embodied practice" (Pottier, 2003:2; Schütz and Luckmann, 1973; Knoblauch, 2005:13), the actors themselves either stimulate or constrain knowledge transfer. It is argued here that this negotiation process is strategically motivated.

It will be shown that, in the case of traditional floodplain fishery, the transfer of knowledge is less delimited and wide and bears no comparison with the picturesque image of the flood being like an open sea.

3. 'No Open Sea' — Limits to Knowledge Sharing

Migratory processes give rise to different people meeting and facilitating new social spaces, in which diverse actors with different livelihoods encounter one another. The flood covering the land with water reminds many people of a wide ocean that can be freely entered by everybody. This metaphor, however, only applies to open space in physical terms. This image of open waterscapes rather describes wide geographical openness. Sharing the same locality inevitably does not mean freely shared ideas, experiences and knowledge. In our case, the strategies of knowledge sharing reflect a highly tuned negotiation process of knowledge flows and capture, based on rational action.

Photo 3: Gone fishing in the flood plains (H.-D. Evers)

3.1 *Knowledge Sharing Strategies*

The human design of the Mekong Delta as a "hydraulic society" (Wittfogel, 1957; Evers and Benedikter, 2009) serves agricultural intensification by means of a complex irrigation system for the purposes of economic development. Channel and dyke water management schemes cut the open land of the delta into grid-like plots to which the floodwaters are guided (see Picture 1). Nevertheless, before the development of irrigation channels and border dykes, the floodwaters spread evenly everywhere,[8] and with this came an abundance of natural fish. In every oral history interview on flood experiences in the past, community elders remember the profusion of fish brought into their homes by the floodwater. In line with other historical accounts, one female respondent vividly recalls her memories from the big flood in 1978, when she and her family were literally sitting on their beds while casting their nets for the fish and snakes in their house.

In accordance with Wittfogel, a local journalist traces back the development of the delta as a hydraulic society. His explanation

[8] For a detailed description of the development of water resource management in the Mekong Delta, see Miller (2003).

tersely represents the popular view on the ecological nature of the Mekong Delta and the changing society-water relationship:

> "The flood is actually called the 'water-rising' season by the local people. The people lived normally with the flood. They lived with the water because at that time there was no dyke. The reason is that at this time the population was little and the cities were not so developed. People lived alongside the channels, rivers and the channels in the fields. Thanks to the water season, they lived by jobs such as catching fish and prawn, making fish sauce and raising fish. They lived completely by nature. [...] The government continued with the work to open the wide land, enlarge production, develop science and technology — they increased rice production because they brought science and technology to the wide land and developed infrastructure. They built factories, services, traffic streets and brought electricity. [...] But the natural water products are used up and limited because so many people catch fish and because of the multiple rice crops that cause environmental pollution — from the pesticides, from the agricultural waste water the water products decrease."

The main aim underlying the current transformation of the human-environment relationship in the Mekong Delta is the strategic use by landless people of water-livelihood knowledge for subsistence. Nevertheless, the motivation to capture or to acquire such knowledge in order to either close off or open up livelihood perspectives might grow as well as experience a new quality in the future. In the future, ecology will enter a new phase as agricultural modernisation is being pushed forward. In the pursuit of modernisation and in order to live up to the country's ambitious rice export orientation, the agricultural policy for the Mekong Delta — the 'rice bowl'[9] of Vietnam — foresees a further shift in land-use patterns from scattering to concentration (Nguyen, 2006:3). This concentration of land, with the ongoing growth of mechanisation, might aggravate the unemployment rate and stabilise the trend of landless households in

---

[9] Nowadays, the region accounts on an annual basis for about 51 percent of Vietnam's total rice production (Nguyen, 2006:1). 90 percent of national rice exports are attributed to this region (Huu Ninh, 2007:8).

rural Vietnam, especially in the Mekong Delta, as observed in the country's post-economic renovation era of the late 1980s (ibid: 37). Small-scale farmers who lose out in this process will increase the numbers of people depending on natural water products as a source of subsistence, leading to great pressure on natural fish resources. This scenario calls for survival strategies from an ever-growing number of landless people.[10] The strategic and controlled transfer of local fishing knowledge is thus an adaptive measure to cope with growing scarcity and competition.

At first glance, competition for natural fish seems to be irrelevant in the case study area due to the exclusive livelihood niches of landed rice farmers on the one hand and landless fishermen on the other. One landless fisherman, originally from the neighbouring province An Giang but now permanently residing in the Northerners' commune, explains:

> "I come from Long Xuyen, An Giang Province. My parents were poor and just gave me the ground for the house but no farming land. With only one ground, you cannot live in An Giang because An Giang does not have so much fish. So we moved here and I saw that the Northerners do not catch fish so much and so I saw that I could easily live here. I live here now since 21 years already thanks to the fish job."

The quote above represents the common situation between the two groups, as encountered in the context of seasonal migration. Paddy farmers show a general lack of interest in fishing, which is due to their stable position as landowners. They tolerate the landless fisherman on their lands during the flood season, whereas, in contrast, by settling down 21 years ago the landless fisherman has turned his knowledge

---

[10] Population growth (1.2 percent natural increase rate in 2008, see GSO 2009: 655) will further increase the intensification of agri- and aquaculture. A decrease in production costs and the introduction of technology in agri- and aquaculture business demands the concentration of land in order to be cost-effective. Industrialisation and mechanized agriculture may not be able to absorb the massive labour externalities of small-scale farmers, who have to sell their plots because their production does not remain competitive.

into a strategic asset because he occupies a new niche within this community, pioneering with exclusive knowledge. The emic perspective of the landed farmers on the stability of their production base explains why they do not consider fishing knowledge as relevant for the pursuit of their own livelihoods. The conceptualisation of local knowledge as integrative neglects the fact that it is also a matter of negotiation as to what kind of knowledge is relevant for one's own purposes. Besides being holistic, local knowledge is backed up with specialised knowledge. Local knowledge as a holistic framework for action is automatically passed on by the socio-cultural embeddedness of a person. Whether access to specialised knowledge is aimed for or not depends on its relevance. Thus, sharing and the reception of knowledge are based on a strategic rationale.

Landless and landowners do not seem to interfere in each other's business. Nevertheless, it will be shown that landless people proactively safeguard their knowledge through the application of certain strategies, even in regards to the 'harmless' paddy farmers. Direct competition mainly exists within the group of landless people itself. How this competition is acted upon needs to be elaborated.

3.2 *Strategies of Non-Sharing and Trust as a Catalyst of Knowledge Transfer*

What was crystallised in the interviews with the Northern Vietnamese rice farmers and landless fishermen were the ways in which the fishermen deal with such growing competition. Knowledge flow is controlled by strategic patterns of action, which serve to proactively obscure livelihood activities in order to keep them secret and to capture the specialised knowledge that forms the basis of everyday life for landless people. In the case of consciously holding back one's skills and knowledge from others, strategies of 'not-telling' or of 'keeping it short' when being asked are applied as evasive forms of communication and interaction. These typical evasive patterns are illustrated nicely in the following interview excerpt. The interview was conducted with a Northern Vietnamese, who himself had acquired certain fishing skills — not through knowledge exchange with the

Southerners but through a very long-term self-instructive process. Within the paddy farmers' commune, this man and his family rather represent an exceptional case since most rice farmers refrain from fishing (see above). However, although for the Northerner this knowledge serves to provide only a small level of extra income, he nonetheless applies the same strategies to actively close off his knowledge from others:

> "They [the Southerners] are very creative and experienced. For each kind such as prawn, abanas, snake-head fish, they have one catching style and one fishing tool — especially the An Giang and Dong Thap people. But they also want to hide their technology. They do not want to show it to all people. [...]. Every day they can sell several 100.000s but they rarely sell the fish here so that the Northerners will see it. They will sell somewhere else, so that the Northerners here do not know that they have so much fish in their region — because, they make a lot of money in the flood season. [...]I am the first man in this commune to successfully learn the job of making bamboo traps. Later my younger and my older brother also learnt it. Now my younger brother is more successful than myself. But also some men around here studied my job and they are around here, not in far places. We have to keep our job a secret, I only tell all the recipes and secrets to my siblings and brothers. [...]When I place the traps I always go in the twilight of the morning so that nobody will see how I make it. Because also for myself, nobody from An Giang or Dong Thap taught me anything. So when I have my career I also want to keep it a secret. One more reason why I want to keep it a secret is that when you place the bamboo trap and pull it up, people see that you have 3 to 4 kilos of snake-head fish inside. They will envy you and try to play tricks on you so that you cannot do your job anymore — by destroying your bamboo trap or by also learning your career."

Selling fish — the Southerner's knowledge and skills-embedded business product — in a far away market to prevent the local population from becoming suspicious is a strategic way of upholding free access to the paddy farmers' lands and the utilisation of its floodwaters as common property. Access to the farmland is a prerequisite for fishing in floodplains, since the seasonal water is managed by ever-more complex irrigation systems (see above) that make access to the water increasingly more dependent on access to the land.[11] By proactively hiding his knowledge-informed success, this man is preventing the Northerners from becoming interested in the potential and the resources of their own lands in the flood season. Although the paddy farmers represent no immediate risk, the fishermen, by their actions, have the strategic foresight to avoid any form of conflict around free access to the flooded lands. This proactivity is referred to as a 'strategy of not-telling'. Another form of direct evasive strategy in regards to the transfer of livelihood knowledge is the strategy of 'keeping it short'. One landless woman, who has come for many years with three other families from An Giang Province to temporarily settle down on the same rice farmer's residential land, articulates the tenor of what was more or less openly indicated in most of the interviews:

> "Sometimes, Northerners come to ask for where and how to place the bamboo traps. Around here, few Northerners place the bamboo traps but when you teach people your career you do not teach them all but very briefly."

Such conscious evasion strategies are based on a fine-tuned deliberation process in which the holder of the knowledge decides with whom to share information with. For the paddy farmer, who never shows off in front of his neighbours and ventures out at "twilight" to place his bamboo traps, distrust certainly is the decisive factor for safeguarding his skills, because he feels that people would "play tricks" on him. It is evident that trust is the dominant factor that facilitates knowledge sharing. Knowledge transfer is bounded in trustworthy, reciprocal relationships — kinship constitutes the basis for such relationships in which knowledge is passed on. In this case, kinship can derive from blood relations as well as being institutionalised through a ceremony that bonds non-biologically-related people together as 'sisters' or 'brothers'. If trust is not guaranteed by blood, then it can grow on the basis of time-consuming investments in personal relationships requiring a strong commitment to mutual reliability (Menkhoff and Gerke., 2002).

---

[11] For a discussion on gender and access to water and land, see Miller (2006).

How much effort needs to be put into trust building and natural bonding between people is succinctly summarised by one respondent, who refers to the experiences of her father who tried to get hold of the fishermen's secrets:

> "[...] my father, although he was very close to the An Giang and Dong Thap people and brought presents to their families many times, they never showed him their secrets and recipes. My father had to dive into the water and touch the bamboo trap to see how it is made although at that time, the water was very high."

As well as trust being the overarching condition for knowledge sharing, other rationalities come into play; economic and cultural motivation regarding the transfer of local knowledge has to be considered. If two people share the same economic status — poor and landless — they will most likely not share their knowledge about where the best fish resources were or the most accommodating landowners who let them stay for free on their compounds. Since the other fisherman's family disposes of quiet the same fishing knowledge they would at least try to keep an economic advantage by hiding their contextualised business knowledge. The woman from An Giang referred to earlier points out:

> "The first time they came here they did not go very spontaneously, they came in advance to observe the field. They saw that the fields here have low water level with lots of grass. Fish usually stays in the grass — as long as the field has grass, the field has fish. So when they saw that this field is perfect, they came together. About this they just talk to their 'brothers' and 'sisters', to the ones they can form groups with. Because when there is only one field but with many bamboo traps, the field does not give enough for all the people. They do not tell it to strangers."

Within this group of fishermen using bamboo traps, the only advantage they have over one another is their 'business secrets'. This is the contextualized knowledge, the 'logistical' know-how that enables them to gain access to the fish resources in the first place, before then applying know-how in the actual exploitation of the fish in the second step.

While economic factors seem to be a strong motivation, the cultural reasoning of knowledge transfer does not seem a predominant pattern. The possible assumption that the Northerners as well as the Southerners in the research site want to keep livelihood knowledge within their respective socio-cultural groups is untenable when contrasted with the empirical data derived from the interviews.

Furthermore, whereas Northerners are not greatly interested in fishing knowledge in the first place, once they possess such livelihood knowledge, being a Northerner alone is obviously not a prerequisite for knowledge sharing with Northerners only. The same applies to Southerners not sharing knowledge exclusively between one another by reference to culture. Although both groups in general — and rather emphatically — articulate cultural differences between one another, e.g. by reference to lifestyle and traits, the 'cultural sameness' seems rather irrelevant for the transfer of knowledge compared to economic motivation.

Both, economic motivation and cultural reference, however, can be annulled by trust. Interviews with the rice farmers and the fishermen showed trust to be the dominant rationale for knowledge sharing. Trust cuts across economic status and culture and is solely based on belief in the 'morality' of the person with whom the knowledge is being shared not to apply this knowledge in any competitively strategic way. Static social group categories such as poor vs. rich, Northerners vs. Southerners, etc. may help as analytical categories, but they do not account for the actual negotiation processes of identifying and crosscutting group membership. Knowledge sharing always follows the strategic considerations of one's own motivations and knowledge by relating to somebody else's goals and knowledge assets on the basis of intersecting, rather than exclusive, social group membership. For example, just because I am female does not mean I share my knowledge only on the basis of gender. Just because I am poor does not mean I only share my knowledge with other poor; therefore, within these strategic considerations of knowledge sharing, trust is an important reason for action.

4. Imitation and Limitation of Knowledge Artefacts

The exchange of knowledge, of course, is not always strategically framed, but can also be restricted by technical limits; the technical limit of sharing is determined by the tacit feature of knowledge. Local knowledge is characterized by its strong components of practical experience and routine. Tacitness defines the non-verbalizable sphere[12] process — knowledge that effectively informs our everyday or specialized knowledge-embedded actions and artefacts. In addition, local knowledge, as knowledge and practices based on everyday experiences, generally endows actors with certain frames of reference for interpretation that enable them to meaningfully act in the social world in which they live (Schütz and Luckmann, 1973). Knowledge can be also embedded in objects (Antweiler, 1995:29), which are here referred to as 'artefacts'. Hand-made bamboo traps in this context would be a typical knowledge artefact created by practical routines and the expertise of putting different materials together efficiently as a first step to obtaining a good catch of fish.

The cases in which these technical limits of knowledge transfer have been indicated directly or indirectly are manifold. The explanation by an interviewed professional bamboo trap maker typically indicates another essential point of tacit knowledge:

> "When my family makes the bamboo traps, many people come to look at us and they also ask questions. I also show them but they cannot do this career because it is very difficult. If they make just one little mistake, the trap cannot work. The most important part of bamboo trap making is the hole where the fish can go into. To make the hole, for example, you need 100 bamboo sticks, but if people use just 99 sticks, the fish will not, or at least be less inclined, to go into the trap. That is why the people around here come here to learn this career, but when they go home to make the traps they cannot sell them."

As directly mentioned or implied in the interviews, the successful replication of knowledge-embedded objects does not work by taking only the right amount of the correct materials or 'ingredients' — this

[12] See Marchand 2003.

Photo 4: Knowledge artefact: Hand-made bamboo trap for wild fishing (H.-D. Evers)

is merely the first step. Besides using the right material in the same quantities (hard facts), the even more decisive factor, however, is getting the exact composition of the different ingredients by following the right step-by-step procedures, which very often involve a strong self-instructive element. The case of a traditional boat building family interviewed at the research site paradigmatically presents the core of tacit knowledge as a typical trait of local knowledge. Becoming a craftsman in the boat business requires long-term (self-) training in the practical aspects boat building through observation and trial and error. Doing things over and over again — repeating the same mistakes, but learning from them — stimulates reflection about the problem and a solution-oriented approach. Over time, the black hole between the problem and the solution becomes accessible, and process-knowledge, which needs to come from the problem to the solution, becomes internalized. It is exactly this process knowledge that constitutes tacitness; it becomes routinized through practical experience. It is very difficult to articulate because it glues together the facts of knowledge-based action; it is the know-how of doing something. This is exemplified by the boat builder, who stated that developing a '*sure eye*' is far more important than using complex mathematical calculations when constructing a steady and long-lasting boat. Marchand (2003) highlights exactly the same phenomenon mentioned by the boat builder in the title of his article "*[...] why the master builder can't explain what he knows*".

Besides the self-instructive process of learning by doing, another factor indicated in the interviews as determining a factor for professionalism is that of becoming a 'master builder', which involves the right amount of creativity, some kind of natural talent and an innate predisposition to develop and professionalize certain skills.

The aforementioned process knowledge, however, is not just a constituent part of knowledge artefacts. Furthermore, it informs the whole procedure of traditional fishing. Hence, catching fish is not done by the bamboo trap alone, no matter how well made the imitation may be. A good catch requires further knowledge on where and when to place the trap, what kind of baits work for what kind of fish, movements of the fish due to different water and weather conditions, etc. The importance of context-bound, situated knowledge is best represented by a fisherman's statement:

> "We [the Southerners] know in which season which kind of fish will move and if they will move in deep water or close to the surface. By eye, we know when the fish goes. So we know where to place the net. The Northerners, in contrast, just look where there is enough space to place a net and there they put it ."

In the interviews with the Northerners on the fishing skills of the landless, niche occupations were given as a rational explanation as to why Southerners are that much more successful in fishing in contrast to the Northerners, who are not into fishing at all. Being without land, the landless have to make a virtue out of the situation in the pursuit of a stable livelihood, while the paddy farmer counts on the products of his land as livelihood collateral. As well as livelihood constraints, natural talent and the creativity of local people, explanations with mysterious, irrational connotations were also given. The professionalism of the landless local people referred back to secret methods, implicating some kind of sacred ecological knowledge on water and its resources etc. Without disregard for the existence of an intuitive and traditional wisdom often assigned to indigenous people in their relationship with nature,[13] this explanation model, nevertheless, would be too shortsighted when considering tacit knowledge. Hence, at least to a degree, the mystery dissolves into the intangible blurriness inherent in tacit knowledge or process knowledge. What might be perceived as secretive or mysterious by 'outsiders' actually is, at least to a certain extent, the mere outcome of a subjective reflection on trial and error in the pursuit of a problem solution. The long-term practical experience gained helps to "develop an eye for something", as indicated in the interview with the boat builder or by the landless fishermen living in the Northerners' commune who "by eye, know where the fish goes". This may then explain that it has less to do with irrational mystery than with rational powers of observation by knowledgeable actors.

Tacit knowledge is knowledge about processes based on experience. It is difficult to verbally articulate; it is internalized in people's actions. It is not standardized in the form of clear-cut verbal instructions on how to get from A to Z, from bamboo string to bamboo trap or from bamboo trap to the big catch. Rather, it is routinized practice but at the same time innovative in adaptation to new situations. Coming back once more to the boat builder referred to earlier, he explains that it would be extremely difficult to become a skillful handicraftsman because "*[...] there are no standards, no basis. It is not like a starter of a motorbike, which you can use for every motorbike alike*".

[13] See Berkes (1999).

The Strategic Turn of Tacit Knowledge

What are referred to above as the "local strategies of knowledge sharing" demonstrate that knowledge can be consciously held back or shared when it comes to coping with livelihood insecurities. Two strategies have been elaborated so far, namely the strategies of 'not telling' and of 'keeping it short'. Furthermore, it has been discussed that the tacit nature of local knowledge in itself has an exclusive property constituted by long-term practical experience, inborn talent and creativity. Since this kind of knowledge is, besides factual knowledge, the precondition for the professionalization of skills, it has a latent strategic element in itself inasmuch as it naturally regulates its access.

The case of the first two strategies refers to actors negotiating access by consciously holding back knowledge. It will now be argued that the technical limit of knowledge sharing can also be turned strategically by actors. This third strategy used to handle the demands of 'outsiders' for knowledge is referred to as 'telling'. In contrast to the first two evasive strategies of communication and interaction, 'telling' serves as a direct form of the communication and proactive strategy of keeping one's face while interacting with others. There are few exceptions to people showing or explaining their skills to a person showing an interest in their career — they follow an 'open' way of dealing with knowledge transfer in the context of growing competition. This third strategy of 'telling' is understandable through its cultural contextualization; this openness in 'telling' about skills has in itself a mechanism of demarcation based on the tacit character of the skills being described. The characteristic of tacitness can back up culturally appropriate ways of communication and interaction by simultaneously upholding the exclusiveness of knowledge — in this respect, tacit knowledge can be strategically turned into a proactive form of communication. If somebody asked you about your formula for fishing success, the only thing you could do is to 'tell' the facts and 'show' him the procedure. Nevertheless, you could not verbalize (see above) process knowledge, and the other person would have to watch your actions closely over a long period and then apply these observations practically in order to successfully internalize process knowledge. Saying nothing when asked makes one lose face due to a complete lack of knowledge. At the same time, the other person can lose face by not paying enough respect when leaving his or her question neglected and the situation uncomfortable. In such a situation, 'telling' only the facts perfectly meets cultural conventions while the completely complementary spectrum of water-livelihood knowledge — as facts and processes — remains exclusive and competitive by definition.

5. Conclusion

It has been shown that local knowledge has adapted to the changing human-water relationship and the growing natural scarcity of fish. It transpires that the strategic element is a constituent part of local knowledge itself (tacitness) as well as inherent in the process of knowledge sharing. 'Not-telling', 'keeping it short' and 'telling' have been discussed as control mechanisms for knowledge flow in the case of seasonal floodplain fishing in the Mekong Delta.

At present, it is rare that Northern paddy farmers in the research commune show an interest in harvesting natural fish. Moreover, those that have acquired such knowledge try actively to withhold it from other paddy farmers in their commune, which certainly indicates competitive behaviour around fishing knowledge as a strategic resource. Possible future scenarios might further stimulate the demand for floodplain fishery knowledge, as growing scarcity makes a resource precious in terms of market value. The decrease of natural fish, for example, could potentially arouse the economic motivation of more affluent people in rural areas to make the strategic decision to acquire these scarce assets. The market price for natural fish products increases proportionately to the amount of cultured fish on the market, the latter being much less valued for its taste than natural fish. The stabilizing trend of landlessness will result in more people looking to diversify their reproductive base — here, fishery knowledge could be a complementary asset in achieving a secure livelihood. The strategic advantage of holding fishery knowledge cannot be foreseen at this stage; however, the potential future conflict around access rights to decreasing natural fish grounds will evidently upgrade such knowledge. Although landless people cannot control the physical accessibility of fish by fencing their lands, they can at least close off the flow and regulate the accessibility of their fishery knowledge through strategic action.

In Vietnamese culture, proverbs play an important role in passing on common knowledge in the form of rules or advice. Proverbs thus can be the reflection of cultural common sense. "*Cho vàng cho bạc chú ai đi buôn*" is a Vietnamese proverb, and can be translated into, "*You can give people gold and silver but you should not show them how to trade*". This proverb vividly relates to the canon throughout this whole text, which advises strategically passing on knowledge by giving only gold and silver — the knowledge artefacts — while holding back knowledge on the procedure, on the 'know-how' to obtain gold and silver in the first place — the contextual process knowledge or the 'business secrets'. Furthermore, it illustrates again the technical limit of knowledge sharing — you can give people some help to start with, but they have to go through the process of trial and error in order to become experienced businessmen. The strategic and technical notions of limits to knowledge sharing are perfectly represented in this saying. In addition, since this advice made it into a proverb, the content certainly is of cultural relevance — knowledge transfer is strategic.

Further to the possibility of being actively and strategically constrained by actors, knowledge transfer in general has less to do with an automatic process of knowledge integration by the recipient than with an active internalization of knowledge as 'embodied practice' (Pottier, 2003:2). Although the existence of local knowledge is universal (Berger and Luckmann, 1980:26–45), local knowledge itself is very much context-specific and to a great extent characterized by process knowledge rather than by verbalizable facts; hence, a great extent of local knowledge lies beyond language (Marchand, 2003). There is more than one universal process knowledge approach to fishing. Some people specialize in fishing with bamboo traps, others specialize in using fishing nets, some use different handmade ingredients for baits, while others use the same ingredients but in different proportions. Thus, even a rather straightforward 'community of practice' such as that represented by the landless fishermen at first glance, becomes highly fragmented by not following inter-subjective standards but a shared, practical personal routine. Due to this very situated nature, one correctly talks of 'knowledge**s**' in plural (Pottier, 2003). While global knowledge can become "localized" (Gerke and Evers, 2006:4), the opposite seems to be rather questionable against the background of this article. Under the paradigm of the "democratisation of development politics" (Nederveen Pieters, 2001:12), development cooperation has accredited the important role of local knowledge and participation for sustainable development since the 1970s (Schröder, 1995:5-7). Development is based on the knowledge on processes (Antweiler, 1995:20), and it is exactly this process knowledge that is difficult to extract and impossible to replant in another context. On the other hand, Sillitoe (1998) stresses the need to develop a "coherent local knowledge intellectual framework" as part of the globalization process. Others such as Neubert and Macamo (2002) argue that local knowledge and scientific knowledge of development have to be combined to achieve participation in development planning. In reality, as this article shows, it seems to be difficult to combine these two knowledge systems. Moreover, time constraints and ambitious output-orientation in development cooperation render local knowledge accounts a quick and simplistic endeavour and reduce it to just another blueprint for development planning.

By postulating local knowledge as flexible and capable of integrating 'new' or 'external' knowledge, the theoretical debate consequently has to assume that this knowledge is freely accessible in the first place — knowledge as an intangible 'take-away'. Nevertheless, it has been shown that the actual processes of knowledge transfer between actors or, for example, between actor and object are prerequisites for any kind of integration. Further, as has been argued in this article, such transfer itself depends on actors' strategies of sharing as well as on inborn talent, creativity and a fair amount of perseverance to vanquish 'tacitness' through long-term self-instructive learning.

What is perfectly described by Mango[14] as the eclectic between global (scientific) knowledge and local knowledge has been well discussed in academia (Long, 2001:73; Pottier *et al.*, 2003). These theoretical insights into global-local knowledge interfaces can also be drawn on for the analysis of inner-'local' negotiation processes on access to knowledge in the context of micro political ecology and knowledge as a strategic resource. This has been the overall aim of this article.

References

Adger, W. N., P. M. Kelly, and N. H. Ninh (2001). Environment, Society and Precipitous Change. In *Living with Environmental Change. Social Vulnerability, adaptation and Resilience in Vietnam*, W. N. Adger, P. M. Kelly, and Nguyen Huu Ninh (Eds.), pp. 3–18. London/New York: Routledge Research Global Environmental Change.

Antweiler, C. (1995). Lokales wissen: Grundlagen, probleme, bibliographie. In *Lokales Wissen und Entwicklung. Zur Relevanz kulturspezifischen Wissens für Entwicklungsprozesse*, S. Honerla and P. Schröder (Eds.), pp. 19–52. Saarbrücken: Verlag für Entwicklungspolitik.

Berger, P. L. and T. Luckmann (1980). *Die gesellschaftliche Konstruktion der Wirklichkeit.* Frankfurt M.: Fischer.

Berkes, F. (1999). *Sacred Ecology. Traditional Ecological Knowledge and Resource Management.* Philadelphia, London: Taylor & Francis.

DFID — Department for International Development (2002). *Poverty and Aquatic Resources in Vietnam: An Assessment of the Role and Potential of Aquatic Resource Management in Poor People's Livelihoods.* www.streaminitiative.org/Library/pdf/DFID/VietnamPovertyReport_23.pdf [accessed 22.11.2007].

Evers, H.-D. (2002). Towards a new sociology of knowledge. Working Paper 15, Department of Sociology, University of Singapore.

Evers, H.-D. and S. Benedikter (2009). *Strategic Group Formation in the Mekong Delta — The Development of a Modern Hydraulic Society.* ZEF Working Paper Series No. 35.

---

[14] Mango argues that "a model that assumes linear knowledge production (outside locality), dissemination (to a locality) and utilization (in locality) is misplaced. Both incoming knowledge as well as certain locally existing bodies of knowledge get transformed, resulting in a broadening of knowledge" (Mango 2002 in Pottier, 2003:25).

GSO — General Statistics Office (2009). *Statistical Yearbook of Vietnam 2008.* Hanoi: Statistical Publishing House.

Gerke, S. and H.-D. Evers (2006). Globalizing local knowledge: Social science research on Southeast Asia, 1970–2000. *Journal of Social Issues in Southeast Asia*, 21(1), 1–21.

Huu Ninh, Nguyen (2007). Flooding in the Mekong River Delta, Viet Nam. United Nations Development Program, Human Development Report 2007/2008, Occasional Paper, *Fighting Climate Change: Human Solidarity in a Divided World.* Human Development Report Office, Occasional Paper 2007/53.

Kelly, P. M., T. V. Lien, H. M. Hien, N. H. Ninh, and W. N. Adger (2001). Managing environmental change in Vietnam. In *Living with Environmental Change. Social Vulnerability, adaptation and Resilience in Vietnam*, W. N. Adger, P. M. Kelly, and Nguyen Huu Ninh (Eds.), pp. 35–58. London/New York: Routledge Research Global Environmental Change.

Knoblauch, H. (2005). *Wissenssoziologie.* Konstanz: UVK Verlagsgesellschaft.

Long, N. (2001). *Development Sociology. Actor Perspectives.* London/New York: Routledge.

Marchand, T. H. J. (2003). A Possible Explanation for the Lack of Explanation; Or, 'Why the Master Builder Can't Explain What He Knows': Introducing Informational Atomism against a 'Definitional' Definition of Concepts, In *Negotiating Local Knowledge. Power and Identity in Development*, J. Pottier, A. Bicker, and P. Sillitoe (Eds.), pp. 30–50. London, Sterling: Pluto Press.

Menkhoff, T. and S. Gerke (Eds.) (2002). *Chinese Entrepreneurship and Asian Business Networks.* London, New York: Routledge.

Miller, F. (2006). Risks, responses and rights: Gender dimensions of water excess and water scarcity in the Mekong Delta, Vietnam. In *Fluid Bonds: Views on Gender and Water*, K. Lahiri-Dutt (ed.), pp. 202–226. Kolkata: Stree.

Miller, F. (2003). *Society-Water Relations in the Mekong Delta: A Political Ecology of Risk.* Division of Geography. Sydney: University of Sydney.

Nederveen Pieterse, J. (2001). *Development Theory. Deconstructions/Reconstructions.* London, Thousand Oaks, New Delhi: Sage.

Nguyen, Ngoc De (2006). *Farmers, Agriculture and Rural Development in the Mekong Delta of Vietnam.* Education Publishing House.

Neubert, D. and E. Macamo (2002). Entwicklungsstrategie zwischen lokalem wissen und globaler wissenschaft. *Geographische Rundschau*, 54(10):12–17.

Pham, Cong Huu, E. Ehlers and Saravanan V. Subramanian (2009). *Dyke System Planning: Theory and Practice in Can Tho City, Vietnam.* ZEF Working Paper Series No. 47.

Pottier, J. (2003). Negotiating local knowledge: An introduction. In *Negotiating Local Knowledge. Power and Identity in Development*, J. Pottier, A. Bicker, and P. Sillitoe (Eds.), pp. 1–29. London, Sterling: Pluto Press.

Rambo, T. A. (2005). *Searching for Vietnam. Selected Writings on Vietnamese Culture and Society.* Kyoto: Kyoto University Press.

Schröder, P. (1995). Lokales Wissen als konstruktives und kritisches Potential für die Entwicklungszusammenarbeit. In *Lokales Wissen und Entwicklung. Zur Relevanz kulturspezifischen Wissens für Entwicklungsprozesse*, S. Honerla and P. Schröder (Eds.), pp.1–15. Saarbrücken: Verlag für Entwicklungspolitik.

Schütz, A. and T. Luckmann (1973). *The Structures of the Life-World.* Translated by R. M. Zander and H. T. Engelhardt. Evanston: Northwestern University Press.

Silitoe, P. (1998). What know natives? Local knowledge in development. *Social Anthropology*, 6(2):203–220.

Tran, Pham Gia and R. Few (2006). Coping with floods in the Mekong Delta, Viet Nam. In *Flood Hazards and Health. Responding to Present and Future Risks*, R. Few and F. Matthies (Eds.), pp. 128–144. London: Earthscan.

UNHCR — Office of the United Nations High Commissioner for Refugees (2000). *The State of the World's Refugees 2000: Fifty Years of Humanitarian Action.* www.unhcr.org/4a4c754a9.html [accessed 08/2009].

Wittfogel, K. A. (1957). *Oriental Despotism — A Comparative Study of Total Power.* Yale University Press.

Chapter

15

Integrating Natural and Social Sciences

Developing a Water Information System for the Mekong Delta

Thomas Menkhoff, Solvay Gerke and Hans-Dieter Evers

Background and Challenge

Integrated watershed management plays an important part in the sustainable use, preservation and analyses of water resources. The term watershed refers to the geographic boundaries of a particular waterbody, its ecosystem and the land that drains to it.[1] Each person is inevitably part of a watershed community, and individual behaviour has a tremendous influence on its balance and sustainability depending on how each person treats watershed resources such as water, air, soil, ground water, plants or animals. If we pollute our own small

[1] http://www.wallkillriver.org/what_is_watershed.htm

watershed, we also negatively affect the larger watersheds downstream which are sometimes referred to as river basins.[2]

A fascinating example of such a river basin is the Mekong, one of world's longest rivers with an estimated length of 4,350 km. The Mekong drains an area of 795,000 km^2 and discharges 475.17 cubic kilometres of water annually. The river originates on the Tibetan Plateau and runs through China's Yunnan province, Myanmar, Laos, Thailand, Cambodia and Vietnam. On March 9, 2010, the Straits Times reported that the "mighty Mekong is drying up".[3] One of the root causes, according to residents in Thai fishing communities along the river, is the construction of dams upstream, causing crops to wither, drinking water levels in wells to fall, and fish stocks to decline. Other challenges at the Mekong's upper reaches include extreme seasonal variations of flow, changing climatic conditions and rapid population increase. All this leads to severe changes in the lower Mekong Delta further south in Vietnam resulting in extreme floods, lack of drinking water, acidification and destruction of habitats.

Towards Effective Integrated Watershed Management

According to development experts, these problems call for effective integrated watershed management which refers to the process of creating and implementing plans, programs, and projects to sustain and enhance watershed functions that affect plant, animal, and human communities within a watershed boundary.[4] Watershed features which agencies seek to manage include water supply, water quality, drainage, stormwater runoff, water rights, and the overall planning and utilization of watersheds. Landowners, land use agencies, stormwater

---

[2] Brüschweiler, S. 2003. Intergrated Water Resources Management — A Way to Sustainability. Info Resource Focus, 1/2003.

[3] Ghosh, Nirmal: "Mighty Mekong is Drying Up". Straits Times, 9 March 2010.

[4] Møller Hansen, J. and Do Hong P. (2005). Integrated water resources management in Vietnam: Present status and future challenges. In *Integrated Water Resources Management in South and South-East Asia*, A.K. Biswas, O. Varis and C. Tortajada (eds.), New Delhi: Oxford University Press.

management experts, environmental specialists, water use purveyors and communities all play an integral part in the management of a watershed. For this purpose, hydrologic, hydraulic, ecologic, and sociological factors must be studied and turned into relevant data points. Inter-agency cooperation among national institutes and authorities at all levels (international, national, regional and local) is also required.

The WISDOM Project

Against this background, the WISDOM project (Evers and Bauer in this volume: Chapter 9; Gerke and Ehlert in this volume: Chapter 14), a joint program between Vietnam and Germany, has been conceptualized to design and implement an actionable knowledge system for the Mekong Delta. The project draws from various disciplines such as hydrology, sociology, information technology and earth observation so as to create an optimized, integrated resource management system.[5] The project planners hope that the integration of various types of data will enable the system's end-users to perform action-oriented analyses on very specific questions aimed at providing them with valuable decision-making tools in supporting regional planning activities. Possible applications include: (a) monitoring of floods and droughts; (b) evaluation of flood and drought risk, damage potential and actual damage; (c) analyses of water quality, pollution and sediment load; (d) improvement of flood prediction via remotely sensed precipitation information; (e) detailed adaptation of surface and sub-surface discharge models; (f) information of land cover and land use changes; (g) observation of settlement development, surface sealing and population growth.

The WISDOM Project's Mandate — Education and Capacity Building

One of the project's key components is education and capacity building. It brings together eight partner organizations from Vietnam

[5] See the following link for more information on the Wisdom project: http://www.wisdom.caf.dlr.de/intro_en.html.

and ten from Germany. Key actors include the Vietnamese Academy of Science and Technology (Division of Remote Sensing and Geographic Information System) and the German Remote Sensing Data Centre of the German Aerospace Centre (DFD-DLR). Dozens of German and Vietnamese scientists are part of the WISDOM project team together with several Ph.D. students from Vietnam and selected European countries. The latter are coordinated by the United Nations University — Environment and Human Security (UNU-EHS) in Bonn in close collaboration with the Bonn Interdisciplinary Graduate School for Development Research BIGS-DR at the Centre for Development Research (ZEF) at Bonn University. Knowledge Management is an important part of the capability development efforts.[6] In November 2009, the authors of this article conducted a training workshop on Knowledge Management at the Mekong Delta Development Research Institute (MDI) at Can Tho University (Vietnam) to enable the participants to understand the power of strategic

Photo 1: Processing fish for export from fish farming in the Mekong Delta

[6] Menkhoff, T., Evers, H.-D. and Chay, Y.W. (eds.) (2010). *Governing and Managing Knowledge in Asia*. Singapore: World Scientific Publishing.

Photo 2: River transport in the Mekong Delta

Photo 3: Fishing boat in the Mekong Delta

Knowledge Management and to appreciate what it takes in terms of leadership, organizational culture, human resource deployment, technology and knowledge process management to create actionable knowledge assets in organizational contexts such as the WISDOM project.

Knowledge Management

Knowledge Management refers to the totality of organisational strategies aimed at creating a smart organisation, which is able to leverage upon its various knowledge assets, to learn from past experiences, whether successful or unsuccessful, and to create new value through knowledge. At the people level, Knowledge Management puts emphasis on the education, learning abilities and competencies of organisational members. At the organisational level, Knowledge Management is concerned about the creation, utilization and development of the collective intelligence of an organisation. Technologically, effective Knowledge Management requires the efficient organisation of a suitable communication and information infrastructure (e.g., intranet) based on relevant taxonomies and knowledge repositories.

Knowledge Management Challenges in the Mekong Delta

In the context of the WISDOM project, the Knowledge Management challenge is to craft and design an information system, which allows absorption of various types of relevant data such as remote sensing data, Geographic Information System data, socio-economic data, digital maps or interpolated point measurements and to organize them in such a way that end-users can query the data depending on the issue they wish to tackle. Key elements include a data entry portal, a database for organized data storage, a visualization tool to display the data and a query mask so that end-users can conduct meaningful analyses with respect to issue-related questions such as: "Which agricultural land areas are most vulnerable to increased flooding?" or "Which socio-economic strata of the population is most affected by river pollution and saline intrusion?"

A key component of the ongoing project work is awareness and capacity building amongst diverse target groups such as government officials and river communities with regard to the tangible benefits of strategic information and knowledge management in the Mekong delta.

As policy-makers, Government officials represent a strategic group in the project landscape who play a key role as knowledge leaders in the ongoing transformation and sustainable development of Vietnam.[7] The resulting information and knowledge management system has to be user-friendly and navigable. This assures sound decision-making and execution with regard to smart watershed-related interventions. Terms such as knowledge management or remote sensing are not always understood by those who can greatly benefit from it. One way of obtaining buy-in amongst potential beneficiaries is to demonstrate how a thoughtful and simple knowledge management approach can help to support decision-making and to tackle calamities such as floods or salt water intrusion.

Towards Climate Change Resilience

While local communities at district and village level often have considerable local knowledge and traditional technologies to effect cost effective and sustainable flood control solutions, they sometimes find it difficult to prepare for flood-induced losses by extending their agricultural production into the dry season through to irrigated crops.[8] Here, thematic knowledge-based communities of practice (COPs) comprising members of local river communities can help to leverage on indigenous know how and innovations.[9] While remote sensing makes it possible to collect all kinds of scientifically interesting data and to perform sophisticated queries, effective knowledge management implies that the resulting information is actionable and in line with strategic imperatives. In the case of the mighty Mekong, one of

---

[7] Evers, H.-D. and S. Benedikter (2009). Hydraulic bureaucracy in a modern hydraulic society — Strategic group formation in the Mekong Delta, Vietnam. *Water Alternative*, 2, 416–439.

[8] Castella, J.-C., Slaats, J., Dang, D.-Q., Geay, F., Nguyen, V.L. and Pham, T.H.T. (2006). Connecting marginal rice farmers to agricultural knowledge and information systems in Vietnam uplands. *Journal of Agricultural Education and Extension*, 12:109–125.

[9] Fforde, A. (2008). Vietnam's informal farmers' groups: Narratives and policy implications. *Südostasien Aktuell*, 1:3–36.

the key challenges is arguably that the river's local woes are trans-boundary and global in nature.

In this respect, the project is ongoing and aimed at achieving a variety of outcomes in areas such as hydrologic modelling, pesticides, endocrine disruptors, knowledge management, socio-economic and vulnerability assessment. Some of its ongoing programs include the following:

Pesticide monitoring

As part of the pesticide monitoring campaign, systematic surface water samples were taken at selected study sites from field discharges and irrigation canals which helped to detect thirteen of fifteen target compounds (buprofezin, butachlor, cypermethrin, difenoconazole, endosulfan sulphate, fenobucarb, fipronil hexaconazole, isoprothiolane, profenofos, pretilachlor, propanil and propiconazole). Fenobucarb, for example, is an insecticide. It is insoluble in water and highly toxic for humans. Once it is detected in rural areas where surface water is often used for drinking, measures can be taken to improve water quality. As such this project component helps to assess the potentially adverse effects of pesticides on humans and aquatic life by comparing measured concentrations with water-quality benchmarks, such as those published by the World Health Organization.

Monitoring of water constituents with remote sensing data

As part of the efforts to monitor water constituents in selected sites around Can Tho, particular types of satellite data are used to map and assess the "turbidity" of water. The term turbidity refers to how clear the water is. Murky water usually implies the existence of particule matter or suspended solids such as clay, silt, and sand from soils, phytoplankton (suspended algae), decaying vegetation, industrial waste and sewage. The extent of turbidity can be mapped with the help of a Modular Inversion and Processing System (MIP). Colours are used to indicate the scaling of turbidity (e.g., light blue colour for low;

pink for high). Measuring and mapping turbidity is an effective way of managing water quality. Respective maps generated with the help of remote sensing data provide officials with important information about the severity of water turbidity issues and enable them to locate the problem so that preventive action can be taken.

Vulnerability assessment research

A Knowledge Management-related objective of the project is to provide data and information about major local vulnerabilities to water-related hazards and climate change. Through the combination of different research methods such as household interviews and wealth rankings, it was found that low-income families who live directly along the river are exposed to various water-related hazards in form of floods, water-level rise and river bank erosion. River bank erosion forces families to rebuild, shift or elevate their homes regularly. This puts a strain on their limited household income. The vulnerability assessment research findings are expected to be instrumental in further increasing the level of sophistication and implementation of planning and flood control institutions aimed at arriving at an integrated approach towards risk and climate change management.

Conclusion

The essential goal of the WISDOM project is to ensure that the Mekong Delta becomes an internationally renowned flagship of innovation and achievement in intelligent, integrated watershed management and climate change resilience. Towards this goal, the project's team members hope that the integration of such data will enable the end-user of the system to perform analyses on very specific questions and thus supply the end-user with a tool supporting regional planning activities. In this regard, the integration of natural and social sciences is of utmost importance for the development of the Water Information System, since it has to show both changes to the natural environment, as well as the socio-economic processes that affect the people living in the Mekong Delta.

References

Brüschweiler, S. (2003). Intergrated water resources management — A way to sustainability. *Info Resource Focus*, 1/2003.

Castella, J.-C., Slaats, J., Dang, D.-Q., Geay, F., Nguyen, V.L. and Pham, T.H.T. (2006). Connecting marginal rice farmers to agricultural knowledge and information systems in Vietnam uplands. *Journal of Agricultural Education and Extension*, 12:109–125.

Evers, H.-D. and S. Benedikter (2009). Hydraulic bureaucracy in a modern hydraulic society — Strategic group formation in the Mekong Delta, Vietnam. *Water Alternative*, 2, 416–439.

Fforde, A. (2008). Vietnam's informal farmers' groups: Narratives and policy implications. *Südostasien Aktuell*, 1:3–36.

Ghosh, Nirmal: "Mighty Mekong is Drying Up". Straits Times, 9 March 2010.

Menkhoff, T., Evers, H.-D. and Chay, Y.W. (eds.) (2010). *Governing and Managing Knowledge in Asia*. Singapore: World Scienticfic Publishing.

Menkhoff, T., Y. W. Chay, H.-D. Evers, and B. Loh (2007). Leadership in knowledge sharing: Creating value through collaboration. *Journal of Asian Business* 23:265–282.

Møller Hansen, J. and Do Hong P. (2005). Integrated water resources management in Vietnam: Present status and future challenges. In *Integrated Water Resources Management in South and South-East Asia*, A.K. Biswas, O. Varis and C. Tortajada (eds.), New Delhi: Oxford University Press.

Tiep, N. X. (2008). Hydraulic works in Vietnam. In *Participatory Irrigation Management and Emerging Issues*, Tiep, N. X. (ed.). Hanoi: Agricultural Publishing House.

World Bank (WB) (2006). Vietnam Environment Monitor 2006. Water Quality in Vietnam with a Focus on the Cau, Nhue-Day and Dong-Nai River Basins. World Bank (WB), Manila.

World Bank (WB) (2008). Vietnam Development Report 2009. Hanoi: World Bank (WB).

Chapter

16

Implementing Knowledge Management in a Geographically Dispersed Organization

The Case of an NGO

Poornima Luthra and Shan L. Pan

1. Introduction

In today's world of globalization and virtualization, geographical dispersedness has become a key characteristic of organizational knowledge (Becker, 2001), and this poses new questions to KM researchers on how organizations integrate knowledge across geographical boundaries, to yield a competitive advantage. Only recently has there been interest in this issue and researchers have indicated that the "conceptual understanding of dispersed teams, too, is still underdeveloped" (Becker, 2001:1039). Boh *et al.* (2007) propose that since the knowledge-based view of the firm at present does not account for how knowledge is distributed across sites, there is a need for new theoretical arguments to understand how managers resolve the challenges of working across geographical boundaries. With this in mind and to contribute towards this emerging area of research, this study aims to understand the implementation of KM programs in geographically dispersed organizations using the case of organization ABC,[1] a non-government cultural relations organization. Since its inception in the 1930s, ABC has grown into an international network of over 7000 people in more than a 100 countries. ABC is hence a *geographically dispersed organization* with the majority of its staff located outside the headquarters in more than 100 countries where the actual operations of the organization take place. Using the case of ABC, this study will provide insights into how geographically dispersed organizations manage their knowledge assets.

2. Theoretical Background

2.1 *The Knowledge-based View of the Firm and Geographical Dispersion*

The business world has undergone a paradigm shift since the early 1990s — from relying on understanding and managing physical goods to focusing on managing intangible assets such as knowledge. Knowledge management (KM) is the deliberate attempt by organizations to "capture, manage and leverage" (Ruppel and Harrington, 2001:37) their knowledge resources to help the organization remain competitive and maintain sustainable competitive advantage (Pan and Leidner, 2003; Zack, 1999). Researchers have highlighted that the four generic processes of KM are knowledge generation, knowledge capturing, knowledge sharing and knowledge utilization (Zheng, 2005). These processes enable an organization to improve their organizational performance and realize immense benefits that include better productivity, improved product quality, and reduced cycle times (Sabherwal and Sabherwal, 2005; Davenport and Prusak, 1998;

[1] The organization's original name has been withheld for confidentiality reasons.

Argot and Ingram, 2000), all leading to improved competitive advantage. According to Sher and Lee (2004), KM leads to competitive advantage in three dimensions: "reduced operating costs, shortened lead-time and product differentiation".

In the knowledge-based view of the firm, KM is seen as an integral element of the competitive strategy of an organization (Earl, 2001) and organizations choose to compete on the basis of knowledge. The integration of knowledge is an important aspect of the knowledge-based view of the firms (Boh *et al.*, 2007) in order for these organizations to gain competitive advantage from managing their knowledge assets. In today's world of globalization and virtualization, geographical dispersedness has become a key characteristic of organizational knowledge (Becker, 2001), and this poses new questions to KM researchers on how organizations integrate knowledge across geographical boundaries, to yield a competitive advantage. Dispersedness of knowledge is an important issue since the creation, storage and transfer of knowledge, particularly tacit knowledge, become significantly more challenging in organizations that are more dispersed. Becker (2001) defines dispersed teams as those "whose members cooperate across geographical boundaries" while Cramton (2001) defines them as "groups of people with a common purpose who carry out interdependent tasks across locations and time, using technology to communicate much more than they use face-to-face meetings". Cummings (2004) argues that geographical locations are a form of structural diversity of an organization that affects the ability of the organization to share knowledge in a manner that would yield performance benefits. According to Becker (2001), the drivers behind the problems arising from dispersed knowledge are due to large numbers resulting in additional resource requirements and the lack of transparency, knowledge asymmetries between the geographically dispersed fragments and structural uncertainty.

Recently, there has been interest in KM issues surrounding geographical dispersion and researchers have indicated that the "conceptual understanding of dispersed teams, too, is still underdeveloped" (Becker, 2001:1039). Boh *et al.* (2007) suggest that two barriers to managing knowledge across geographical boundaries are coordination costs and the power of local ties. They argue that distance reduces the spontaneity of informal knowledge sharing making it difficult for employees in such organizations to build a common ground, social relationships, maintain awareness and focus on the project. In addition, local workers and managers tend to build relationships with people within their own geographical location, making it difficult to share knowledge across geographical boundaries. These barriers, and the prevalence of geographically dispersed organizations, make the management of knowledge in such organizations an important issue for organizations. In Becker's (2001) work, he develops and evaluates strategies for dealing with dispersed knowledge. Becker (2001) believes that "the dispersedness of knowledge has important consequences for organizations and causes particular management problems", a view supported by Boh *et al.* (2007).

Boh *et al.* (2007) propose that since the knowledge-based view of the firm at present does not account for how knowledge is distributed across sites and given that the barriers to collaboration (such as coordination costs and the pull of local ties) can outweigh the advantages in dispersed organizations, there is a need for new theoretical arguments to understand how managers resolve the challenges of working across geographical boundaries. Boh *et al.* (2007) suggest that future research should extend the knowledge-based view of the firm by "examining how managerial decision making is related to integrating knowledge in organizations" that are geographically dispersed. Hence, to contribute towards this emerging area of KM research and the work done by researchers like Becker (2001) and Boh *et al.* (2007), some important questions that need to be addressed, e.g., how do geographically dispersed organizations implement their KM programs?

3. Research Methodology

The objective of this paper is to understand how geographically dispersed organisations manage knowledge. According to Eisenhardt (1989), the case study methodology enables us to focus on understanding the dynamics present within single settings and hence, we use an in-depth interpretive case study approach to bring to surface these research issues, as well as enable us to investigate a contemporary phenomenon within its real-life context (Yin, 1994). Interpretive researchers argue that access to reality is contingent upon social attributes such as language, shared meanings and artifacts, as well as the context in which events and work practices occur (Klein and Myers, 1999; Orlikowski and Baroudi, 1991; Walsham, 1993, 1995a, 1995b). In addition, the interpretive stance does not require any propositions or hypothesis to be formulated prior to the field studies (Kaplan and Maxwell, 1994; Klein and Myers, 1995).

We conducted a case study at organization ABC and studied their knowledge management (KM) strategy. The primary technique for collecting data in this research was the interview method which was conducted in the headquarters and one regional office in 2006 and 2007 (primarily face-to-face interviews with the exception of a few video-conferences) — 23 interviews were conducted in the headquarters over a ten-day period in 2007, and 26 interviews were conducted in the regional office over a 2 month period in 2006. The interviews were semi-structured to obtain a breadth of information, opinion and experience (Fontana and Frey, 1994). By asking questions that enable open-ended answers, participants were better able to express their ideas and opinions. The interviews, which lasted anywhere between 45 minutes to 90 minutes were recorded with the organization's and participants' permission. The interviewees were with top and senior management, middle management and language educators who ranged from various levels of the hierarchy.

The interviews were retrospective in nature. The researcher drew on the perceptions of the interviewees as revealed in their comments during the interviews. The majority of ABC staff interviewed (approximately 65 percent) had been with the organization for more than 8 years and hence were able to provide their views on the KM program over the period from the late 1990s to 2007. About 20 percent of the ABC staff interviewed had been with the organization for between 15 to 25 years. Historical reconstruction of the incidents was subsequently performed by the field researchers. The responses from the interviewees were triangulated with data collected from other interviewees and sources to ensure its accuracy. "Triangulation" of data (Stake, 1994), or using multiple sources of data collection, allows an investigator to address a broader range of historical, attitudinal, and behavioral issues. By using the narratives from one subject to confirm or contradict others through the process of triangulation, inter-subject reliability was enhanced (Miles and Huberman, 1994). No preferential treatment was given to any particular interviewee's comments. In addition, the researcher made special efforts to detect whether the interviewee was attempting to conceal or withhold any details.

In addition to the interview data collected, a number of secondary resources were also used that included strategy papers, staff survey reports, corporate brochures and external consultant reports.

4. The Case

ABC[2], a non-government organization, was established in the 1930s as a cultural relations organization which aims to foster mutually beneficial relationships between people in the country it represents and other countries, as well as to increase appreciation of its own country's ideas and achievements. ABC conducts work in six sectors: arts, education and training, language teaching, governance, information, and science and health. The work done in these six areas supports and complements the strategy of the government of the country it represents. In January 1999, ABC, a geographically dispersed organization, began its KM journey which included a variety of components.

5. ABC's KM Journey

5.1 *Create Positions in the Organizational Structure to Guide and Manage KM*

ABC created the posts of Knowledge Manager, Chief Knowledge Officer (CKO), Intranet Manager and Intranet Assistant in 1999. In addition, a "Knowledge Strategy Group" was established which began

[2] The organization's original name has been withheld for confidentiality reasons.

developing a strategy to manage the organization's knowledge. Influenced by the work of the World Bank, it was decided that the strategy would focus on knowledge sharing (KS) with the aim that that would encourage collective empowerment rather than external management. The group carried out a consultation process to promote the need for KM within the organization, as well as to obtain information of the needs of ABC staff. The discussions as part of this consultation process focused on the following themes: making the most of technology, building teams, access to information, improving quality, putting ideas into practice, and values and behaviors that support KS. The consultation sessions highlighted the need for a greater sense of engagement and inclusion of staff in the organization, and this provided an internal pressure for a KM strategy. The KS strategy paper explained:

> "They (Internal employees) want to know about the important things that are happening. But they also want to feel that their experience and knowledge is valued. They require the support, motivation and tools to share knowledge more effectively with others, starting with their own immediate work group."

5.2 *Create a KM Strategy That Is Based on an Understanding of the Needs of Employees*

Having an understanding of the needs of its employees through the consultation sessions and with the manpower in place to manage knowledge, ABC created its KM strategy to guide the organization in managing its knowledge assets. ABC was very careful to ensure that KM was not implemented as a "fad" in the organization and to achieve this, they made efforts to ensure that the concepts of KM were in line with the problems the organization was dealing with at that time.

5.3 *Build an IT Infrastructure for KM and Tools That Facilitate Knowledge Sharing*

Following the creation of the strategy, ABC adopted an IT-led approach to managing the organization's knowledge. ABC began with the roll-out of the new global infrastructure for the organization with a renewed focus that would provide a reliable basis for business applications software, sharing knowledge and the use of web technology to access ABC's audiences. Such an investment of resources into building an IT infrastructure is crucial to KM.

Knowledge and learning centers (KLCs) were set up to encourage KS and learning within the council. These KLCs, which were communities of practice, helped bring ABC staff with similar expertise or interests and provide a forum for KS. In addition, the pilot of three databases was carried out. In 2001, an online chat facility was launched to encourage ABC staff to communicate with one another. However, the use of primarily technological tools that did not address the needs of the users who comprised of the staff of ABC, and the structural silos inherent in the organizational structure at the time prevented these tools from gaining the buy-in that it hoped to obtain. The former global knowledge facilitator expressed her view:

> "In the early stages, one of the challenges was that the initial response was that it was just a matter of having the right software and sourcing and that would be the solution..."

5.4 *Create Senior Positions and Departments to Guide and Manage KM Activities and Improve the Awareness of KM within the Organization*

In 2002, ABC recruited a Director of KM from outside the organization at a time when there was a need to translate the theory of managing knowledge into practice. Her approach to KM was to encourage collaboration through the use of tools such as communities of practice. To enable this to happen, a KM team was setup. The role of the KM team was to guide and manage KM activities, and in doing so increase the awareness of KM and its benefits. At the same time, ABC's "Web team" launched an improved 'ABC World' intranet homepage that was designed to communicate better to their staff.

5.5 *Conduct a Knowledge Audit to Identify the KM-Needs of the Organization*

In 2003, the Director of KM conducted a knowledge audit that researched the business needs of the organization in relation to KM and found that ABC needed a corporate intranet, improved records management and avenues for collaboration. The most important thing that this exercise revealed was that there was a significant amount of duplication of work and "reinvention of the wheel". It was found that with the various countries in which ABC were represented conducting their own projects (with strategic guidance from the headquarters) in the various sectors, there was a lot of repetition of projects. An ABC staff explained:

> "We are one of the most intelligent organizations...And we do immense amounts of work driven by intelligent people having very good ideas which in itself has led to enormous inefficiencies. But we duplicate efforts; we don't share expertise very well. Although we are an intelligent organization, we don't seem to be a learning organization so the driver [for KM] is that it would simply save us huge amounts of money and time."

In addition, the audit also showed that the structure of the organization promoted silos that prevented effective KS across the geographically dispersed organization. Since its establishment, ABC adopted a "hub and spoke" organizational structure — the hub being the headquarters and the spokes being the overseas country offices. This type of structure encouraged communication with headquarters rather than direct communication between the overseas offices. Within this structure, the regional directors were located in the headquarters while the country directors for each region were located in their respective countries and these country directors would report directly to the regional directors. The nature of its structure created an "us and them" type feeling, also known as structural silos, between the headquarters of ABC and its overseas offices. The former global KM facilitator for ABC commented:

> "I think generally in the organization, because it's been quite a traditional hierarchy and it's been a hub and spoke model, there has always been a

> perception of us and them. Us overseas and them in 'country X' who were "doing it to us"."

5.6 *Pitch KM at Senior Management to Gain Leadership Buy-in*

Armed with the results of the knowledge-audit, the Director of KM began promoting KS as an enabler of the achieving the goals of the organization to the senior management team (SMT). She pitched KS as a key enabler of achieving the effectiveness and efficiency and therefore gained the buy-in of the SMT. A former global knowledge facilitator expressed her opinion:

> "In order to get buy-in from the senior management here, the top of the organization, she [the Director of KM] had to make sure that it tied in with the objectives of [ABC's new Corporate Strategy] so it was expressed in such a way that it was quite high-level so she would get buy in from senior managers but it was not the best way to get buy-in from other staff and that's why we have had to put in so much effort in explaining what we are talking about."

5.7 *Modify the KM Strategy to Adopt a Collaborative Approach to KM with Technology as an Enabler*

At that time, the SMT was re-evaluating their current corporate strategy and creating the new strategy. Based on the results of the knowledge-audit and with the support of senior management, the Director of KM re-introduced a modified Global Knowledge Sharing (GKS) Strategy. The GKS Strategy's objectives were to enable ABC to (1) develop world class products and (2) deliver excellent services to its customers by effectively sharing and utilizing its collective knowledge. The scope of the KS program was to provide knowledge and information services to serve the needs of the over 7000 ABC staff worldwide. While the previous strategy emphasized a technology-driven approach, the new strategy adopted a collaborative approach.

ABC's strategy was based on 'practical' KM and aims to help real people solve real business problems by providing the right information to the right employees at the right time. At the individual level, KM at ABC is about helping employees to get their work done. At the country or regional level, KM helps to obtain information, knowledge and experiences on what other countries and regions are doing with regards to a project or the plan for the country, for example. At the organizational level, ABC needs to evaluate their performance and this requires information on the impact of projects on customers, success stories and testimonials, as well as an accurate and complete picture of all the projects and events carried out in each country.

Some of the key areas of focus of the KS Strategy include: implementing projects to build capability in designing and delivering products and services for external customers, connecting people, aligning with the commissioning process and organizational values, investing in user-friendly technologies, as well as fulfilling government requirements for information/data handling policies. To do these, ABC's GKS strategy adopts a collaborative approach to KS that promotes and facilitates internal networking by giving time and space for ABC staff to share ideas, experiences and work practices in addition to collaborative tools to stay connected. In addition, the new KS strategy adopted a bottom-up approach to KS by identifying and nurturing existing internal networks and build on ABC staff's creativity, passion and enthusiasm to work together. This is based on the understanding that knowledge can be conscripted, and can only be volunteered. In terms of culture, the KS strategy works towards promoting an organizational culture that celebrates success from failures, promotes engagement with staff, ensures that individuals take responsibility to seek information before designing products, and that they adopt good communication practices in sharing their experience with others.

In approving the strategy, ABC's SMT endorsed the principle that improving KS, fostering exploitation of ABC's collective knowledge, and supporting collaboration were critical enablers of ABC's new strategy. These enablers align specifically with:

- Delivering what clients and customers want — by the sharing of marketing experience, preventing reinvention of the wheel, and

the co-operative development of regional and global products and services;
- Releasing creativity and potential — by supporting flexible team working approaches across geographical and hierarchical boundaries and the underpinning cultural changes required to achieve this;
- The extension of ABC's global reach through building external on-line communities to engage in new ways with its target audiences.

Table 1 depicts the strategic knowledge types that were central to ABC's KS Program:

The KS program identifies the following locations in which the above knowledge resides:

- In written documents, reports and document repositories
- In conversation and narratives

Table 1: Strategic knowledge types of ABC's knowledge sharing program

Know ABC	Know ABC's Customers	Know ABC's Product Offerings	Know Why (and Why Not)	Know Who
• Knowledge about corporate priorities, policies and standards	• Knowledge about conducting marketing research and applying the findings to design products	• Knowledge about our products and services offerings worldwide	• Knowledge about good practices and lessons learnt to design, implement and deliver products and services	• Knowledge about who the internal staff are and their area of expertise
	• Knowledge about who ABC's clients, customers and stakeholders are and ABC's relationship with them	• Knowledge about UK ideas, achievements, news, content, photos		• Knowledge about who the external experts are

- In people (i.e., in the heads of colleagues with experience in specific areas).

5.8 *Dedicate Resources Towards pilot KM projects*

Following the approval of the KS Strategy, ABC introduced a number of pilot projects that included the introduction of KS awards to encourage and reward good KS behavior, held the first KM conference to increase awareness of KS in the Council, launched communities of practice and training to build these communities, began piloting communities and collaboration sites to encourage collaborative work, and launched regular teleconference sessions for the interested knowledge champions in the regions. The KS awards were introduced to create a culture in which colleagues trust and respect one another, in which the organization rewards achievement and success, in which staff are willing to share their mistakes and learn from one another. The awards celebrate cross-team working that has been effective in bringing together colleagues in different countries and departments.

To enable ABC's employees to share knowledge, ABC's KS program encourages the use of the following tools: teleconference, videoconference, burning a CD, Online communication, Electronic discussion lists, Real-time discussions with a facilitator and Weblogging (a tool that enables staff to publish information and make it available to colleagues or the network instantly).

5.9 *Identify KM as a Key Enabler of the Corporate Strategy of the Organization*

In ABC's new strategy, the organization once again reiterates that to be effective, they need to be able to tap into what they know as an organization and facilitate collaboration, especially among geographically dispersed teams. By managing their own knowledge and experience, ABC aims to enable the whole ABC network to benefit from innovation and developments in any part. This would also ensure consistency in standards and skills globally.

5.10 *Restructure the Organization to Require Collaboration*

ABC is a unique organization in which most of its employees are geographically dispersed making KM both essential and more difficult to implement. The way the organization was arranged prior to the reorganization had created silos between the various country offices and the headquarters resulting in immense duplication of products and services. Realizing that the structure was playing a significant role in hindering KS, ABC restructured the organization. Instead of having over 100 country offices, the new structure had just 13 regions. These regions were given regional budgets for products and services thereby forcing the countries in each region to work together. The aim of regionalization was to reduce the significant duplication of products and services by encouraging the countries in a region to work together. Each region is working towards its own individual change programs in its countries to enable KS, the release of creativity, better understanding of customers and greater efficiency. As a result of regionalization, the regional directors have been relocated away from the headquarters to a country in their regional jurisdiction. With regionalization, ABC staff overseas was forced to think and work regionally.

With regionalization, the KM team needed to create awareness of the KS as an enabler of regional collaboration and ABC staff needed tools to enable them to work regionally. In response to these needs, the KM team held a global KM conference, published KS booklets and conducted a KS workshop for overseas KM project managers that had been chosen. At the same time, the organization hired a knowledge manager. This raised the profile of KS and helped align it with the corporate strategic changes taking place.

5.11 *Introduce Easily Understood Corporate Values That Best Represent the Organization*

Underlying the elements of the new strategy is a set of values guiding ABC which were created to address the problems of the previous organizational values that were not well-understood. To help people understand the new values, ABC encouraged discussion to translate the values into terms that made sense for everyone. In addition to these values, ABC introduced the dual-key working, which is as much about culture changes as it is a mechanism. Dual-key working brings the headquarters and overseas knowledge about capacity and demand together. It involves making decisions about strategy and products and services to meet identified needs and expectations, and ensuring that demand is balanced with ABC's own ambition. This new way of working helps build a KS culture between the regions and the headquarters.

6. Conclusion

In this paper, the KM journey of a geographically dispersed organization has been described. The early proponents of KM in ABC had vision of a knowledge organization which has definitely become a reality and the organization has made significant progress in implementing their KM program. In moving forward, the organization faces two main challenges. One of the biggest challenges facing ABC in the coming years is in ensuring that the new structure of the geographically dispersed organization in the headquarters and overseas results in KS and a reduction in the duplication of projects. For ABC staff to really partake in KS to make it a part of the fabric of organizational life requires a "knowledge-friendly" organization culture. In such a culture, staff has a positive orientation to knowledge. Employees are bright, intellectually curious, willing and free to explore, and executives encourage their knowledge creation and use. Also, staff is not inhibited in sharing knowledge. They are not alienated or resentful of the company and don't fear that sharing knowledge will cost them their jobs. ABC will need to work towards creating that culture across its widely dispersed organization.

References

Argote, L. and P. Ingram (2000). Knowledge transfer: A basis for competitive advantage in firms. *Organizational Behaviour and Human Decision Processes*, 82:150–169.

Becker, M. C. (2001). Managing dispersed knowledge: Organizational problems, managerial strategies and their effectiveness. *Journal of Management Studies*, 38(7):1037–1051.

Boh, W. F., Y. Ren, S. Kiesler, and R. Bussjaeger (2007). Experstise and collaboration in the geographically dispersed organization. *Organization Science* 18(4):595–612.

Cramton, C. D. (2001). The mutual knowledge problem and its consequences for dispersed collaboration. *Organization Science*, 12(3):346–371.

Cummings, J. N. (2004). Work groups, structural diversity, and knowledge sharing in a global organization. *Management Science*, 50(3):352–364.

Davenport, T. H. and L. Prusak (1998). *Working Knowledge*. Boston: Harvard Business School Press.

Earl, M. (2001). Knowledge management strategies: Toward a taxonomy. *Journal of Management Information Systems*, 18(1):215–233.

Eisenhardt, K. M. (1989). Building theories from case study research. *Academy of Management Review*, 14:532–550.

Kaplan, B., and J. A. Maxwell (1994). Qualitative research methods for evaluating computer information systems. In *Evaluating Health Care Information Systems: Methods and Applications*, J. G. Anderson, C. R. Aydin and S. J. Jay, J. (Eds.), pp. 45–68. Thousands Oaks: Sage.

Klein, H. K. and M. D. Myers (1999). A set of principles for conducting and evaluating interpretive field studies in information systems. *MIS Quarterly*, 23:67–94.

Orlikowski, W. J. and J. J. Baroudi (1991). Studying information technology in organizations: Research approaches and assumptions. *Information Systems Research*, 2:1–28.

Pan, S. L. and D. E. Leidner (2003). Bridging communities of practice with information technology in pursuit of global knowledge sharing. *Journal of Strategic Information Systems*, 12(1):71–88.

Ruppel, C. and S. Harrington (2001). Sharing knowledge through intranets. *IEEE Transactions on Professional Communication*, 44:37–52.

Sabherwal, R. and S. Sabherwal (2005). Knowledge management using information technology: Determinants of short-term impact on firm value. *Decision Sciences*, 36:531–567.

Sher, P. J. and V. C. Lee (2004). Information technology as a facilitator for enhancing dynamic capabilities through knowledge management. *Information and Management*, 41:933–945.

Walsham, G. (1993). *Interpreting Information Systems in Organizations*. Chichester: Wiley.

Walsham, G. (1995a). The emergence of interpretivism in IS research. *Information Systems Research*, 6:376–394.

Walsham, G. (1995b). Interpretive case studies in IS research: Nature and method. *European Journal of Information Systems*, 4:74–81.

Yin, R. (1994). *Case Study Research: Design and Methods.* Beverly Hills, CA: Sage.

Zack, M. H. (1999). Developing a knowledge strategy. *California Management Review*, 41(3):125–145.

Zheng, W. (2005). A conceptualization of the relationship between organizational culture and knowledge management. *Journal of Information and Knowledge Management*, 4(2):113–124.

Index